Crystal Field Effects in Metals and Alloys

Crystal Field Effects in Metals and Alloys

Edited by A. Furrer
Institute for Reactor Technique
Swiss Federal Institute of Technology
Würenlingen, Switzerland

Plenum Press · New York and London

Library of Congress Cataloging in Publication Data

International Conference on Crystal Field Effects in Metals and Alloys, 2d, Zürich, 1976.
Crystal field effects in metals and alloys.

"Proceedings of the Second International Conference on Crystal Field Effects in Metals and Alloys held at Zürich, Switzerland, September 1-4, 1976."
Includes indexes.
1. Crystal field theory—Congresses. I. Furrer, A. II. Title.
QD475.I52 1976 546'.3 76-55802
ISBN 0-306-31008-2

Proceedings of the Second International Conference on Crystal Field Effects in Metals and Alloys held at Zürich, Switzerland, September 1–4, 1976

A Division of Plenum Publishing Corporation
227 West 17th Street, New York, N.Y. 10011

Printed in the United States of America

CONFERENCE ORGANIZATION

Sponsorship:

Swiss Federal Institute of Technology (ETH)
Zürich, Switzerland

Chairman:

W. Hälg, Zürich, Switzerland

Scientific Secretary:

A. Furrer, Würenlingen, Switzerland

Administrative Secretary:

Mrs. V. Stämpfli, Würenlingen, Switzerland

International Advisory Committee:

E. Bucher, Konstanz, BRD
K. H. J. Buschow, Eindhoven, The Netherlands
W. J. L. Buyers, Chalk River, Canada
D. R. Cooper, Morgantown, USA

Preface

The idea of this conference grew out of the rapidly increasing volume of experimental facts and theoretical concepts related to the problem of crystal-field effects in metals and alloys. The crystal field plays an important role in the understanding of the energetic level structure of ions in condensed matter. In particular, the magnetic properties of rare earth metals and alloys are strongly influenced by the crystal field. In the phenomenological theory the crystal field successfully describes the static and dynamic magnetic properties of these systems. On the other hand the microscopic origin of the crystal field in metals is not yet fully understood. However, recent years have seen some of the areas of crystal-field effects mature to the point that they should be summarized and brought to the active notice of a larger audience. In addition, a number of exciting developments have occured which deserve attention.

This book contains 13 invited and 45 contributed papers presented at the 2nd international conference on crystal-field effects in metals and alloys held at Zürich, Switzerland, September 1-4, 1976. Emphasis was placed on the following specific categories of interest: spin waves and excitons, soft modes and critical effects, magnetic properties, physical properties influenced by crystal-field effects, actinides and valency. Because the conference was relatively small, about 120 participants, and because the topic was relatively narrow, recent work in the field could be treated thoroughly and the present state of knowledge assessed comprehensively. Only papers actually presented at the conference are published. All papers have been reviewed for scientific and technical content. The papers are grouped by session number for convenient gathering of the same topic in the same place.

I should like to thank the many people who helped to make the conference a success. I am indebted to the conference chairman, W. Hälg, for continuous support, and to the members of the advisory committee, E. Bucher, K.H.J. Buschow, W.J.L. Buyers, and B.R. Cooper, for advice and comments on the scientific program. I thank all the experts who acted as session chairmen, and particularly B.R. Cooper who was given the difficult task of summarizing the papers and discussions of the conference. It is a pleasure to express appreciation to the ETH Zürich for providing the facilities and for financial support. I owe gratitude to our indefatigable secretary, Mrs. V. Stämpfli, for all her labors in cheerfully shepherding participants, authors and referees before, during and after the conference. Finally I acknowledge the efforts of the publisher towards a rapid publication of these proceedings.

A. Furrer

Contents

RECONCILING LOCALIZED MOMENTS WITH LATTICE PERIODICITY

K. W. H. Stevens

Department of Physics, University of Nottingham
University Park, Nottingham NG7 2RD, England

ABSTRACT

The original crystal field model is incompatible with lattice periodicity. Yet experiment provides many examples of localized moments and conduction electrons. The problem of beginning with a very basic Hamiltonian, and recasting it to show localized moments and conduction properties is considered using degenerate perturbation theory. The result is a spin-Hamiltonian of typical form for the localized moments, with conduction electrons as well. At no point is crystal field theory invoked.

***

Crystal field theory has been very successful in describing the properties of rare earth insulators and conductors, so it may surprise you that I have chosen to discuss a theory which does not use crystal field ideas. The reason is that although crystal field theory provides a conceptually simple framework in which to interpret experimental results, it has serious limitations of a mathematical nature. In avoiding these I find that I increasingly want to drop crystal field ideas, while retaining the spin-Hamiltonian descriptions.

For simplicity I shall consider a rigid periodic crystal containing rare earth ions. It may be an insulator or a conductor. The Hamiltonian is

$$\mathcal{H} = \sum_i \left|\frac{p^2}{2m}\right|_i - \sum_{i,M} \left|\frac{Z_M e^2}{r_{iM}}\right| + \tfrac{1}{2} \sum_{i\neq j} \frac{e^2}{r_{ij}} + \ldots$$

where ... includes the spin-orbit, Zeeman and any other interactions. The Hamiltonian is invariant under interchange of electrons and under the symmetry operations of the space group. It does not contain any explicit crystal field energies. To progress, approximations are necessary. A common one is to replace the mutual coulomb repulsions of the electrons by an average potential, giving an unperturbed Hamiltonian

$$\mathcal{H}_o = \sum_i \left|\frac{p^2}{2m}\right|_i - \sum_{i,M} \left|\frac{Z_M e^2}{r_{iM}}\right| + \sum_i V(r_i)$$

where V(r) is a periodic potential. $\mathcal{H}_o$ describes independent electrons moving in Bloch functions which extend throughout the crystal. They are filled to a Fermi level. There are no local moments at this stage. If they are to emerge they will have to arise from the perturbation $\mathcal{H} - \mathcal{H}_o$, which is then not obviously small. So although $\mathcal{H}_o$ retains all the symmetries of $\mathcal{H}$ it is not a very promising unperturbed Hamiltonian for a system which shows localized moments.

A more promising unperturbed Hamiltonian would seem to be one which has a rare earth ion, perturbed by a crystal field, at each rare earth site. That is, one might take

$$\mathcal{H}_o = \sum_A h_A^{xtl}$$

where

$$h_A^{xtl} = \sum_i \left|\frac{p^2}{2m} - \frac{Z_A e^2}{r_{iA}} + V_A(x,y,z)\right|_i + \tfrac{1}{2} \sum_{i \neq j} \left|\frac{e^2}{r_{ij}}\right| .$$

h_A^{xtl} is the crystal field Hamiltonian for site A, and the summation in it is over the electrons at site A. For a conductor we might expect an extra term of the form

$$h_{cond} = \sum_{\text{conduction electrons}} \left|\frac{p^2}{2m^*}\right| .$$

The apparatus of crystal field theory can be used to deal with each h_A^{xtl}, to give local moments at each site, and it seems obvious that the perturbation $(\mathcal{H} - \mathcal{H}_o)$ will produce the various interactions between localized moments, and with the conduction electrons. In fact such a perturbation theory is difficult to carry out, and in spite of its superficial attractions I believe it should be rejected, for the following reasons. In constructing each h_A^{xtl} it is necessary to allocate particular electrons to particular sites, thus introducing distinctions. Also $\mathcal{H}_o$ does not have the space group symmetries of $\mathcal{H}$, for the co-ordinates of a given electron are measured from the nucleus to which it has been allocated. $\mathscr{H}_0$ therefore has much less symmetry than $\mathscr{H}$. We have the unusual feature that the perturbation $(\mathscr{H} - \mathscr{H}_0)$ substantially *increases* the symmetry. In the perturbation theory of the states it seems that one will be attempting to express high symmetry states in a power series of states of lower symmetry! The method I shall outline avoids these problems. Its application to insulators is described in Stevens (1976).

I believe the unperturbed Hamiltonian should have the following properties. It should not distinguish between electrons, it should have the space group symmetries of $\mathscr{H}$, it should contain local moments and conduction electrons, the perturbation $(\mathscr{H} - \mathscr{H}_0)$ should be small and the perturbation theory should be tractable.

It is remarkable how much information we have about wavefunctions of electrons in solids without having solved any relevant Schrodinger equations. For example, given a rare earth compound of known composition and structure one can usually predict quite a lot about its magnetic properties, using the crystal field theory of an isolated ion. No doubt we shall have some examples in due course! I suggest we use this information to construct the unperturbed Hamiltonian, for we know $\mathscr{H}$ and $\mathscr{H}_0$ is only an auxiliary quantity to enable us to do perturbation theory. So why not choose it to suit ourselves? It should have, as eigenstates, the states which we have reason to believe are close to being eigenstates of $\mathscr{H}$. As we shall probably want it to be Hermitian we should ensure that these states are mutually orthogonal. Normally when we speak of orbitals of ions in crystals we use non-orthogonal orbitals. This is a nuisance in the present context, so I shall insist that all the electronic orbitals are mutually orthogonal. One first imagines that a band structure calculation has been made. All the band states are mutually orthogonal. They can be used to construct mutually orthogonal and localized Wannier functions (Wannier 1937, Kohn 1973). When in future I speak of atomic-like orbitals I shall mean their Wannier analogues (which are less well-localized). For rare earth conductors there is no point in using Wannier functions for the conduction electrons. I shall assume their orbitals are typical Bloch functions. They are orthogonal to the Wannier functions of other bands.

For a given rare earth crystal we can describe a chosen and approximate ground state by specifying the one-electron states which are occupied. In general there will be many possible low-lying states because of the degeneracies from the unfilled 4f-shells and from the conduction band. But with orthogonal one-electron orbitals all such states will be mutually orthogonal. Similarly we can construct an infinity of mutually orthogonal excited states. Of course, when we are allocating electrons to orbitals we have to ensure that the exclusion principle is obeyed. A straightforward procedure is to allocate them, with specified spin orientations, as if they are distinguishable, and then change to Slater determinants. A more compact procedure is to use the method of second quantization, introducing creation and annihilation operators for the various orbitals plus spin. Any desired combination of Slater determinants is then obtained by operating on the vacuum with a suitable expression in creation operators.

The states which are obtained by such a procedure are mutually orthogonal, so they could be the eigenstates of a Hermitian operator. Given any such state, $|n\rangle$, we may form its projection operator, $P_n = |n\rangle\langle n|$. Then

$$\sum_n \lambda_n P_n$$

has each state $|n\rangle$ as an eigenstate, with eigenvalue λ_n. The λ_n are therefore at our disposal. In the original version of this theory I thought a good choice for λ_n would be the expectation value of $\mathscr{H}$ taken for the state $|n\rangle$. That is, $\lambda_n = \langle n|\mathscr{H}|n\rangle$, for if $|n\rangle$ approximates to an eigenstate of $\mathscr{H}$, then $\langle n|\mathscr{H}|n\rangle$ should approximate to an eigenvalue. Thus

$$\sum_n \langle n|\mathscr{H}|n\rangle |n\rangle\langle n|$$

should be a good unperturbed Hamiltonian. I now think a slightly different choice for λ_n is preferable. But to show why it is convenient to go a little further with $\lambda_n = \langle n|\mathscr{H}|n\rangle$. Since we know the states $|n\rangle$ and we know $\mathscr{H}$ it is a reasonably straightforward exercise to simplify $\langle n|\mathscr{H}|n\rangle$. The steps are similar to those in the theory of many-electron atoms (Griffith 1971). The main difference is that we have vastly many more electrons, and we have a lattice of nuclei. As with the atom the $\langle n|\mathscr{H}|n\rangle$ reduce to sums of one- and two-electron matrix elements, examples being

$$\langle\phi_A|\frac{p^2}{2m}|\phi_A\rangle \ , \quad \langle\phi_A\phi_B|\frac{e^2}{r_{12}}|\phi_A\phi_B\rangle \quad , \quad \text{etc.}$$

Also, as with the atom, most of them can be given physical meanings. For example the $\langle n|\mathscr{H}|n\rangle$ for a given $|n\rangle$ contains the kinetic energies of all the electrons of state $|n\rangle$, added together of course. Similarly it contains the coulomb energies of the electrons in the fields of the nuclei. In an atom all the electrons are attached to the same nucleus, so this energy is part of the binding energy. In our case an electron in a given 4f orbit has a binding energy to its own nucleus, and a crystal-field-like energy to all the other nuclei. Similarly the coulomb repulsions between electrons give binding energies for electrons in orbitals on the same nucleus, but crystal field and multipole type energies when the orbitals are on different nuclei. Somewhat similar considerations enter for the conduction electrons. But the important points are that $\langle n|\mathcal{H}|n\rangle$ contains all the main energy terms and although no specific crystal field has been introduced the energies typically associated with it can be recognized as present. (It should be noted though that our effective crystal field must be evaluated using orthogonal one-electron orbitals.)

Once the $\langle n|\mathcal{H}|n\rangle$ have been evaluated it becomes easier to see that in spite of the nice energy properties,

$$\sum_n \langle n|\mathcal{H}|n\rangle |n\rangle\langle n|$$

does not have the rotational symmetries of $\mathcal{H}$, though it does have the translational and indistinguishability requirements. The difficulty stems from the fact that in defining the second quantized operators directions of quantization for orbit and spin are introduced. Fortunately the problem is readily circumvented, and then one finds that there is a bonus! The various $|n\rangle$ can be arranged into families. For example, the states in which each rare earth site is $(4f)^r$, with all the conduction electrons in the lowest conduction band can be regarded as a family. Another would be that in which, say, two sites have $(4f)^{r-1}$, the rest have $(4f)^r$, and the displaced 4f-electrons are accommodated in the lowest conduction band. If one examines the states in a given family, the $\langle n|\mathcal{H}|n\rangle$ are not all equal. (For a given $(4f)^r$ the terms of different L,S have quite different energies. Also the electrons in the conduction band can be re-arranged to have different kinetic energies.) I could use more sophisticated separations into families, but this would lead to unnecessary complication here. I therefore use about the simplest concept of family. For a given family one can evaluate the mean $\langle n|\mathcal{H}|n\rangle$, say $\overline{\langle n|\mathcal{H}|n\rangle}$. Then the operator

$$\mathcal{H}_o = \sum_n \overline{\langle n|\mathcal{H}|n\rangle} |n\rangle\langle n| = \sum_{\text{families}} \{\overline{\langle n|\mathcal{H}|n\rangle}\, P_n\}$$

where P_n denotes the projection operator for the whole family to which $|n\rangle$ belongs, is found to have all the symmetry properties of $\mathcal{H}$. It is therefore a much better choice for the unperturbed Hamiltonian, at least on symmetry considerations. Its use has the effect that, in the unperturbed Hamiltonian the conduction band has zero width, with a finite gap to the next band. Now comes the bonus. There is a very convenient and powerful perturbation theory for degenerate manifolds, which does not work at all well with quasi-degenerate manifolds (Messiah 1962). The trick of replacing $\langle n|\mathcal{H}|n\rangle$ by $\overline{\langle n|\mathcal{H}|n\rangle}$ has therefore both improved the symmetry aspects of the theory and opened the way to the use of a powerful perturbation formalism. It is also worth noting that the dimensions of the ground manifold, $\mathcal{H}_o$, are those associated with having each rare earth site possess a definite number of 4f electrons with a definite number of conduction electrons and double-occupied other orbitals.

The spin-Hamiltonian theory of isolated ions arises because the perturbation theory used there gives an effective Hamiltonian within a specific manifold of states. A similar thing happens with general perturbation theory. The low-lying eigenvalues of $\mathscr{H}$ are separated from the rest and they are given as the eigenvalues of an effective operator acting within P_o. But since P_o has the characteristic feature that all its states have definite numbers of electrons in the various shells, the form of the effective Hamiltonian can be deduced without going through the detailed perturbation theory, though this is necessary to get expressions for the parameters it contains. The second quantized formalism is now particularly convenient, for the conservation on number is easily expressed. Denote the rare earth sites by A, B, ... and the creation operators at site A for 4f orbitals plus spin by a_p^*, a_q^*, ..., and similarly for B, etc. Similarly denote the conduction band creation operators by c_k^*, where k includes spin. Then $\mathscr{H}_{eff}$ can be expected to include combinations of the forms:

$$a_p^* a_q \quad , \quad b_p^* b_q \quad , \quad a_p^* b_q^* b_r a_s \quad , \quad c_{k_1}^* c_{k_2} \quad , \quad c_{k_1}^* a_p^* a_q c_{k_2}$$

where every a* is accompanied by an a, to preserve the electron number at a site. But every a*a commutes with every b*b, etc. Let us therefore neglect, for the moment, all combinations which couple sites to one another or to conduction electrons. That is we retain only combinations of the form

$$a^*a \quad , \quad b^*b \quad , \quad a^*a^*aa \quad , \quad c_{k_1}^* c_{k_2} \quad , \quad c_{k_1}^* c_{k_2}^* c_{k_3} c_{k_4} \quad .$$

Then, all those which pertain to site A can be collected together into h_A. Similarly, those for site B, in h_B, etc. and those pertaining to the conduction band into h_{cond}. The result is

$$h_A + h_B + \ldots + h_{cond} \quad .$$

Every h_A commutes with every h_B, etc. Also a particular h_A contains annihilation and creation operators for 4f electrons at A in such a way that although the combinations may change the states they do not change the number of electrons. But this is precisely the thing which angular momentum operators do, and it is a comparatively simple matter to write a given h_A as equivalent to an angular momentum operator defined at site A. Similarly h_B can be written as equivalent to an angular momentum operator defined at site B, etc. Thus

$$h_A + h_B + \dots + h_{cond}$$

can be replaced by an equivalent operator of type

$$h_A(\ell_A, s_A) + h_B(\ell_B, s_B) + \dots h_{cond}$$

The two forms are not identical, for the states of the former are not simple product states of h_A, h_B, etc. for h_A, h_B, etc. act in the same Hilbert space, whereas the states of the latter are direct products. Further, if we examine the forms of the localized angular momentum operators we find, because the point symmetry has been preserved throughout, that it has the form which one would obtain from a crystal field theory. The main difference is that its parameters have to be obtained from perturbation treatment of $(\mathscr{H} - \mathscr{H}_0)$, rather than from a crystal field theory. h_{cond} is not replaced by an angular momentum form. It also produces a width to the conduction band.

There remains the terms in $\mathscr{H}_{eff}$ which couple sites together, and sites to conduction electrons. The a*a, b*b, etc. are simply replaced by their equivalent angular momentum operators.

I shall not go further with this problem, for I have essentially finished what I had aimed to achieve; to give a justification for the form of the Hamiltonian (localized angular momentum operators, coupled to one another and to conduction electrons) which is the starting point for most discussions of rare earth conductors. However, unlike these I have at no point introduced crystal field theory, but rather I have tried to show that by means of a suitably chosen perturbation theory, starting from a very general Hamiltonian, it is possible to demonstrate that the problem of reconciling localized moments with periodicity, which has divided magneticians for many years, can be resolved. I hope, too, that these ideas will help resolve some of the difficulties which have arisen over the significance of some experimental 'crystal field' parameters, by giving new meanings to them.

REFERENCES

Griffith, J. S., Theory of Transition Metal Ions (Cambridge: University Press) (1971).

Kohn, W., Phys. Rev. B, 7, 4388 (1973).

Messiah, A., Quantum Mechanics, Vol. II, Chap. XVI, Section 17 (Amsterdam: North-Holland) (1962).

Stevens, K. W. H., Phys. Rep. C, 24, 1 (1976).

Wannier, G. H., Phys. Rev., 52, 191 (1937).

SPIN WAVES IN ANISOTROPIC MAGNETS

Per-Anker Lindgård

Research Establishment Risø
DK-4000 Roskilde, Denmark

ABSTRACT

The theory of excitations in anisotropic magnets is reviewed. It will be shown, by discussion of a simple model system with Heisenberg interaction plus planar single ion anisotropy, that the conventional spinwave theory is treating the single ion anisotropy very poorly. A new Bose operator expansion of tensor operators which is different from the Holstein Primakoff transformation will be discussed. The spin waves fulfil the Goldstone theorem exactly for planar anisotropy. An infinite order perturbation approach shows that in general only renormalized parameters are obtainable from spin wave data of anisotropic magnets. The approach shows clearly the general structure of the ground state corrections due to crystal field effects. The Er data for the cone phase can be analyzed successfully with single ion anisotropy and isotropic exchange interaction, J(q). The Tb data can be understood on the basis of the same model.

INTRODUCTION

The theory of magnetic excitations in strongly anisotropic magnets has been in constant development during the past decade. The development followed closely the neutron scattering results in more and more crystal field dominated systems. In particular the rare earth metals and compounds served as examples for which is was clear that the conventional (C) spin wave theory, developed for the Heisenberg magnet, was unsatisfactory and other methods i.e. the pseudo Boson[1] and later the standard basis operator technique[2] had to be employed. Although these techniques can be applied for any ratio of the crystal field, V_c, relative to the exchange interaction, H_{ex}, they have in general only been used for crystal field dominated systems. The reason is that they suffer from the disadvantage that the crystal field must be accurately known from other sources in order to diagonalize the single site Hamiltonian before a treatment of the collective excitations can be initiated. The simple concept of anisotropy, which makes the spin value a simple variable, also breaks down and each compound must be treated separately. The heavy rare earth metals were thought to be strongly anisotropic yet sufficiently dominated by the exchange interaction that the conventional spin wave theory should work. However, Egami and Brooks[3] strongly advocated that groundstate corrections due to crystal field effects were significant in order to obtain quantitative agreement between anisotropy parameters obtained from spin wave measurements and other sources, i.e. susceptibility measurements. Lindgård and Danielsen[4] developed a systematic Bose operator expansion for any tensor operator and showed that the renormalization of the anisotropy parameters assumes a particular simple form not found by Egami and Brooks[3]. It was found that the anisotropy renormalizes less for the spin oscillations in the hardest direction, say ω_x, and more for the more easy direction, say ω_y (z is in the easiest direction). The renormalization could be described by the two parameters $m(T)\pm\frac{1}{2}b(T)$, where $m(T) = \langle a^+a\rangle/S$ is the spin deviation parameter and $b = \langle aa\rangle/S$ is the socalled ellipticity parameter, which is zero if $\omega_x = \omega_y$. It was later shown that this holds for any wave vector q.[9] Since it is well known from the antiferromagnetic ground state problem[5] that the Hartree Fock approximation treats the ground state corrections rather crudely a more accurate method is desirable. We shall here present such a method which combines some of the virtues of the pseudo boson (spin) and the conventional theory. The method, which we shall denote the matching of matrix element (MME) method, is shortly as follows. The original Hamiltonian which has a non-diagonal single site term is transformed into a new Hamiltonian with only a diagonal single site part. This is achieved by expanding each spin (tensor) operator in an infinite expansion of boson or spin operators. The advantage is that the new Hamiltonian is identical in form to that for an isotropic antiferromagnet, which has been intensively studied[5]. In a Hartree Fock sense the ground state corrections for this Hamiltonian is considerably reduced relative to the original one, simply because a partial diagonalization has been performed. The expense is that an effective two-ion anisotropy is introduced. We shall demonstrate that it is very difficult to distinguish this from a possible genuine two-ion anisotropy. Tables of Bose operator expansions for any tensor operator for cubic and hexagonal crystal fields have been developed[6] to first order in V_c/H_{ex}. For the spin operator expansion the expansion has been generalized to infinite order[7] by a canonical transform method and explicit expressions have been given to second order.

SIMPLE MODEL SYSTEM

Let us begin by considering a planar ferromagnet and compare the results of the MME theory and the conventional spin wave theory, improved by the Hartree Fock correction. The Hamiltonian is

$$H = \sum_i H_i^s + H_{int} + \text{const} \qquad (1)$$

where

$$H_i^s = - H_{ex} S_i^z + D(S_i^x)^2 \;, \quad H_{ex} = 2S\sum_j J_{ij} \qquad (2)$$

$$H_{int} = - \sum_{ij} J_{ij} \vec{s}_i \cdot \vec{s}_j \;, \quad s_i^\alpha = S_i^\alpha - \delta_{\alpha z} S \qquad (3)$$

We shall assume that $D > 0$ and that $H_{ex} > D(S-\frac{1}{2})$ so that the ground state of H_i^s is not a singlet. In the Holstein Primakoff[8] (HP) transformation the considered states to which Bose states are equivalated are the pure states $|S,m\rangle$, eigenfunctions for $H_o = - H_{ex} S_i^z$ and with $|S,S\rangle$ assumed as the ground state. A finite anisotropy, D, perturbs these states to $\psi_m = \sum_n a_{mn} |S,n\rangle$. However, let us follow the general HP idea and asign a Bose state $|m\rangle$ to each of the new states ψ_m and require that a well ordered Bose operator expansion (WOBE) has the same matrix elements as a given operator O_p we wish to expand (we call this the matching of matrix element method)

$$<\psi_{\dot{m}'}|O_p|\psi_n> = <m'|WOBE|n> \quad (4)$$

In this way we obtain by standard perturbation theory the generalized HP transformation[6]. For example to first order in D/H_{ex}

$$S_z = S - a^+a + \frac{D}{2H_{ex}} \sqrt{S(S-\frac{1}{2})}\,(a^2+a^{+2}) + .. \quad (5)$$

$$S^+ = \sqrt{2S}\,\{a - \frac{D}{2H_{ex}}(S-\tfrac{1}{2})a^+ + \ldots\}\ ,\ S^- = (S^+)^+ \quad (6)$$

The new feature is that a S^+ operator can both annihilate and create a Bose state. When substituted back into (2) the last term in (5) clearly cancels the off-diagonal anisotropy term thereby diagonalizing H_i^s. Extending this result to higher order of perturbation is very cumbersome using the standard perturbation technique. However, for the simple Hamiltonian H_o, the method of canonical transformation is found very useful[7]. Let us seek a unitary operator exp(U) which diagonalizes the single site Hamiltonian H^s (2).

$$\tilde{H}^s = e^U H^s e^{-U} = \sum_{n=0}^{\infty} \frac{1}{n!}[U,\ldots[U,H^s]\ldots] \quad (7)$$

Because of the simple commutation relation $[O_{\ell m},H_o] = O_{\ell m}{*}\,mH_{ex}$ for any tensor operator $O_{\ell m}$, U can easily be found for a general crystal field to any order of perturbation in V_c/H_{ex}. If we transform the full Hamiltonian (1) (or a more general one) it assumes the form of an effective two-ion anisotropic one with a diagonal single site part.

$$\tilde{H} = -\sum_{ij} J_{\ell m}^{\ell' m'}(ij)\,O_{\ell m}(i)\,O_{\ell' m'}(j) \quad (8)$$

for which the conventional methods (i.e. HP) can be used. Assuming the exchange interaction to be dominant the two lowest lying states of $\tilde{H}^s$ are |S,S-1> and |S,S>. Using the MME method[4] we can then find a spin or Bose operator expansion of (8). Likewise the operators generating the transverse modes can be expanded as follows

$$\tilde{S}_x = S_x(u+v)+w; \quad \tilde{S}_y = S_y(u-v)+w \quad (9)$$

Explicitly to second order in $d = D/H_{ex}$ we find

$$u = 1+\frac{1}{8}(S-\tfrac{1}{2})(S-\frac{3}{2})d^2 \quad , \; v = -\tfrac{1}{2}(S-\tfrac{1}{2})d+\tfrac{1}{2}(S-\tfrac{1}{2})(S-1)d^2 \tag{10}$$

The operator expansions (9) are distinct from those encountered in the pseudo spin theories with respect to the conserved spin length $S = \tilde{S}$, the inclusion of the well ordered higher order spin terms, w, and the perturbation expansion of the coefficients, u and v, which makes a direct diagonalization unnecessary.

The bilinear part of the transformed Hamiltonian $\tilde{H}$ is identical to the antiferromagnetic Hamiltonian.

$$\tilde{H} = \text{const.}+ \frac{1}{2S} \sum_q \{A_q S_q^+ S_q^- + \tfrac{1}{2}B_q (S_q^+ S_{-q}^+ + S_{-q}^- S_q^-)\} + w \tag{11}$$

$$= \text{const.}+ \sum_q \{A_q a_q^+ a_q + \tfrac{1}{2}B_q (a_q^+ a_{-q}^+ + a_{-q} a_q)\} + wb \tag{12}$$

When additional ground state corrections (AGC), present when $B_q \neq 0$, are neglected the spin wave energy is

$$E_q = \sqrt{A_q^2 - B_q^2} \tag{13}$$

For the planar ferromagnet (1) we find

$$A_q^{MME} = (u^2+v^2)\,\omega_q - 2uvH_{ex} = \omega_q + D(S-\tfrac{1}{2})+\ldots$$

$$B_q^{MME} = 2uv(\omega_q - H_{ex}) = D(S-\tfrac{1}{2})J(q)/J(0)+\ldots \tag{14}$$

where $\omega_q = 2S(J(0)-J(q))$, $H_{ex} = 2SJ(0)$ and $J(q)=\Sigma_R\, e^{iqR} J_R$. It is clear that the Goldstone theorem i.e. $E_{q\to 0} \to 0$ is exactly fulfilled with the functions (14) in (13). This is not the case using the conventional (HP) theory which also assumes the form (12), say (C12), but with

$$A_q^c = \omega_q + D(S-\tfrac{1}{2})$$

$$B_q^c = D\sqrt{S(S-\tfrac{1}{2})} \tag{15}$$

It was shown[9] that if the higher order terms wb in (C12) are treated in the Hartree Fock (HF) approximation the Goldstone theorem is fulfilled to first order in 1/S. Assuming nearest neighbour interaction one can express the functions in the HF approximation as

$$A_q^{CHF} = \omega_q(1-m+m') + D(S-\tfrac{1}{2})(1-2m-b)$$
$$B_q^{CHF} = \omega_q b' + D(S-\tfrac{1}{2})(1-2m-b) \qquad (16)$$

in terms of the general HF correction factors

$$m = \sum_q m_q/N,\ m' = \sum_q m_q\gamma_q/N,\ m_q = \frac{1}{S}\{\frac{A_q(T)}{E_q(T)}(n_q+\tfrac{1}{2})-\tfrac{1}{2}\}$$
$$b = \sum_q b_q/N,\ b' = \sum_q b_q\gamma_q/N,\ b_q = \frac{-1}{S}\ \frac{B_q(T)}{E_q(T)}(n_q+\tfrac{1}{2}) \qquad (17)$$

where $n_q = \{\exp(E_q(T)/kT)-1\}^{-1}$ is the spin wave population factor and $\gamma_q = J(q)/J(0)$. We are now able to compare the various results. The most significant feature of the MME approach is that the effective two ion anisotropy (8) introduces a q-dependence of B_q^{MME} (15). As can be seen from (17) this reduces the AGC for (11) or (12) relative to that for (C11) or (C12) as we would expect since a partial diagonalization was performed in arriving af (8). The CHF result (16) approaches the MME result (14) by introducing a q-dependence of B_q^{CHF}. However, by comparing the HF approximation results to leading order in d one finds

$$B_q^{MME,HF} = D(S-\tfrac{1}{2})\{1-(1+\delta)\omega_q/2S\} \qquad (18)$$

$$B_q^{C,HF} = D(S-\tfrac{1}{2})\{1-\delta'\omega_q/2S\} \qquad (19)$$

where $\delta = \frac{1}{N}\Sigma_q\gamma_q^2/E_q > 0$ and $\delta' = \frac{1}{N}\Sigma_q\gamma_q/E_q > 0$. The AGC introduces $\delta > 0$ and therefore further enhances the wave vector dependence of B_q. The reason for the difference between (18) and (19) is that the single ion anisotropy is not treated to the same level of accuracy by the C and the MME approach, the latter being the potentially most accurate. Expanded to the same level of accuracy they give identical results. For further discussion we shall now turn to a more rich example than the planer ferromagnet.

THE GENERAL CASE

Let us consider the transverse spin wave dispersion for a general cone structure, which contains the spiral and the ferromagnetic structure as limiting cases. We now also consider a general crystal field $V_c = \lambda\Sigma_\ell B_{\ell m} O_{\ell m}$ (local z-axis). λ is a formal expansion parameter ($=1$ in the result). After the usual Fourier transformation and neglecting the higher order spin terms we can write the transformed Hamiltonian as (11)

$$\tilde{H} = \text{const.} + \frac{1}{2S}\sum_q\{A_q S_q^+ S_q^- + \tfrac{1}{2}B_q(S_q^+ S_{-q}^+ + S_{-q}^- S_q^-)\} \qquad (20)$$

with

$$A_q = H_{ex}(1+\gamma) - \Delta_q^{ex}(u^2-v^2) - A_q^{ex}(u^2+v^2) - 2uvB_q^{ex}$$
$$B_q = -B_q^{ex}(u^2+v^2) - 2uvA_q^{ex} \qquad (21)$$

The parameters γ, u and v are infinite expansions in $\lambda B_{\ell m}/H_{ex}$, γ is accounting for the diagonal single site anisotropy. The index, ex, indicates known linear functions[10] of the isotropic exchange interaction. The spin wave energy is, neglecting additional ground state and temperature dependent corrections (AGC), $E_q = \Delta_q^{ex}(u^2-v^2) + (A_q-B_q)^{\frac{1}{2}}$. Eliminating one parameter by the Goldstone theorem and using a similar notation as Keffer[5] we can write the energy in the illustrative form

$$E_q = \Delta_q^{ex}M^{+-} + \{\omega_q^x\omega_q^y M_{xx}M_{yy} - 2R\omega_q^y M_{yy}\}^{\frac{1}{2}} \qquad (22)$$

where $\omega_q^x = (A_o^{ex}+B_o^{ex}) - (A_q^{ex}+B_q^{ex})$ and $\omega_q^y = (A_o^{ex}-B_o^{ex}) - (A_q^{ex}-B_q^{ex})$; for the planar ferromagnet $\omega_q^x = \omega_q^y = \omega_q$ and $\Delta_q^{ex} = B_q^{ex} = 0$. R is the effective anisotropy parameter defined below in (27) for $\tilde{m} = \tilde{b} = 0$. $M^{\alpha\beta}$ are the reduced matrix elements for the spin operators $\tilde{S}_q^\alpha\tilde{S}_q^\beta$ in the true ground state 1>. If we put $M^{\alpha\beta} = 1$ (22) reduces to the conventional expression derived by Cooper et al.[11]

$$E_q^{conv.} = \Delta_q^{ex} + \{\omega_q^x\omega_q^y - 2R_o\omega_q^y\}^{\frac{1}{2}} \qquad (23)$$

Using the eigenstates of H_o as a trial ground state, $1>_0$, the above transformations (9) give

$$2SM^{+-} = \tfrac{1}{2}\langle \tilde{S}_q^+ \tilde{S}_q^- - \tilde{S}_q^- \tilde{S}_q^+ \rangle = (u^2-v^2)\langle S_q^+ S_q^- \rangle_0$$

$$SM_{xx} = \langle \tilde{S}_q^x \tilde{S}_q^x \rangle = (u+v)^2 \langle S_q^x S_q^x \rangle_0 \qquad (24)$$

$$SM_{yy} = \langle \tilde{S}_q^y \tilde{S}_q^y \rangle = (u-v)^2 \langle S_q^y S_q^y \rangle_0$$

If we define $\langle S_q^+ S_q^- \rangle_0 \equiv 2S(1-\tilde{m})$, $\langle S_q^x S_q^x \rangle_0 = S(1-\tilde{m}+\tilde{b})$ $\langle S_q^y S_q^y \rangle_0 = S(1-\tilde{m}-\tilde{b})$, then $\tilde{m}$ and $\tilde{b} \equiv \langle S_q^+ S_{-q}^+ \rangle_0 / 2S$ are zero at zero temperature. In this case $(M^{+-})^2 = M_{xx}M_{yy}$ and (22) reduces to the conventional expression (23), when we introduce a renormalized exchange interaction $J(q)^{RPA} = (u^2-v^2)\ J(q) = \langle \tilde{S}^z_{Sw} \rangle / S\ J(q)$. The MME approach, treating the single site part of the Hamiltonian correctly, therefore yields the RPA result for the exchange interaction to all orders in λ. Furthermore it renormalizes the anisotropy consistently to a given order in $\lambda B_{\ell m}/H_{ex}$. This is not easy to obtain by a Greens function approach on the original Hamiltonian using the RPA decoupling.

However, the matching of matrix element method allows us to go a step further by generalizing (24). When B_q is large and comparable to A_q, as for the antiferromagnetic problem, the fully aligned eigenstate of H_0 is a poor trial ground state. Let us instead choose the final spin wave eigenstates of (20) as $|\rangle_0$ to calculate the matrix elements $M^{\alpha\beta}$ (24). In this case $\tilde{m}$ and $\tilde{b}$ may be large and non-zero at zero temperature. The resulting expression for the transverse spin wave energy is then

$$E_q = \Delta_q^r + \{\omega^x_{q,r}\omega^y_{q,r}[1-(\frac{\tilde{b}}{1-\tilde{m}})^2] - 2R_r\omega^y_{q,r}\}^{\frac{1}{2}} \qquad (25)$$

in terms of the further renormalized exchange integral

$$J(q)^r = (u^2-v^2)(1-\tilde{m})\ J(\vec{q}) \qquad (26)$$

and the renormalized anisotropy parameter

$$R_r = (1-\frac{\tilde{b}}{1-\tilde{m}})\{2uv[(1-\tilde{m})A_0^{ex}+\tilde{b}B_0^{ex}]+(u^2+v^2)[\tilde{b}A_0^{ex} + (1-\tilde{m})B_0^{ex}]\} \qquad (27)$$

the functional form of (25) is different from the conventional expression (23). This allows a different fit to experimentally determined spin wave energies for a cone structure. The (AGC) may in general additionally change the functional form, whereas the higher order terms, w in (9), to a large extent can be incorporated in effective u and v. The spin wave theory estimates $\tilde{m} = Sm$ and $\tilde{b} = \frac{1}{2}Sb$ in the Hartree-Fock approximation given in (17). Also the simpler "ground state correction" due to a perturbation of the cone structure by a hexagonal crystal field can be incorporated in $\tilde{m}$ and $\tilde{b}$.

We can define two renormalized anisotropy parameters in (20)

$$D_0 = [1+\gamma-(u^2+v^2)]H_{ex}, \quad D_2 = -2uvH_{ex} \tag{28}$$

To first order in λ D_m reduce to the relatively simple expression given by Lindgård[10] being linear combinations of $B_{\ell m}$. If the expansion parameters $B_{\ell m}/H_{ex}$ are not << 1 the expansion in (28) must be carried to the appropriate order in λ and the simple relations no longer hold.

The absence of a giant two-ion anisotropy in the heavy rare earth metals was recently concluded[10] on the basis of a first order analysis of spin wave data on Er[12]. The formula used for the spin wave energy was identical in form to (25), which has here been shown to be the simplest expression, valid in general when the parameters are interpreted as being renormalized. It was found that an excellent fit could be obtained on the basis of (25) with the single extra parameter $\varepsilon = \frac{\tilde{b}}{1-\tilde{m}}$. A fit to (23)[12] can only be obtained by introducing an extra wave vector dependent function (of unknown origin and with many parameters). We interpret this to mean that the ground state corrections (either due to the hexagonal perturbation of the cone structure or to quantum mechanical effects) for the cone structure in Er are indeed very significant and eliminate the need for introducing a giant two-ion anisotropy. For Tb the ground state corrections invalidate the simple relation between the field derivative of E_q^2 and A_q.[13] A priliminary analysis indicates that the results can be understood on the basis of single ion anisotropic and isotropic exchange interaction. It is difficult to estimate the magnitude of the ground state corrections. The Hartree Fock approximation gives a quite small lower limit. Conventional estimates of a possible genuine two-ion anisotropy gives likewise very small magnitudes. However, in this case one expects symmetry breaking effects which can be measured directly; they are found to be small for Tb.[14]

The conclusion is that an analysis of transverse spin wave data in anisotropic systems at most allows a determination of a renormalized exchange interaction $J^r(q)$, and of two renormalized anisotropy parameters D_o and D_2. The parameters are in general not simply related to the original ones in the Hamiltonian. Additional information about the parameters may be obtained by varying external parameters, such as magnetic field or temperature. A genuine two-ion anisotropy, which is not symmetry breaking, can only be uniquely determined if its q-dependence differs from that of the isotropic interaction and more than one q-dependent quantity has been measured Finally we remark that the relation between the present theory and the pseudo spin theory was demonstrated and that the present approach is found to be advantageous when the crystal field is incompletely known.

REFERENCES

1. G.T. Trammell, J. Appl. Phys. 31, 3625 (1960) and B. Grover, Phys. Rev. 140A, 1944 (1965)
2. W.J.L. Buyers, T.M. Holden, E.C. Swensson, R.A. Cowley and M.T. Hutchings, J. Phys. C. 4, 2139 (1971) and A.B. Haley and P. Erdös, Phys. Rev. B5, 1106 (1972)
3. M.S.S. Brooks and T. Egami, J. Phys. C, 6, 531 and 3719 (1973) and 7, 979 (1974); M.S.S. Brooks, Phys. Rev. B1, 2257 (1970) and T. Egami, J. Phys. C5, L85 (1972)
4. P.-A. Lindgård and O. Danielsen, J. Phys. C7, 1523 (1973)
5. F. Keffer, Handbuch der Physik, XVIII/2, 1 (1966)
6. P.-A. Lindgård and A. Kowalska, J. Phys. C9, 2081 (1976)
7. P.-A. Lindgård and J.F. Cooke, to be published
8. T. Holstein and H. Primakoff, Phys. Rev. 58, 1098 (1940)
9. J. Jensen, J. Phys. C8, 2769 (1975)
10. P.-A. Lindgård, Phys. Rev. Lett., 36, 385 (1976)
11. B.R. Cooper, R.J. Elliott, S.J. Nettel and H. Suhl, Phys. Rev. 127, 57 (1962)
12. R.M. Nicklow, N. Wakabayaski, M.K. Wilkinson and R.E. Reed, Phys. Rev. Lett. 27, 334 (1971)
13. J. Jensen, J.G. Houmann and H. Bjerrum Møller, Phys. Rev. B12, 303 (1975)
14. P.-A. Lindgård and J.G. Houmann, in Proceedings of Conf. on Rare Earths, Durham, England (Ed. E.W. Lee, Institute of Physics, London), (1972)

SPIN WAVES IN HoZn

B. Hennion*, J. Pierre+, D. Schmitt+ and P. Morin+

* D.Ph. G.S.R.M., Orme les Merisiers,
91190 - Gif-sur-Yvette, France
+ Laboratoire de Magnétisme, C.N.R.S.,
166X, 38042-Grenoble-Cedex, France

ABSTRACT

The spin wave spectrum of HoZn (CsCl-structure) has been studied at 4.2 K. Two branches were followed along the principal symmetry directions. The data have been analysed following a Green's functions formalism to obtain the Fourier transform of interactions. The dispersion curves are well reproduced, but their precise energy position requires the addition of second order terms in the starting Hamiltonian.

INTRODUCTION

The spin wave spectra for pure rare earth metals have been extensively studied, but a few results are available for their intermetallics compounds (Pr_3Tl[1], HoP, $TbAl_2$, $HoFe_2$,...). The behaviour of dispersion curves, as well as the energy gaps for q=0, were shown to depend on the crystal field (CEF) parameters through the level scheme and matrix elements between levels. Besides one-ion anisotropy and the long range Heisenberg interaction, it has been sometimes found necessary to introduce anisotropic exchange terms or magnetoelastic interactions[2].

EXPERIMENTS

The cubic compound HoZn orders ferromagnetically at 75 K along a threefold axis ; however below 23 K, the magnetization axis is a twofold one[3]. The moment reaches 8.45 μ_B at 4.2 K and follows at low temperatures a $T^{3/2}\ e^{-\Delta/T}$ law (Δ = 18 K). The spin wave spectrum was studied at 4.2 K by inelastic neutron scattering on the H4 triple axis spectrometer at the Nuclear Center of Saclay. The spectrometer was operated in the constant q mode.

The single crystal was a 2 cm^3 platelet cut perpendicularly to a binary axis. Another binary axis was mounted vertically, giving a <110> diffraction plane. Thus the principal symmetry directions in the Brillouin zone (insert of figure 3), except XM, could be investigated. Two branches were observed in these directions and in the [112] direction, with energy gaps at the zone center Γ of 1.9 and 8.7 meV (figure 1). The value of the first one (21 K) is in agreement with magnetization data.

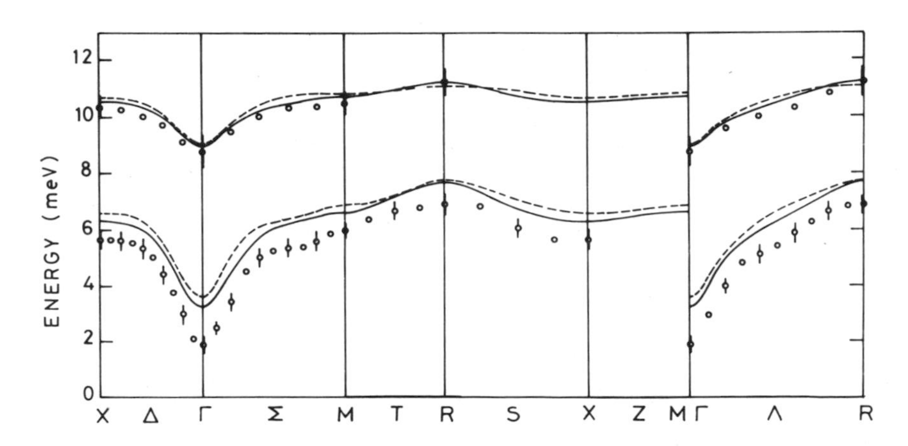

Fig. 1 : Dispersion curves of HoZn at 4.2 K. Dashed curves : W = 0.36 K, x = 0.08, continuous curves with biquadratic exchange B = 0.001 K.

For the first branch, the dispersion law is initially quadratic in scattering vector and isotropic (up to $|q| = 0.4\ \Pi/a$), but becomes anisotropic for larger q. The second branch displays similar features, but with a smaller dispersion. Particularly the "knees" along ΓM and ΓR directions are found on both branches.

INTERPRETATION

The semi-classical theory of spin waves - isotropic moment with anisotropy of energy - was found inadequate due to the anisotropic reduction of the moment by crystal field. It would lead, with the parameters deduced from neutron spectroscopy (W = 0.36 K, x = 0.08)[4] to a gap of around 12 meV for q = 0. Instead, we used a Green functions formalism as described by Buyers et al[1]. The starting Hamiltonian is divided in a single ion part including a molecular field interaction

$$2 \sum_{i \neq j} I_{ij} \langle S_{z_i} \rangle S_{z_j}$$

and an interionic part. The single ion term is first diagonalized to obtain eigenvalues ω_n and eigenfunctions $|n\rangle$. The splitting by exchange of the holmium multiplet is represented on the figure 2.

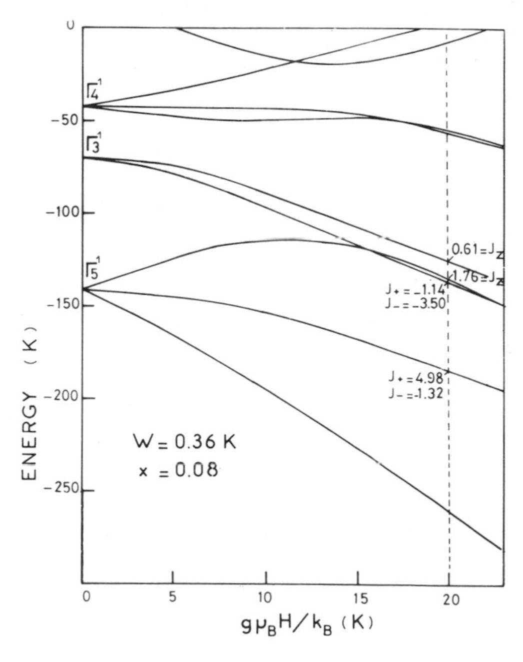

Fig. 2 : Splitting of Ho multiplet by the exchange field in HoZn. J_z, J_+ and J_- represent the matrix elements S_{z0n}, S_{+0n}, S_{-0n} (see text) for the exchange field at 4.2 K.

The single site dynamical susceptibility is defined by

$$g^{\alpha\beta}(\omega) = \sum_{m,n} \frac{S_{\alpha mn} \; S_{\beta nm} \; (f_m - f_n)}{\omega - \omega_n + \omega_m}$$

where α, β equal +, - or z, $S_{\alpha mn} = \langle m| S_\alpha |n\rangle$ and f_n is the Boltzman factor. Contrarily to the case of tetragonal or rhomboedral magnetic symmetries, the orthorhombic symmetry leads to the simultaneous occurrence of S_{+mn} and S_{-mn}. Thus, one defines one longitudinal generalized susceptibility

$$G^{zz}(q, \omega) = g^{zz}/(1 - 2J(q)\; g^{zz}),$$

where $J(q) = \sum_j I_{ij} \exp(i\vec{q}.\vec{R}_{ij})$ is the Fourier transform of interactions, but four transverse coupled susceptibilities

$$G^{++}(q, \omega) = G^{--}(q, \omega) = g^{++}/D$$

$$G^{\overset{+-}{-+}}(q, \omega) = \left[g^{\overset{+-}{-+}} + J(q)(g^{++2} - g^{+-}\; g^{-+}) \right] /D$$

with $D = (1 - g^{+-}\; J(q))(1 - g^{-+}\; J(q)) - g^{++2}\; J(q)^2$

The magnon energies for a given q value correspond to the ω-poles of the susceptibility functions. The analysis of the two observed branches leads to interpret them as transverse excitations toward the two first excited levels, corresponding to the largest matrix elements, (fig. 2) and gives, for the above CEF parameters, similar variations of J(q). The longitudinal excitation toward the third excited level is expected in the range 9.7 - 10.7 meV with a smaller intensity and dispersion. It could perhaps been separated from the second observed branch with an improved resolution.

The interionic interactions I_n may be computed by the inverse Fourier transform. However, $I_o = a^3/8\Pi \int J(q)\, d^3q$ integrated over the Brillouin zone does not vanish, that means an incorrect description of the anisotropy. Deleting I_o, we obtain the theoretical dispersion curves (dashed line of fig. 1), which are shifted from their experimental position. The fact that the difference between calculated and observed spin-waves energies is almost constant over the whole Brillouin zone suggests that this difference results from single-ion properties.

The one-ion CEF parameters have been determined[4] at low temperatures (15 and 68 K) on the paramagnetic dilute $Ho_{0.15}Y_{0.85}Zn$ compound. These parameters are not expected to change much with the concentration or temperature, but may be modified by the magnetic order. However, slight changes of the starting parameters do not improve the calculated energies. Similarly to the interpretation of magnetization curves[4], we are led to introduce second order terms in the Hamiltonian. As no magnetostriction is observed along two-fold axis, we write a biquadratic exchange term :

$$\mathcal{H}_b = B(1/4<O_2^0 + 3\ O_2^2> (O_2^0 + 3\ O_2^2) - 12<O_2^1>\ O_2^1)$$

The value $B = 10^{-3}$ K which describes well the magnetization curves leads to the plot of J(q) drawn in figure 3 and improves slightly the fit of dispersion curves (continuous line on figure 1). The value of J(q=0) is then in rough agreement with the value obtained from ordering temperature (-1.5 K).

CONCLUSION

The energy location of dispersion curves is improved by including anisotropic exchange, but further work is needed to precise its exact expression. The study of the temperature dependence of excitations, particularly near the rotation temperature, will perhaps give valuable informations on this point. The interactions I_n (Table I) with first and third neighbourgs are dominant and negative (as expected for a ferromagnet in this formalism), the determination of the last ones being rather uncertain. However, interactions up to the 6th neighbourgs are necessary to allow a correct description of the form of the branches. Their variation with distance is rather far from the predictions of the Ruderman-Kittel theory.

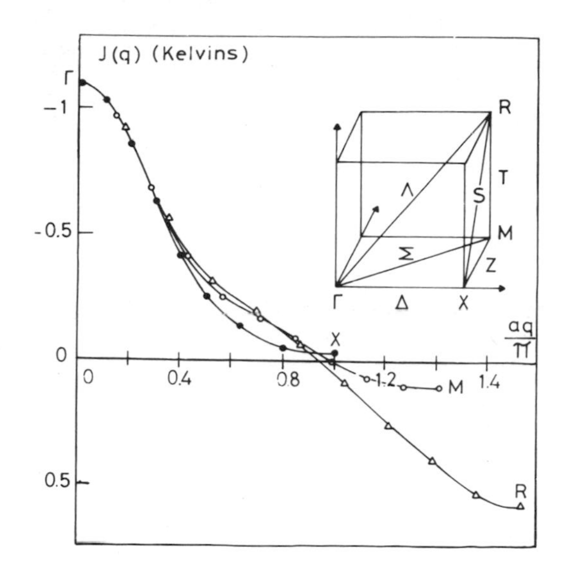

Fig. 3 : Fourier transform of exchange interactions. As insert symmetry directions in the Brillouin zone.

Table I : Values of I_n deduced from experiment

Neighbour n	1	2	3	4	5	6	7	8	⩾9
I_n (10^{-4} K)	-835	-74	-149	54	-35	-51	-44	-28	<20

REFERENCES

1. T.M. Holden, W.J. Buyers, Phys. Rev. B 9, 3797 (1974)
T.M. Holden, W.J. Buyers, A. Perreault, Phys. Rev. B 11, 266 (1975)

2. J. Jensen, J.G. Houmann, H.B. Møller, Phys. Rev. B 12, 303 (1975)

3. P. Morin and J. Pierre, Sol. Stat. Commun., 13, 537 (1973)

4. D. Schmitt, P. Morin, J. Pierre, Proc. of the 2nd Intern. Conf. on Crystal Field Effects in Metals and Alloys, Zurich (1976).

GROUND- AND EXCITED-STATE SPIN WAVES IN HOLMIUM PHOSPHIDE

A. Furrer*, P.M. Levy**, and E. Kaldis***

* Institut für Reaktortechnik ETHZ, EIR, CH-5303 Würenlingen, Switzerland

** New York University, Department of Physics, New York, N.Y. 10003, USA

*** Laboratorium für Festkörperphysik, ETH, CH-8093 Zürich, Switzerland

ABSTRACT

The magnetic excitation spectrum of HoP (flopside spin structure, T_c=5.5 K) has been studied by neutron inelastic scattering. At 1.5 K five spin-wave branches have been found which are almost independent of wave vector. The results are analysed in terms of a single-ion Hamiltonian containing a crystal-field and the effects of bilinear and quadrupolar pair interactions. At 4.2 K excited-state spin waves have been observed which correspond to transitions from the upper member of the ground-state doublet $\Gamma_3^{(2)}$. The stability of the unusual flopside spin structure is discussed.

INTRODUCTION

Holmium phosphide belongs to the large class of NaCl-structured rare earth monopnictides. Below T_c=5.5 K a magnetically ordered flopside spin structure has been observed (1). In this structure there are ferromagnetic (111) planes with the moment directions along the cube edges, and the direction of the moments in an adjacent plane are perpendicular, i.e., along another edge of the cube. Below T_c, HoP is tetragonally distorted, but the distortions are small and could not be precisely measured (2). The splitting of the ground-state multiplet $^5I_8$ due to the octahedral crystal field is known from neutron spectroscopy (3,4): the crystal-field parameters are W=-0.025 meV and x=0.75 in the notation of Lea et al. (5). The effective exchange field can be roughly estimated from the ordered magnetic moment which has been determined to be 8.8 μ_B at 1.3 K (1). More detailed information on the nature of the exchange interaction, however, requires neutron inelastic scattering experiments on single crystals.

In the present work we report on neutron inelastic scattering measurements of magnetic excitations in single-crystal HoP and give a theoretical interpretation of the observed spectrum. In a previous experiment (4) an intense excitation from the ground state to the sixth excited state has been observed, while other ground state transitions appeared as unresolved shoulders. Additional measurements have been performed with high energy resolution. At 1.5 K five spin-wave branches have been resolved. At 4.2 K the observed spectra demonstrate the presence of excited-state spin waves. All spin-wave branches turn out to be almost independent of wave vector.

To obtain a tractable model we based our analysis of the observed spectra on a single-ion Hamiltonian containing the crystal-field and the effects of bilinear and quadrupolar pair interactions. We conclude from our results that the appearance of the flopside spin structure is due to the quadrupolar pair interaction.

EXPERIMENTAL

The neutron inelastic scattering experiments were performed on a triple-axis spectrometer at the reactor Diorit, Würenlingen. The measurements have been carried out in the constant-$\vec{Q}$ mode of operation with neutron energy loss. The analyser energy was kept fixed at 5 meV. A cooled beryllium filter has been used to avoid higher-order contamination. The measurements were made at 1.5 K and 4.2 K for wave vectors along the three principal symmetry directions around the (111) and $(\frac{3}{2}\ \frac{3}{2}\ \frac{3}{2})$ reciprocal-lattice points. The sample was a cylindrical single crystal of 0.5 cm^3 volume. For details of the single-crystal growth see Ref. (4) and the references cited therein.

Examples of the distributions of the scattered neutrons are shown in Figure 1. At 1.5 K there are four resolved lines at 0.8 meV, 1.3 meV, 2.0 meV, and 2.5 meV and one intense peak at 10.4 meV with an unresolved shoulder at the high energy side. As the temperature is raised to 4.2 K the line at 10.4 meV remains, while the peaks in the low energy part are considerably shifted. Measurements have been performed for different wave vectors, but the observed energy spectra

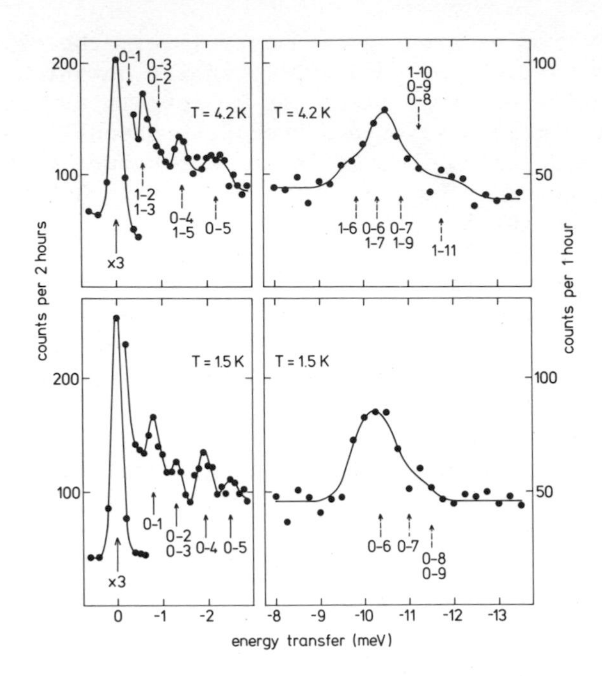

Figure 1: Energy distributions of neutrons scattered from HoP at the centers of the (111) and $(\frac{3}{2}\ \frac{3}{2}\ \frac{3}{2})$ zones (low and high energy transfer, respectively). The solid arrows denote the transition energies to which the model parameters have been fitted. The broken arrows point to the transition energies calculated from the model. The identification of the transitions is given by the numbers connected with each arrow (0: ground state, 1: first-excited state, etc.). The solid lines are drawn as a guide to the eye.

turned out to be identical to those shown in Figure 1 within the statistical spread of the data.

THEORY

We describe the magnetic behaviour of HoP by a Hamiltonian which contains crystal-field, bilinear and quadrupolar pair interactions:

$$\mathcal{H} = \mathcal{H}_{cf} - \sum_{i>j} \left(J_{ij}\, \vec{S}_i \cdot \vec{S}_j + K_{ij}\, (\tfrac{1}{3} Q_i Q_j + P_i P_j) \right) \quad , \tag{1}$$

where $Q_i = 3S_z^2(i) - S(S+1)$ and $P_i = S_x^2(i) - S_y^2(i)$.

To obtain a tractable model of the magnetic behaviour of HoP, we write Eq. (1) in the mean field approximation:

$$\mathcal{H}_{A,B} = \mathcal{H}_{cf} - (\lambda_1 \langle\vec{S}\rangle_{A,B} + \lambda_2 \langle\vec{S}\rangle_{B,A}) \cdot \vec{S} - \frac{1}{3}(\lambda_3 \langle Q\rangle_{A,B} + \lambda_4 \langle Q\rangle_{B,A}) Q$$
$$- (\lambda_3 \langle P\rangle_{A,B} + \lambda_4 \langle P\rangle_{B,A}) P \quad . \tag{2}$$

Here, the indices A,B denote the two sublattices of the Ho ions in HoP, and $\lambda_1(\lambda_3)$ and $\lambda_2(\lambda_4)$ are the bilinear (quadrupolar) intra- and inter-sublattice coupling parameters, respectively. The Hamiltonian (2) comprises four parameters λ_i and ten expectation values which have to be determined self-consistently. For actual calculations, however, Eq. (2) needs some simplifications. First we set the spin components along the y-axis equal to zero: $\langle S_y\rangle_A = \langle S_y\rangle_B = 0$. In a second specialization we require the two sublattices to be equivalent. This leads to the following relations (6):

$$\begin{aligned} \langle S_x\rangle_B &= \langle S_z\rangle_A \quad , \\ \langle S_z\rangle_B &= \langle S_x\rangle_A \quad , \\ \langle Q\rangle_B &= \tfrac{3}{2}\langle P\rangle_A - \tfrac{1}{2}\langle Q\rangle_A \quad , \\ \langle P\rangle_B &= \tfrac{1}{2}\left(\langle Q\rangle_A + \langle P\rangle_A\right) \quad . \end{aligned} \tag{3}$$

Combining Eqs. (2) and (3) yields

$$\mathcal{H}_A = \mathcal{H}_{cf} - (\lambda_1 \langle S_x\rangle_A + \lambda_2 \langle S_z\rangle_A) S_x - (\lambda_1 \langle S_z\rangle_A + \lambda_2 \langle S_x\rangle_A) S_z$$
$$- \left(\tfrac{1}{3}(\lambda_3 - \tfrac{1}{2}\lambda_4)\langle Q\rangle_A + \tfrac{1}{2}\lambda_4 \langle P\rangle_A\right) Q \tag{4}$$
$$- \left(\tfrac{1}{2}\lambda_4 \langle Q\rangle_A + (\tfrac{1}{2}\lambda_4 + \lambda_3)\langle P\rangle_A\right) P \quad ,$$

and a similar expression for $\mathcal{H}_B$. The problem is now simplified to solve the one-sublattice Hamiltonian (4) with four parameters λ_i and four expectation values. Finally, if we make a third specialization that the local distortions in the crystal are such as to leave it cubic on the average, we obtain (6)

$$\mathcal{H}_A = \mathcal{H}_{cf} - (\lambda_1 \langle S_x\rangle_A + \lambda_2 \langle S_z\rangle_A) S_x - (\lambda_1 \langle S_z\rangle_A + \lambda_2 \langle S_x\rangle_A) S_z$$
$$- \tfrac{1}{3}(\lambda_3 - \lambda_4)\langle Q\rangle_A (Q - P) \quad . \tag{5}$$

Thus we have three parameters $(\lambda_1, \lambda_2, \lambda_3 - \lambda_4)$ and three expectation values to determine self-consistently.

ANALYSIS

The observed energy spectra have been analysed on the basis of Eq. (5). From the crystal-field parameters of HoP (3,4) we conclude that the ground state is nearly six-fold degenerate, while the excited crystal-field states are at least 100 K above the ground state. Since the energies of the magnetic interactions are of the order of T_c=5.5 K, the dynamics of the spin system of HoP is essentially determined by the six-fold degenerate ground state, and consequently we use in our fitting procedure only the four low-energy transitions observed at 1.5 K (see Fig. 1). The starting values of the coupling parameters have been estimated on the basis of the "cubic model" developed by Kim et al. (7). In fact, the cubic model in which the magnetic interactions are projected onto the six-fold degenerate ground-state manifold gives a very good description of the magnetic properties of HoP; in particular, the coupling parameters derived from the cubic model agree within 5 % with the parameters determined from the full Hamiltonian (5).

In a first step we neglect the quadrupolar pair interactions. The least-squares fit to the four low-energy transitions at 1.5 K yields $\lambda_1=(2.11\pm0.17)\times10^{-2}$ meV and $\lambda_2=(\pm0.74\pm0.06)\times10^{-2}$ meV with a fitting error $\chi^2=2.7$. These parameters alone, however, cannot stabilize the flopside spin structure. Trammell (8) has shown that the flopside spin structure is due to magnetic dipole forces. With our parameters, however, a ferromagnetic spin arrangement has the lowest energy for $\lambda_2>0$, while an antiferromangetic spin arrangement is favoured for $\lambda_2<0$. Thus we conclude that the flopside spin structure of HoP cannot be explained as being due to dipole forces. We will show that quadrupolar pair interactions are responsible for the flopside structure as pointed out by Levy and Chen (9).

The final least-squares fitting procedure was based upon the full Hamiltonian (5). We obtained

$$\lambda_1 = (1.74\pm0.14)\times10^{-2} \text{ meV} ,$$

$$\lambda_2 = (\pm0.91\pm0.08)\times10^{-2} \text{ meV} ,$$

$$\lambda_3-\lambda_4 = (1.2 \pm0.2)\times10^{-5} \text{ meV} ,$$

with a fitting error χ^2=1.2. The temperature dependence of the order parameters is shown in Figure 2. The transitions predicted by this model are indicated in Figure 1. The overall agreement with the observed energy spectra is rather good. Of particular interest are the transitions between the first excited state and higher excited states in the 4.2 K data. These excited-state spin-wave intensities result directly from the large transition matrix elements and the population factor.

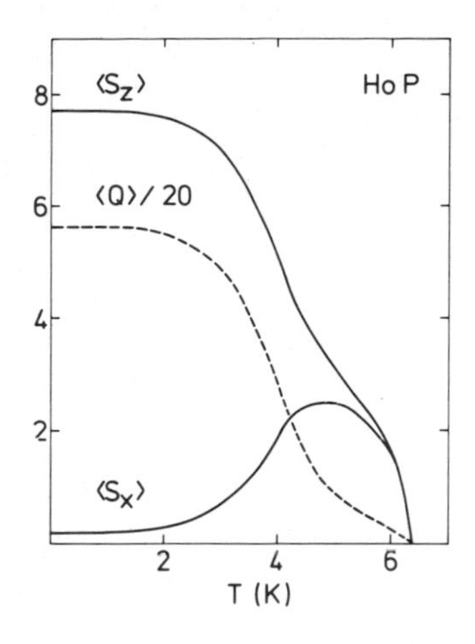

Figure 2:
Temperature dependence of the order parameters of Ho^{3+} in HoP.

DISCUSSION

The parameters determined from the model Hamiltonian (5) give a quantitatively correct description of the magnetic excitation spectrum of HoP. The model predicts a second-order phase transition at a temperature of 6.3 K which is in reasonable agreement with the observed T_c=5.5 K, and a low-temperature moment which is slightly higher than the observed value [1].

The model equation (5) whose results we presented in this paper does not give a detailed description of the quadrupole interaction. It is likely, however, that the important quadrupole coupling is the inter-sublattice coupling λ_4. Thus, with $(\lambda_3-\lambda_4)>0$ from our fit to the magnetic excitations we have $\lambda_4<0$, i.e. we obtain a mutually perpendicular array of quadrupoles [9]. Moreover, with $(\lambda_3-\lambda_4)>0$ the quadrupole interaction lowers the energy of the spin system and stabilizes the flopside spin structure of HoP.

Although our model gives a good description of the overall magnetic behaviour of HoP, there are still some important questions which have to be further investigated. We cannot decide from our spin-wave experiments whether the bilinear inter-sublattice coupling is ferromagnetic ($\lambda_2>0$) or antiferromagnetic ($\lambda_2<0$). If $\lambda_2>0$ ($\lambda_2<0$) the angle between the spin directions in adjacent (111) planes should be smaller (greater) than 90° for $0\leq T<T_c$. Stevens and Pytte [10] have suggested a distortion pattern to explain the flopside spin structure. They suppose that the phosphorus ions are alternately displaced along the +x and -x directions, so that half the Ho ions are in compressed octahedra of their neighbours and half in extended octahedra. In this model the two sublattices are not equivalent. A similar distortion has recently been found in UO_2 [11].

Neutron diffraction experiments on single crystals would be highly desirable to further elucidate these questions as well as to check the temperature dependence of the order parameters and the nature of the phase transition.

REFERENCES

1. H.R. Child, M.K. Wilkinson, J.W. Cable, W.C. Koehler, and E.O. Wollan, Phys. Rev. 131, 922 (1963)
2. F. Lévy, Phys. Kondens. Mater. 10, 85 (1969)
3. R.J. Birgeneau, E. Bucher, J.P. Maita, L. Passell, and K.C. Turberfield, Phys. Rev. B8, 5345 (1973)
4. A. Furrer and E. Kaldis, Magnetism and Magnetic Materials - 1975, AIP Conf. Proc. Vol. 29 (1976), pp. 264
5. K.R. Lea, M.J.M. Leask, and W.P. Wolf, J. Phys. Chem. Solids 23, 1381 (1962)
6. A. Furrer and P.M. Levy, to be published
7. D. Kim, P.M. Levy, and L.F. Uffer, Phys. Rev. B12, 989 (1975)
8. G.T. Trammell, Phys. Rev. 131, 932 (1963)
9. P.M. Levy and H.H. Chen. Phys. Rev. Lett. 27, 1385 (1971)
10. K.W.H. Stevens and E. Pytte, Solid State Commun. 13, 101 (1973)
11. J. Faber, Jr., G.H. Lander, and B.R. Cooper, Phys. Rev. Lett. 35, 1770 (1975)

NEUTRON CRYSTAL-FIELD SPECTROSCOPY IN $PrSn_3$

A. Furrer*, A. Murasik*+, and Z. Kletowski°

* Institut für Reaktortechnik ETHZ, EIR,
CH-5303 Würenlingen, Switzerland
\+ Permanent Address: Institute of Nuclear Research,
Swierk Research Establishment, Otwock, Poland
° Institute of Low Temperature and Structure Research,
Polish Academy of Science, Wrocław, Poland

ABSTRACT

Neutron inelastic scattering experiments on polycrystalline $PrSn_3$ have been performed. A least-squares fitting procedure including exchange fluctuations yields the crystal-field parameters $x=0.79\pm0.03$ and $W=0.157\pm0.005$ meV, which correspond to a Γ_5 ground state with an overall splitting of 74 K, and the paramagnetic molecular field parameter $\lambda_p=(1.8\pm0.3)\times10^{23}$ Oe^2/erg. The crystal-field parameters are consistent with susceptibility and zero-field magnetization data. The anomalous temperature dependence of the crystal-field transition line widths is interpreted in terms of spin-electron relaxation effects.

INTRODUCTION

The intermetallic compounds $La_{1-c}Pr_cSn_3$ (where $0\leq c\leq1$) have been extensively studied in the last few years. They crystallize in the cubic $AuCu_3$ structure. Below ∿8.5 K $PrSn_3$ is a type A antiferromagnet (1-4). The particular interest, however, arises from the superconducting properties for low Pr concentration (5). Moreover, resistivity measurements (6-7) suggest that these compounds are Kondo systems. For a quantitative understanding of all these effects information on the crystal-field levels is indispensible.

In the present work we report on the crystalline electric field levels of $PrSn_3$ determined by neutron spectroscopy. Exchange fluctuations have been taken into account in the data analysis. The unusual temperature dependence of the crystal-field transition line widths is qualitatively discussed and related to the magnetic resistivity.

EXPERIMENT AND ANALYSIS

The inelastic neutron scattering experiments were carried out on a triple-axis spectrometer at the reactor Saphir at Würenlingen, Switzerland. The spectrometer was operated in the neutron energy gain configuration and in the constant-Q mode. The incident neutrons filtered by pyrolythic graphite had a constant energy of 14.96 meV. The measurements have been performed at various temperatures, T, between 78 K and 293 K and scattering vectors between 2.1 $Å^{-1}$ and 2.9 $Å^{-1}$. By measuring at different scattering vectors, $\vec{Q}$, undesired phonon scattering can be identified. In particular, with increasing scattering vector $\vec{Q}$ the phonon intensity increases with Q^2, whereas the crystal-field intensity decreases according to $f^2(Q)$ where $f(Q)$ is the form factor. Details of the sample preparation are given in Ref. (2).

Observed energy spectra for Q=2.1 $Å^{-1}$ are shown in Fig. 1. The spectra are characterized by a large quasielastic scattering contribution and a broad peak centered at around 6.5 meV. By studying the Q-dependence these scattering contributions can be identified as crystal-field transitions. In particular, when Q is increased from 2.1 $Å^{-1}$ to 2.9 $Å^{-1}$ the peak at 6.5 meV decreases by 14 % in agreement with the reduction predicted by the form factor. Therefore, we conclude that the phonon contributions in $PrSn_3$ are weak and do not give rise to a peaked structure, which is in agreement with the phonon picture observed for the isostructural compound $PrPd_3$ (8).

Following the work of Lea et al. (9), the crystal field of cubic symmetry splits the ground state multiplet $^3H_4$ into a singlet Γ_1, a doublet Γ_3 and two triplets Γ_4 and Γ_5. Since the spectra do not exhibit well resolved crystal-field transition lines we applied a least-squares fitting procedure in which both the energies and the intensities of the crystal-field transitions are simultaneously adjusted to the observed energy spectra. We have shown in a previous publication (10) that exchange fluctuations considerably change the transition probabilities of Pr^{3+} in $PrSn_3$ with respect to those calculated in the limit of non-interacting Pr ions. Therefore, in order to base our fitting procedure on a correct intensity input,

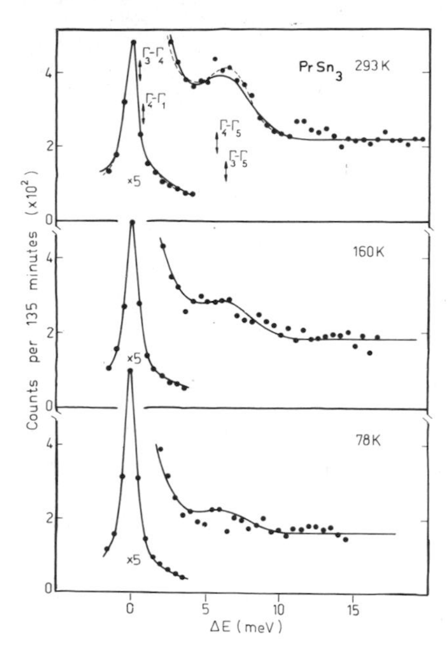

Figure 1: Energy spectra of neutrons scattered from polycrystalline $PrSn_3$ for Q=2.1 $Å^{-1}$. The full lines are the results of the least-squares fits for non-interacting Pr ions. The broken line takes into account exchange fluctuations.

exchange fluctuations have been taken into account in the dynamic effective field approximation (11). The background has been assumed to be constant, and both the elastic peak and the crystal-field transition peaks have been approximated by Gaussians. The resulting fit parameters are:

$$W = 0.157\pm0.005 \text{ meV},$$
$$x = 0.79 \pm0.03,$$
$$\lambda_p = (1.8\pm0.3)\times10^{23} \text{ Oe}^2/\text{erg},$$
$$\gamma_{78\text{ K}} = 3.6\pm0.4 \text{ meV},$$
$$\gamma_{160\text{ K}} = 3.1\pm0.4 \text{ meV},$$
$$\gamma_{293\text{ K}} = 1.3\pm0.3 \text{ meV},$$

where W and x are the crystal-field parameters in the notation of Lea et al. (9) and λ_p is the paramagnetic molecular field parameter. The widths of the crystal-field transition lines, γ, have been obtained from the fitted Gaussians by taking account of the instrumental resolution. The crystal-field parameters of $PrSn_3$ correspond to a Γ_5 ground state separated by 60-74 K from the excited states Γ_1, Γ_3 and Γ_4 in agreement with the energetic scheme of the crystal-field levels determined for the isostructural system $PrPd_3$ (8).

DISCUSSION

Crystal-field effects can significantly influence the magnetic properties. This may serve as an experimental check for the crystal-field parameters obtained from our least-squares fitting procedure. The calculated susceptibility is in good agreement with the experimental data obtained by Czopnik et al. (2). A more sensitive check is possible from magnetization measurements. Figure 2 shows the results of a neutron diffraction experiment performed by Bytonski et al. (4) on the same sample which has been used in the present study. The magnetic moment at 4.2 K has been determined to be 2.08±0.20 μ_B. Using the molecular field parameter λ_m=44.5×10^{23}Oe^2/erg which predicts the correct Néel temperature the calculated moment at 4.2 K is 2.01 μ_B. This value as well as the calculated temperature dependence of the magnetic peak intensity (see Fig. 2) are in good agreement with the experimental data.

The present experiments clearly demonstrate that the crystal-field level scheme of $PrSn_3$ markedly departs from that determined for the diluted system $La_{1-c}Pr_cSn_3$ ($c \leq 0.1$) from susceptibility and specific heat data (5) (Γ_1 ground state, overall splitting 25 K).

Of particular interest is the anomalous temperature-dependence of the line widths γ, which decrease with increasing temperature. The energy widths of crystal-field levels are due to a number of effects, but relaxation effects are generally the most important broadening mechanisms. The line widths γ obtained from our fit correspond to the total intrinsic widths minus exchange broadening contributions which are implicitly taken into account in our analysis. Following Heer et al. (12) we estimate the line-width contri-

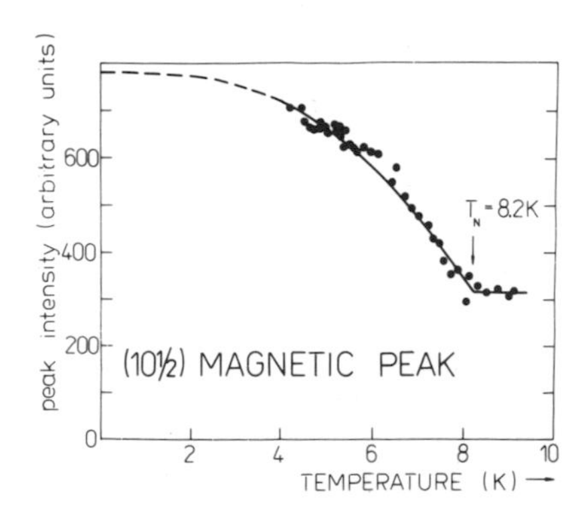

Figure 2: Magnetic peak intensity versus temperature. The solid line represents the calculated values scaled to the experimental data.

butions due to spin-lattice relaxation effects to be smaller than 0.5 meV at room temperature, i.e. they are small compared to the line widths obtained from our fit. We suggest that these line widths are the result of spin-electron relaxation effects. The spin-electron relaxation mechanism can be described by a characteristic relaxation time, τ, which is related to the line width by Heisenberg's uncertainty relation:

$$\gamma(T) \sim \frac{\hbar}{2\tau(T)} \quad . \qquad (1)$$

Following Korringa [13] we expect

$$\frac{1}{\tau(T)} \sim T \quad , \qquad (2)$$

which is in contradiction to our observations. This discrepancy is most likely due to the Kondo behaviour of $PrSn_3$. Indeed, the magnetic resistivity of $PrSn_3$ [6,7,14] cannot be explained by calculating the spin-disorder resistivity in the first Born approximation. Qualitative agreement, however, is obtained by extending the calculation to the second Born approximation which yields an additional relaxation time. We may have measured this relaxation time in the present experiments.

REFERENCES

[1] T. Tsuchida and W.E. Wallace, J. Chem. Phys. 43, 3811 (1965)

[2] A. Czopnik, Z. Kletowski, and B. Staliński, phys. stat. sol. (a) 3, K263 (1970)

[3] P. Lethuillier, J. Pierre, G. Fillion, and B. Barbara, phys. stat. sol. (a) 15, 613 (1973)

[4] R. Bytonski, A. Murasik, and H. Bąk-Ptasiewicz, Institute of Nuclear Research, Swierk, Poland (unpublished)

[5] R.W. McCallum, W.A. Fertig, C.A. Luengo, M.B. Maple, E. Bucher, J.P. Maita, A.R. Sweedler, L. Mattix, P. Fulde, and J. Keller, Phys. Rev. Lett. 34, 1620 (1975)

[6] A.I. Abou Aly, S. Bakanowski, N.F. Berk, J.E. Crow, and T. Mihalisin, Phys. Rev. Lett. 35, 1387 (1975)

[7] P. Lethuillier and P. Haen, Phys. Rev. Lett. 35, 1391 (1975)

[8] A. Furrer and H.-G. Purwins, J. Phys. C: Solid State Phys. 9, 1491 (1976)

(9) K.R. Lea, M.J.M. Leask, and W.P. Wolf, J. Phys. Chem. Solids 23, 1381 (1962)

(10) A. Murasik, A. Furrer, and Z. Kletowski, Solid State Commun. 19, 65 (1976)

(11) A. Furrer and H. Heer, Phys. Rev. Lett. 31, 1351 (1973)
A. Furrer, Solid State Commun. 16, 839 (1975)

(12) H. Heer, A. Furrer, W. Hälg, and O. Vogt, these Proceedings

(13) J. Korringa, Physica 16, 601 (1950)

(14) B. Stalinski, Z. Kletowski, and Z. Henkie, phys. stat. sol. (a) 19, K165 (1973)

CRYSTAL FIELDS IN PrX_3(X=In, Tl, Pb, Sn) INTERMETALLIC COMPOUNDS WITH Cu_3Au STRUCTURE °

W. Groß[+], K. Knorr[++], A. P. Murani[*], K. H. J. Buschow[**]

[+] Physikal. Inst. d. Universität Frankfurt
[++] Physikal. Inst. d. Universität Mainz
[*] Institute Max von Laue-Paul Langevin, Grenoble (ILL)
[**] Philips' Research Laboratories, Eindhoven

ABSTRACT

We have determined the crystal field splittings of Pr^{3+} in $PrIn_3$, $PrTl_3$, $PrPb_3$, and $PrSn_3$ by inelastic neutron scattering. The values of the deduced crystal field parameters are found to depend strongly on the constituent X, though all systems have nonmagnetic ground-states. The line widths in the different systems are compared.

The systems investigated in the present work crystallize in the Cu_3Au structure as do the corresponding dilute systems $(La, Pr)X_3$. A considerable amount of work has been done on these systems concerning the specific heat, magnetic and superconducting properties[1, 2, 3, 4, 5], which are all influenced by the crystal-field (CF) splitting of the 4 f states of the magnetic ions and therefore yield some information about the CF parameters. It was thought to be of considerable interest to determine the CF splittings more directly by inelastic neutron scattering.

The samples were prepared by enclosing stoichiometric amounts of the components between specially shaped Ta-sheets inside a sealed iron mold which was heated to between 1150 and 1250°C depending on the compound. Thus evaporation of the high vapour pressure components was avoided. The $PrSn_3$ sample was melted in an argon arc furnace and homogenized in a levitation furnace. The crystallographic phases were verified by neutron diffraction. The neutron scattering experiments were carried out on the time-of-flight spectrometers IN7, IN4, and IN5 at the ILL Grenoble. The results for $PrIn_3$ are given in fig. 1 together with some relevant experimental parameters. Because of lack of space the observed spectra for the other compounds are not presented but the results are briefly described here. Thus for $PrTl_3$ at 5K one upward excitation could be seen at 3.3 meV. At elevated temperatures 25K and 50K additional transitions were observed at 2.42 and 1.19 meV. $PrPb_3$ exhibited two well-resolved peaks at low temperature (2K) occuring at 1.63 and 2.63 meV, which became less clearly separated, producing a wide structure between 1.6 and 2.4 meV at 20K. Finally in $PrSn_3$, apart from an excitation occuring at certain q-values at an energy of around 5, 5 to 6, 5 meV which will later be discussed in more detail, no resolved CF transition was observed above the Neel-temperature of 8.6 K. In the ordered phase, however, a magnetic excitation at 1.5 meV was found.

The cubic CF is known to split the ground multiplet of Pr^{3+} into 4 levels Γ_1, Γ_3, Γ_4, Γ_5 whose arrangement can be determined by two CF-parameters x and W or equivalently B_4 and B_6 [6]. Magnetic dipol transitions occur between Γ_1- Γ_4, Γ_3- Γ_4, Γ_3- Γ_5, Γ_4- Γ_5 with transition probabilities as given by Birgeneau[7]. The extraction of the CF-level schemes shown in fig. 2 from the results described in the previous section was straightforward and was obtained by fitting the data to the formula of neutron scattering cross-section with x, W, and the line width $\Delta E/E$ of CF-transitions as the most important parameters. These parameters together with the effective point charges (see below) are listed in the table. Only $PrSn_3$ needs further comment. As mentioned above, no resolved CF transition was found in the paramagnetic phase whereas a clear magnetic excitation at 1.5 meV was present at lower temperatures. Therefore we tried to explain this result with the two alternative CF splittings derived from specific heat measurements on (La, Pr) Sn_3 by Bucher et al.[2] and a molecular field in the [100] direction consistent with an ordered moment of $2.65 \pm 0.3\ \mu_B$ found by Lethuillier et al.[8] A molecular field calculation shows that the present result is only compatible with Bucher's second scheme, reproduced in fig. 2, and a mean field of about 80 kG.

In the following we compare our results with the work of other authors. From the measurements of the susceptibilities and specific heats of $PrIn_3$ [3,5], $PrTl_3$ [1], and $PrPb_3$ [1,4,9] the respective authors concluded that the Pr^{3+} ground-states in these three systems

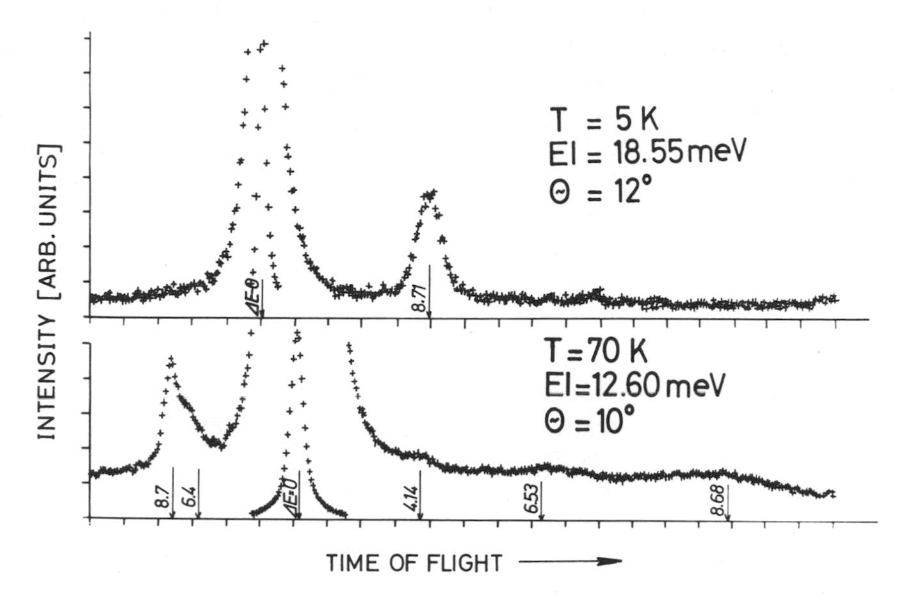

Fig. 1: Time-of-flight spectra measured on $PrIn_3$, T giving the temperature, E_i the neutron incident energy and Θ the total scattering angle. The numbers on the arrows are the neutron energy transfers in meV.

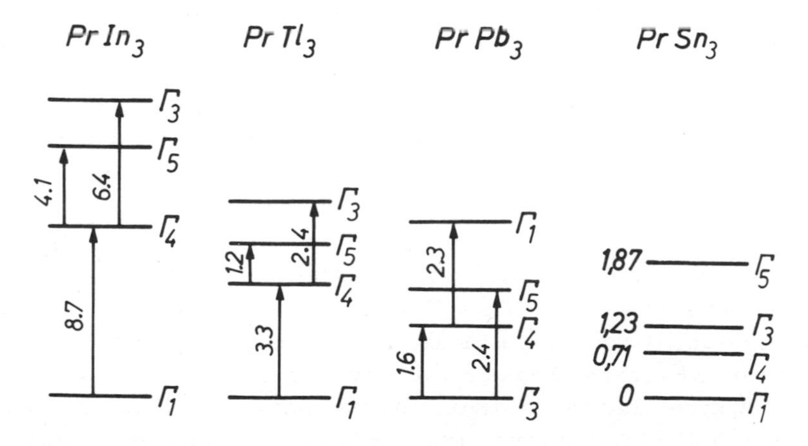

Fig. 2: Energy level schemes of the four systems. The arrows indicate the transitions actually observed. The numbers are the energy transfers in meV. The scheme for $PrSn_3$ is taken from Bucher et al.[2]

Table: CF parameters x and W and line widths $\Delta E/E$ extracted from the spectra. q_1 and q_2 are the effective point charges of nearest and second nearest neighbors respectively.

System	x	W [meV]	$\Delta E/E$	q_1 [|e|]	q_2 [|e|]
$PrIn_3$	-0.66	0.229	0.08	6.0	11.6
$PrTl_3$	-0.62	0.081	0.10	3.0	6.4
$PrPb_3$	0.09	-0.0303	0.42	-2.9	-8.3
$PrSn_3$	-0.95	0.044	-	0.1	-1.2

are Γ_1 in $PrIn_3$ and $PrTl_3$ and Γ_3 in $PrPb_3$. The first excited state was assumed to be Γ_4 for all three systems at an energy of 100K, 37K, and 18K above the ground-state, respectively. These findings are in excellent agreement with the splittings deduced from the present neutron data although the actual values of the CF-parameters x and W are different from theirs. This is perhaps not surprising since our CF-parameters were determined by at least two transitions for each system whereas those derived [10] from the magnetic and specific heat properties cited above are mainly sensitive to the low lying levels. The dependence of the susceptibility on the Pr-concentration in (La, Pr) Pb_3 and its lack of concentration dependence in (La, Pr) In_3 and (La, Pr) Tl_3 was interpreted as indicating that $PrIn_3$ and $PrTl_3$ are pure CF-Van Vleck paramagnets whereas in $PrPb_3$ exchange effects play an important role. This finding is supported by the line widths of the neutron spectra listed in the table, which increase with decreasing separation of the first magnetic excited state (Γ_4) from the nonmagnetic ground-state. Also the deviation of depression of the superconducting transition temperature in $(La, Pr)X_3$ (X=Tl, Pb) from an Abrikosov-Gorkov behaviour is in line with the ground-states being nonmagnetic. The situation is less clear for $PrSn_3$. In the paramagnetic phase only quasielastic magnetic scattering of a width of 1.0 meV was observed. This line width is probably caused partially by the CF-splitting and partially by exchange effects. A similar absence of CF-transitions in the paramagnetic phase was found in TbSb [11] which also is an overcritical induced antiferromagnet. A peak at about 6 meV energy transfer in neutron energy gain, which lead Murasik et al. [15] to suggest a CF-splitting with a Γ_1 ground-state and the Γ_3, Γ_4, Γ_5 states grouped together at about 6 meV, could be reproduced but was identified as a phonon effect by its dependence on temperature and q-transfer. As quoted above our result for the ordered phase is compatible with Bucher's second scheme for (La, Pr)Sn_3 but not with the first one, which was however used to explain some superconducting properties of this system in terms of the theory of Keller et al.[12]. From the complex resistivity behaviour of $PrSn_3$ [13] one might conclude strong s-f-scattering in this compound. Clearly more investigations are needed here. Finally we calculated the effective charges of the first (q_1) and second (q_2) neighbors as given in the table from B_4 and B_6 and nonrelativistic $\langle r^4 \rangle$ and $\langle r^6 \rangle$. One notices a significant change in the value of the charges and even of their polarity within the series but except for Sn the signs of the charges are identical to those given by Bucher et al.[14]

REFERENCES

1. E.Bucher, K.Andres, J.P.Maita, and G.W.Hull Jr., Helv.Phys.Acta 41, 723 (1968)
2. E.Bucher, J.P.Maita, G.W.Hull Jr., J.Sierro, C.W.Chu, and B.Lüthi in Devine RAB(Ed.)Proc.1st Conf.on Crystalline Electric Field Effects in Metals and Alloys (Montreal 1974)p.221
3. K.H.J.Buschow, H.W.de Wijn, and A.M. van Diepen, J.Chem.Phys. 50, 137 (1969)
4. P.Lethuillier, G.Quezel-Ambrunaz, A.Percheron, Solid State Commun. 12, 105 (1973)
5. A.M.van Diepen, R.S.Craig, W.E. Wallace, J.Phys. Chem.Solids 32, 1867 (1971)
6. K.R.Lea, M.J.M.Leask, and W.P.Wolf, J.Phys.Chem. Solids, 23, 1381 (1962)
7. A.J.Birgeneau, J.Phys.Chem.Solids 33, 59 (1972)
8. P.Lethuillier, J.Pierre, G.Fillion, and B.Barbara, phys.stat.sol. (a) 15, 613 (1973)
9. E.Bucher, K.Andres, A.C.Gossard, and J.P.Maita, Proc. 13th Int.Natl.Low Temp.Conf., Boulder (1972)
10. P.Lethuillier and J.Chaussy, J.dePhys.(France)37, 123(1976)
11. T.M.Holden, E.C.Svensson, W.J.L.Buyers, and O.Vogt, Phys.Rev. B 10, 3864 (1974)
12. J.Keller and P.Fulde, J.of Low Temp.Phys. 4, 289 (1971)
13. A.I.Abou Aly, S.Bakanowski, N.F.Berk, J.E.Crow, and T.Mihalisin, Phys.Rev.Lett. 35, 1387 (1975)
14. E.Bucher and J.P.Maita, Solid State Commun.13, 215 (1973)
15. A.Murasik, A.Furrer, Z.Kletowski, Solid State Commun. 19, 65 (1976)

o This work was supported by the Sonderforschungsbereich 65 "Festkörperspektroskopie" Darmstadt/Frankfurt, financed from special fonds of the Deutsche Forschungsgemeinschaft.

CRYSTAL FIELDS IN Pr-HYDRIDES

K. Knorr[+] and B.E.F. Fender[++]

[+]Institut für Physik, Universität Mainz, Germany

[++]Inorganic Chemistry Lab., University of Oxford, England

ABSTRACT

Inelastic neutron scattering was used to determine the crystal field splittings of PrD_2 and $PrD_{2.5}$. For the PrD_2 crystal field at the Pr-site is cubic and can be explained by a negative charge of the hydrogen. The splitting observed for $PrD_{2.5}$ can be described assuming a well defined short range order caused by a mer-configuration of the half filled octahedral hydrogen sites leading to an orthorhombic crystal field. The structure and the paramagnetic susceptibility of the Pr-hydrides are discussed on this basis.

The rare earth metals ideally form two stoichiometric hydrides, the dihydrides with CaF_2-structure where only the tetrahedral sites are occupied by hydrogen and the trihydrides with BiF_3-structure where the octahedral sites are filled, too. For intermediate H-concentrations ($2 < y < 3$ in RE H_y) the octahedral sites are partially occupied [1] (fig 1). The electric conductivity usually decreases with increasing y. The change is most spectacular for CeH_y where a metal-semiconductor transition was observed around $y = 2.75$[2]. This effect was explained by an anionic state of the H-ions. The magnetic susceptibility of PrH_y showed unusual changes when y is varied[3]. The results on PrH_2 were interpreted by crystal field effects whereas $PrH_{2.5}$ was

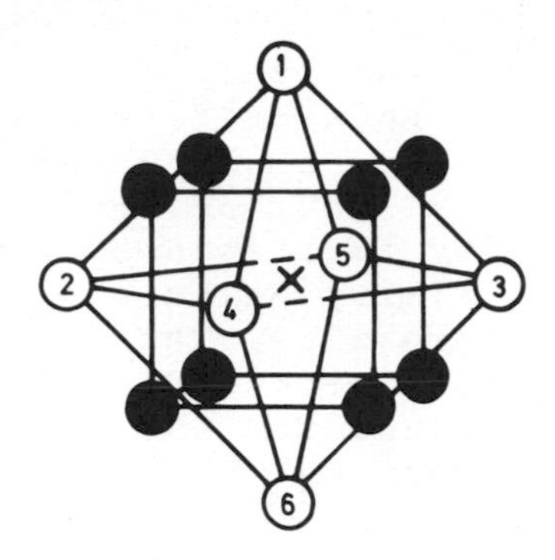

Fig. 1: The first hydrogen neighbours in tetrahedral sites (●) and the second hydrogen neighbours in octahedral sites (○) of Pr (✖) in PrH_y ($2 < y < 3$).

thought to show a change of the valence of the Pr-ions from +3 to +4 at about 30 K.

We did neutron crystal field spectroscopy measurements on PrD_2 and $PrD_{2.5}$ in order to get information on the magnetic behaviour, the arrangement of the hydrogen in the octahedral sites for y = 2.5 and the charge of the H-ions. A more detailed description of our work will be given elsewhere.

The inelastic neutron scattering experiments were performed on the time-of-flight multidetector spectrometer IN4 of the Institut Laue-Langevin at Grenoble. The data were taken on powdered samples at temperatures from 10 to 300 K. For PrD_2 a single crystal field transition was observed at an energy transfer of 21meV, for $PrD_{2.5}$ five crystal field transitions could be detected at 12, 16.5, 24.5, 26.5, and 29 meV (fig 2). These findings could be seen over the whole temperature range though less clearly at higher temperatures because of a higher background which, at least for $PrD_{2.5}$, was probably due to additional weak transitions from excited crystal field levels. Signals of optical phonons were observed in both samples but will not be discussed here.

For PrD_2 the crystal field at the Pr^{3+}-site is cubic and mainly generated by the 8 nearest hydrogen neighbours in tetrahedral sites. The corresponding hamiltonian H^{tet} and the point charge expressions for the parameters B_4^{tet} and B_6^{tet} can be found in ref. 4. The experimental result demands B_4^{tet} = 0.0082 meV, B_6^{tet} = 0.000069 meV, i.e. a $_5$ ground state and Γ_1, Γ_3, Γ_4 states around 21 meV, or the same splitting reversed. The ambiguity is lifted by a comparison of the magnetic susceptibility calculated for either case with the experimental result[3]. The first solution (Γ_5 lowest) gives a much better fit. The point charge model then leads to values for the D-ions of -0.64 from B_4^{tet} and -0.80 from B_6^{tet}.

$PrD_{2.5}$ needs some more considerations. The mere fact that the splitting in the neutron spectrum stayed constant with temperature rules out a change of the valence state of the Pr-ion. We therefore tried to explain the results on $PrD_{2.5}$ by a crystal field hamiltonian acting on the J = 4 ground multiplet of Pr^{3+} which takes the general form $H = H^{tet} + \sum_i w_i H_i^{oct}$ where H_i^{oct} describes the crystal field generated

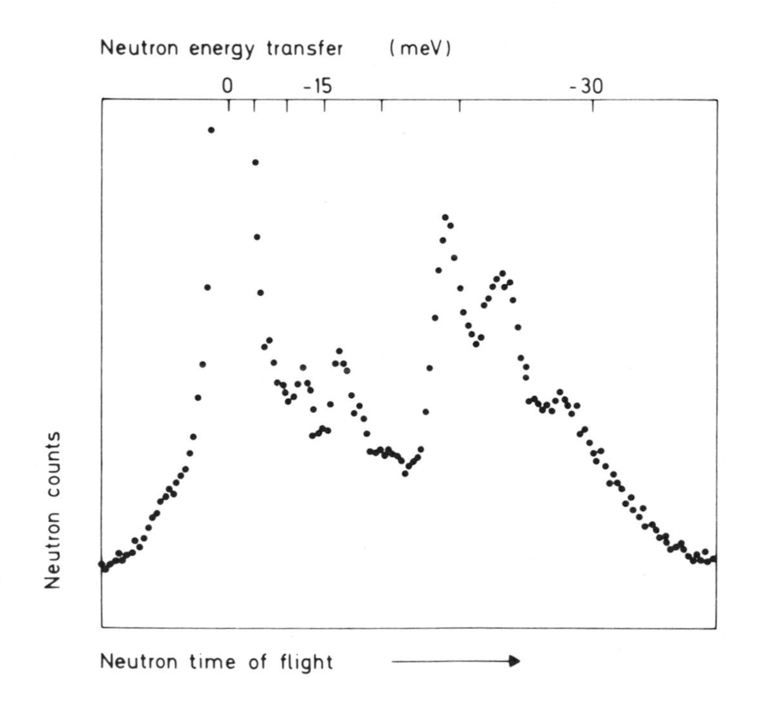

Fig. 2: The neutron time of flight spectrum for $PrD_{2.5}$ at 50 K.

by the different configurations i with the statistical weights w_i of the second nearest neighbours in the octahedral sites (fig 1). The operators H_i^{oct} are generally non-cubic and depend on three crystal field parameters of 2nd, 4th and 6th order. In order to be able to fit the five transitions observed with this model we had to make a number of assumptions: H^{tet} is unchanged with respect to PrD_2. All five transitions are transitions from the ground state since they were already present at low temperatures. The magnetic susceptibility [3] below 80 K is basically Van-Vleck-like i.e. the ground state is nonmagnetic and well separated from the excited states. (The deviation from the Van-Vleck-behaviour at very low temperatures would be then most likely due to a configuration with small statistical weight which leads to a magnetic ground state and whose crystal field transitions were however too weak to be seen in the neutron experiment.)
One finds that the results can be fitted by the crystal field of a single configuration of the octahedral sites, namely the configuration mer-XA_3 where the sites 1,2 and 3 are occupied (fig 1). The hamiltonian generated by this arrangement is orthorhombic and of the form

$$H^{oct}(\text{mer-}XA_3) = B_2\, 3\hat{O}_2^2 + B_4\left[\tfrac{7}{4}\hat{O}_4^0 - 5\hat{O}_4^2 + \tfrac{35}{4}\hat{O}_4^4\right] + B_6\left[\tfrac{6}{16}\hat{O}_6^0 + \tfrac{105}{16}\hat{O}_6^2 - \tfrac{63}{8}\hat{O}_6^4 + \tfrac{231}{16}\hat{O}_6^6\right]$$

with B_2 = -0.1 meV, B_4 = -0.00675 meV, B_6 = 0.0003 meV. A point charge calculation with charges of -0.7 taken from PrD_2 yields the values -2, -0.0016, 0.0002 respectively. The signs agree with the figures from the fit, again supporting the concept of anionic hydrogen. As usual the 2nd order term is grossly overestimated by the point charge model.

On these grounds we propose to regard PrH_y ($2<y<3$) in the following, certainly somewhat idealized way: Besides the dihydride and the trihydride there exists another stable phase with the formula $PrH_{2.5}$ which is characterized by a well defined at least short range order of the octahedral sites, namely the mer-XA_3 configuration. Hydrides with $2<y<2.5$ are mixtures of PrH_2 and $PrH_{2.5}$. This model would explain the variation of the magnetic susceptibility with y.

REFERENCES

1. C.G. Titcomb, A.K. Cheetham, B.E.F. Fender, J. Phys. C 7, 2409 (1974)

2. G.G. Libowitz, J.G. Pack, W.P. Binnie, Phys. Rev. B6, 4540 (1972)

3. W.E. Wallace, K.H. Mader, J. Chem.Phys. 48, 84 (1968)

4. K.R. Lea, M.J.M. Leask, W.P. Wolf, J.Phys.Chem. Solids 23, 1381 (1962)

NEUTRON SPECTROSCOPY OF THE CRYSTAL FIELD SPLITTING IN CERIUM HYDRIDE

P. Vorderwisch*, S. Hautecler+, J.B. Suck° and H. Dachs*

* Hahn-Meitner-Institut für Kernforschung Berlin GmbH, D-1000 Berlin 39, Germany
+ C.E.N./S.C.K., B-2400 Mol, Belgium
° Institut Laue-Langevin, F-38042 Grenoble, France

ABSTRACT

The crystal field splitting of the ground state $^2F_{5/2}$ of Ce^{3+} in fcc cerium hydride has been investigated by inelastic neutron scattering using deuterated samples CeD_x with concentrations between x = 1.94 and x = 2.75. For all samples investigated three crystal field transitions are observed which cannot be explained without taking into account a non-cubic surrounding of part of the Ce ions.

INTRODUCTION

Cerium forms a hydride phase CeH_x with a wide range of non-stoichiometry between $x \sim 1.8$ and $x = 3$. According to recent neutron diffraction measurements $CeD_{2.00}$ should have the ideal CaF_2 structure[1], i.e., that hydrogens fill all the tetrahedral interstitial sites of the fcc cerium lattice. For $x > 2.00$ it was found that a few percent of the tetrahedral sites are vacant, the excess hydrogens entering octahedral sites[1,2]. Long range order has been observed at room temperature[2] for compositions between $2.15 \leq x \leq 2.40$ and at x = 2.75 at low temperature[1]. With respect to the electronic properties, CeH_2 has a metallic character, whilst CeH_3 is an insulator; moreover, in $CeH_{2.75}$ a metal to semiconductor transition occurs at ~ 250 K[3]. Therefore we thought that a study of the crystalline electric field (CEF) in cerium hydride would be worthwhile for the understanding of these interesting features. Up to now two CEF investigations on this system have been performed on $CeH_{2.00}$ [4] and $CeH_{1.98}$ [5]. In both cases the interpretation was done assuming a crystal field of cubic symmetry, which would split the $^2F_{5/2}$ ground state of Ce^{3+} into a quartet Γ_8 and a doublet Γ_7. Specific heat data[4] gave a splitting energy of 13.4 meV and Γ_8 as the lower level. In a neutron inelastic scattering (INS) study[5] a peak at an energy transfer of 45 meV was assigned to the transition between both levels. We have done additional INS measurements on deuterated samples for $1.94 \leq x \leq 2.75$ to investigate the influence of the occupation number of interstitial sites on the CEF splitting.

MEASUREMENTS

Our INS measurements have been performed at the time-of-flight spectrometer IN4 of the Institut Laue-Langevin.

Fig. 1 gives an example of the results obtained in the upscattering experiment on $CeD_{2.01}$. Under these experimental conditions, comparable to those used in ref. 5, the spectra are dominated by phonon scattering. However, clearly no peak can be seen at an energy transfer Δ = 45 meV. On the other side, at Δ = 18.5 meV and around Δ = 6 meV maxima in the scattered intensity have the correct momentum-transfer behaviour and may therefore be assigned to a CEF transition. The peak at Δ = 54 meV must be attributed to localized vibrations of octahedral hydrogens[6], which indicates that in our sample some tetrahedral vacancies are present. This is in contradiction with the $CeD_{2.00}$ structure determination[1], but in agreement with the structure derived for $LaD_{2.02}$ [7].

Fig. 2 shows results obtained in downscattering on $CeD_{1.94}$ at 70 K. For comparison results on $LaD_{1.94}$ at the same temperature are added, which give an indication for the phonon contribution in the $CeD_{1.94}$ spectra. It shows up that in the spectrum of the lower scattering angle three peaks may uniquely be assigned to CEF transitions. These three peaks are also identified in the corresponding spectra, for the same sample composition at 4.6 K as well as for $CeD_{2.42}$ at 4.6 K. But in the $CeD_{2.74}$ spectra they are so broadened that their structure is nearly lost.

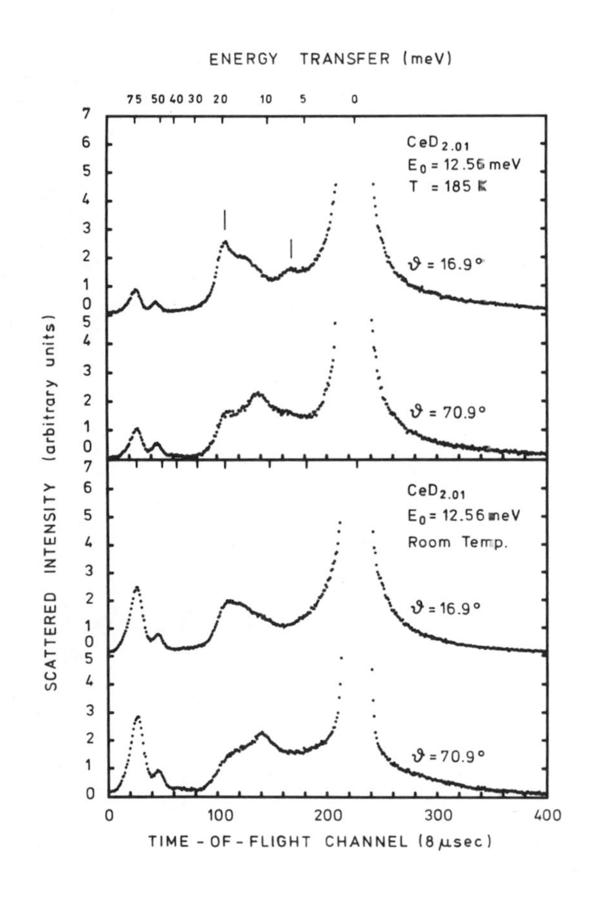

Fig. 1 Time-of-Flight Spectra for Neutron Upscattering by $CeD_{2.01}$

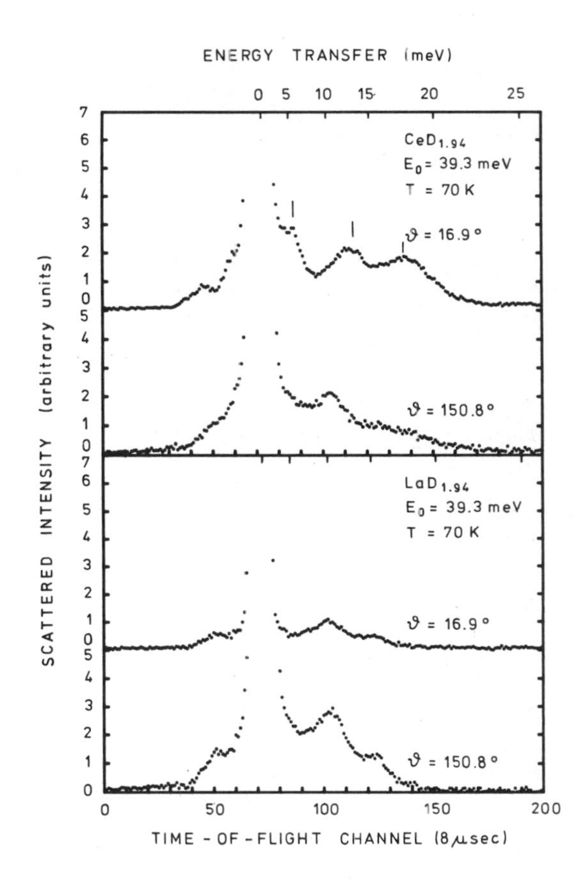

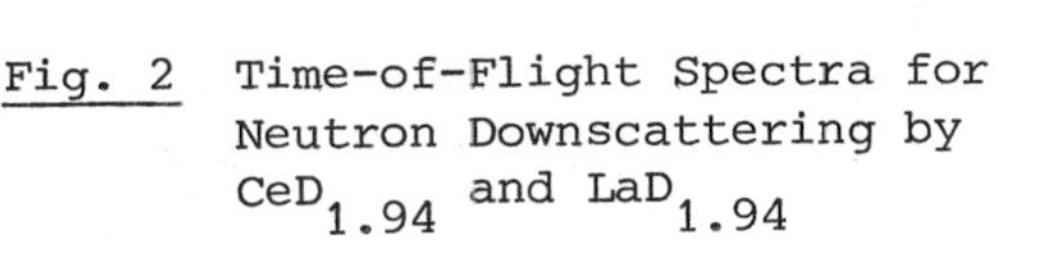

Fig. 2 Time-of-Flight Spectra for Neutron Downscattering by $CeD_{1.94}$ and $LaD_{1.94}$

ANALYSIS AND DISCUSSION

Our results for $CeD_{1.94}$ and $CeD_{2.01}$ can obviously not be explained within the assumption of a crystal field of cubic symmetry. In a non-cubic CEF the level scheme consists of three doublets, between which at low temperature in downscattering experiments only two transitions should occur. Assuming that 7% of the tetrahedral sites are vacant, as found for $LaD_{2.02}$ [7], a qualitative description of the $CeD_{1.94}$ results is possible: the configuration $CeD^{t}_{1.86}D^{o}_{0.08}$ leads to 44% Ce^{3+} ions surrounded by 8 D^{t}, while the remaining Ce^{3+} ions are in a non-cubic surrounding. This gives a superposition of two level schemes from which we can get 3 transitions under the experimental conditions. Furthermore, the influence of the octahedral hydrogens brings a broadening of the levels, which is in agreement with the results on the samples with higher concentrations. A detailed numerical analysis is at present under study.

ACKNOWLEDGEMENT

The authors wish to thank Prof. Wenzl, Mr. A. Carl and Mr. K.H. Klatt for their support in sample preparation at KFA Jülich, Germany.

REFERENCES

1. A.K. Cheetham and B.E.F. Fender, J. Phys. C: Solid St. Phys. 5, L30 (1972)

2. C.G. Titcomb, A.K. Cheetham and B.E.F. Fender, J. Phys. C: Solid St. Phys. 7, 2409 (1974)

3. G.G. Libowitz, J.G. Pack and W.P. Binnie, Proc. 8th Rare Earth Research Conf., Reno, Nevada, Vol. I, p. 310 (1970)

4. Z. Biegański, Bull. Acad. Polon. Sci., Sér. Sci. Chim. 19, 581 (1971)

5. D.G. Hunt and D.K. Ross, J. Less-Common Met. 45, 229 (1976)

6. P. Vorderwisch and S. Hautecler, phys. stat. sol. (b) 66, 595 (1974)

7. A.F. Wright, Chemistry Information Meeting, ILL Grenoble, p. 29 (1975)

EPR STUDIES OF EXCITED STATE EXCHANGE AND CRYSTAL-FIELD EFFECTS IN RARE EARTH COMPOUNDS

C. Y. Huang,* K. Sugawara,† and B. R. Cooper**

*Los Alamos Scientific Laboratory, Los Alamos, NM 87545
†Physics Department,+ Case Western Reserve University, Cleveland, OH 44106
**Department of Physics, West Virginia University, Morgantown, WV 26506 U.S.A.

ABSTRACT

EPR studies[1-4] in excited crystal-field states of Tm^{3+}, Pr^{3+}, and Tb^{3+} in singlet-ground-state systems[5] and[6] in the excited state of Ce^{3+} in CeP are reviewed. Because one is looking at a crystal-field excited state resonance, the exchange, even if isotropic, does not act as a secular perturbation. This means that one obtains different effects and has access to more information about the dynamic effects of exchange than in conventional paramagnetic resonance experiments. The Tm and Pr monopnictides studies are paramagnetic at all temperatures. The most striking feature[1,2] of the behavior of the $\Gamma_5^{(2)}$ EPR in the Tm compounds is the presence of an anomalous maximum in the temperature dependence of the g-factor. We discuss the relationship of this effect to anisotropic exchange. The results of the EPR of the excited $\Gamma_5^{(2)}$ level of Tb^{3+} (g-factor becomes very large at T_N) in antiferromagnetic TbX (X=P, As, Sb) and that of the excited Γ_8 level of Ce^{3+} in antiferromagnetic CeP will also be reported. For sufficient dilution of the Tb^{3+} in the terbium monopnictides, the systems become paramagnetic (Van Vleck paramagnets) down to 0°K. We have studied[4] the Tb^{3+} excited state resonance EPR in $Tb_{0.1}La_{0.9}P$ as an example of behavior in such systems.

*Work performed under the auspices of U.S.E.R.D.A.
†Present address: NASA Lewis Research Center, Cleveland, OH
+Supported in part by U.S. NSF Grant No. DMR 74-08033.

INTRODUCTION

We have observed[2] the resonance in the excited $\Gamma_5^{(2)}$ level of Tm^{3+} in TmP between 20°K and 300°K. The high temperature g-factor smoothly approaches the crystal-field-only value expected for a predominantly fourth-order crystal field. The shift from that crystal-field-only value has the 1/T dependence expected for an exchange shift proportional to the susceptibility, and the fact that the approach to the asymptotic value is from below indicates small antiferromagnetic exchange as expected. On the other hand, the anomalous peak at about 40°K cannot be explained on the basis of small isotropic exchange. Similar anomalous behavior has been observed by us[4] in TmAs and several years ago by Cooper et.al.[1] in TmN. We relate these experimental observations to crystal-field and exchange effects, particularly emphasizing the information obtained[4] about anisotropic exchange.

The EPR of the excited $\Gamma_5^{(2)}$ level of Tb^{3+} in antiferromagnetic TbX (X=P, As, Sb) has been studied.[3] The "g-shift" from the crystal-field-only g-factor diverges as T approaches T_N, and can be expressed as $A/(T-T_N)^n$ for TbP and TbAs with $n=1.3 \pm 0.3$.

To complement these experiments in singlet-ground-state Tb^{3+} antiferromagnets, we have also studied[6] the excited state EPR for the Kramers ion Ce^{3+} in antiferromagnetic CeP (T_N=9°K) from 77°K to 300°K. From the g-factor at high temperature and the temperature dependence of the EPR intensity, the resonance has been identified as originating from the excited Γ_8 quartet which lies about 170°K above the ground state Γ_7 doublet.

EXCITED STATE EPR IN PARAMAGNETIC SINGLET-GROUND-STATE SYSTEMS

A. Background

The first observation of excited state EPR was made by Cooper et.al.[1] in TmN, and their results and our own succeeding results[2,4] for excited state EPR in the other thulium monopnictides of NaCl-structure, and especially those for TmP, provide probably the most striking results of excited state EPR. The crystal-field-only level scheme for Tm^{3+} in the pertinent predominantly fourth-order octahedral crystal-field is shown in the inset of Fig. 1. The original motivation for the experiments of Cooper et.al. lies in the fact that since there are two Γ_5 levels, as shown in Fig. 1, the EPR g-values for these states would be expected to vary strongly with the x parameter determining the ratio of fourth-to-sixth-order crystal-field anisotropy. Thus, observation of the $\Gamma_5^{(2)}$ resonance would determine the x parameter. The results of the TmN experiments, as described below, while providing that information, actually provided information of much wider impact than just that result and provided the basic motivation for our own more recent and wider ranging studies.

As shown in Fig. 1, the high and low g Γ_5 states interchange at about x=-0.8, and this is just the region of x that applies to the thulium monopnictides. As discussed below, our results[2,4] indicate that the x parameter is to the right of the g crossing for thulium nitride and phosphide, and to the left of the crossing for the arsenide and antimonide. It should be noted that the Γ_4 paramagnetic resonance has not been observed, presumably because the crystal-field-only g value is so low.

In the original observations in TmN and our own subsequent investigations, exchange effects, in fact, play an important role. Indeed the interplay between exchange and crystal-field effects and the evidence on the nature of the exchange interaction is a particularly interesting part of the results. Exchange effects give a temperature-dependent g-shift from the temperature independent crystal-field-only value. When the exchange is isotropic and much smaller than the crystal-field, theory[1] predicts a g-shift proportional to a quantity differing only slightly from the susceptibility, i.e., going approximately as 1/T for the higher temperature range. This indeed is the behavior observed for the higher temperature part of the range studied in TmN and TmP, giving a measure of the overall exchange strength. However, g anomalies at lower temperature are indicative[4] of significant anisotropic exchange.

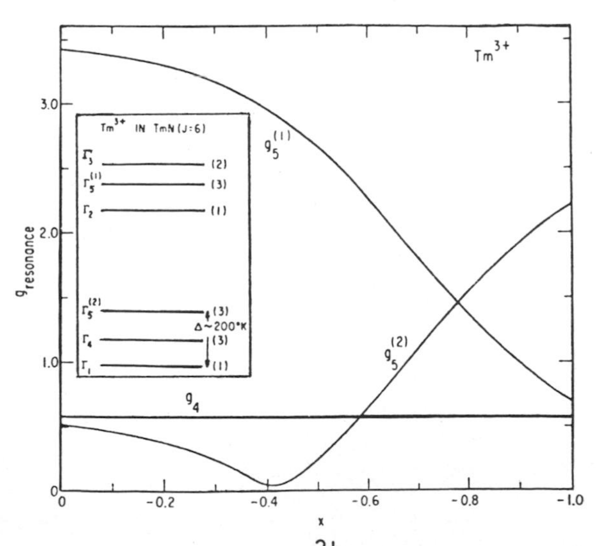

Fig. 1: Resonance g-values for Tm^{3+} in TmN as a function of x. The inset shows the crystal field energy level scheme for Tm^{3+} in TmN.

B. Experimental Results

In this section we review the results for powder samples investigated at ∿ 9.2 GHz. We will review the results for TmN, TmP, and PrP. We will also discuss briefly our preliminary results for TmAs, TmSb, and $Tb_{0.1}La_{0.9}P$. All of these systems have the NaCl crystal structure and are known to be paramagnetic down to ∿ 1°K. The lack of magnetic ordering and Van Vleck-type susceptibility for these systems can be understood when crystal-field effects are dominant over exchange, and the crystal-field ground state is a singlet.

(a) TmN and TmP. Figure 2(a) shows the observed[1] g-behavior for TmN while Fig. 2(b) shows the EPR intensity behavior indicating the excited state origin of the observed effects. This intensity behavior indicates that the EPR is originating from an excited state at about 200°K above the Γ_1 ground state, while the high temperature g behavior is consistent with an x value close to x = -1. This shows that the EPR is originating from the $\Gamma_5^{(2)}$ excited state. The most striking feature of these results is the lower temperature g behavior, going through a minimum at about 100°K, and then rising sharply with decreasing temperature. The results of Cooper et.al.[1] showed a corresponding temperature variation of the linewidth in the same temperature range.

Figures 3(a) and 3(b) depict the observed[2] g-values and linewidths of TmP measured at 9 GHz, respectively. The temperature variation of the intensity, shown in Fig. 4, shows that the resonance originates from an excited level at 60°K above the Γ_1 groundstate, which we identify as the $\Gamma_5^{(2)}$ level. This energy splitting is consistent with that indicated by neutron scattering[7] and the Gd^{3+} probe EPR technique.[8] The g-value of TmP, just as that of TmN, smoothly approaches 2.20 with increasing temperature indicating

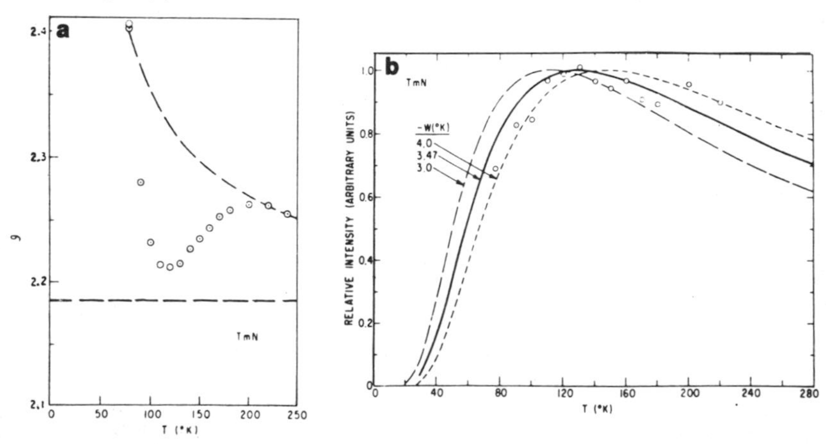

Fig. 2: (a)Temperature dependence of resonance g-value in TmN. Curve of 1/T dependence shown with asymptotic limit at g=2.186. (b)Temperature dependence of intensity of $\Gamma_5^{(2)}$ resonance of Tm^{3+} in TmN.

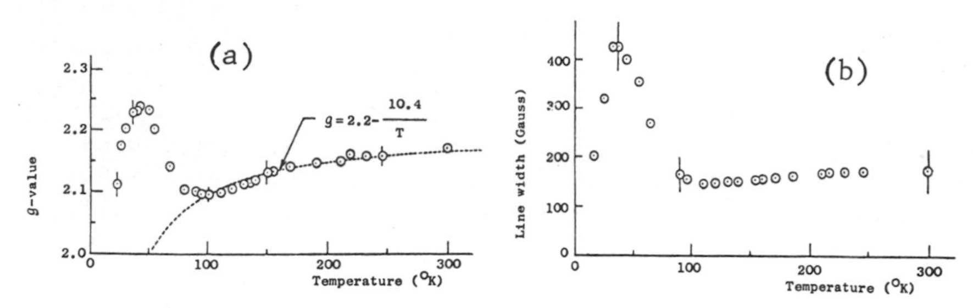

Fig. 3: (a)Temperature dependence of the g-value of the excited $\Gamma_5^{(2)}$ level of Tm^{3+} in TmP. (b)Temperature dependence of the EPR line width of the excited $\Gamma_5^{(2)}$ level of Tm^{3+} in TmP.

that the crystal-field anisotropy is very close to the fourth-order only limit. The experimentally observed g-values above 100°K can be fitted to the expression g = 2.2 - 10.4/T, as shown by the dashed curve in Fig. 3(a). The experimental g-value approaches the high temperature limit from below, which is consistent with the expectation that any exchange in TmP will be antiferromagnetic, as compared to the ferromagnetic exchange expected for TmN.

As in TmN, the g-behavior at lower temperatures for TmP departs strongly from the 1/T higher temperature dependence. We were able to follow the behavior to lower temperatures than in TmN, and to fully define the anomaly as a maximum centered at about 50°K as shown in Fig. 3(a). The line width, shown in Fig. 3(b), has its maximum at about the same temperature.

(b) TmAs. Figure 5 shows the g-behavior for the excited triplet state resonance[4] in TmAs. The behavior is qualitatively similar to that for TmN and TmP, but the anomalous behavior is not so well defined. The data for T > 150°K can be fitted to g = 2.21 - 19/T. From the signal intensity (very weak), we concluded that the resonance originates from an excited level about 250°K above Γ_1. This energy splitting is too high for $\Gamma_5^{(2)}$, since

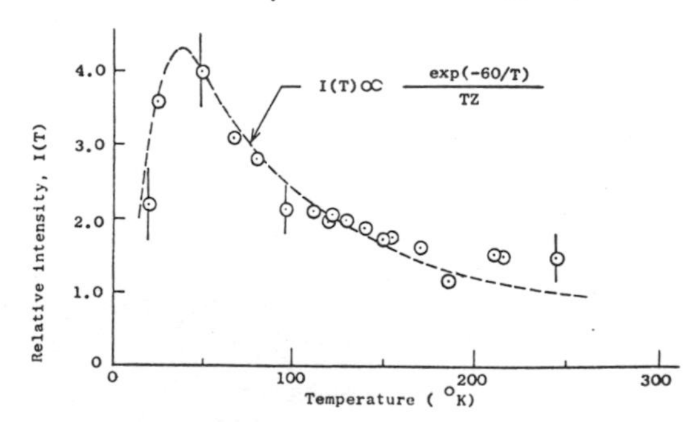

Fig.4: Temperature dependence of the EPR intensity of the $\Gamma_5^{(2)}$ level of Tm^{3+} in TmP.

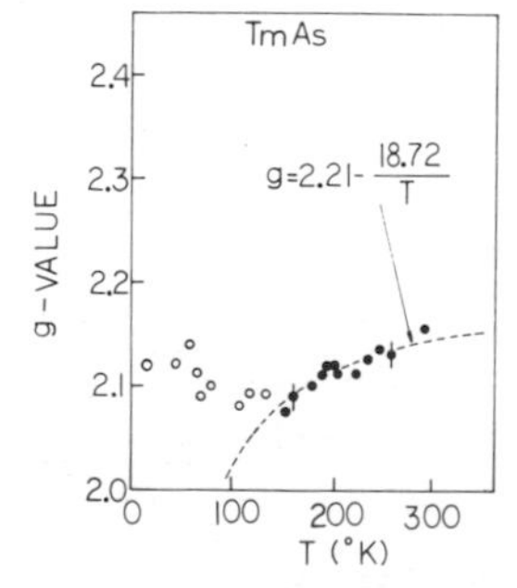

Fig.5: Temperature dependence of the g-value of Tm^{3+} in TmAs.

our Gd^{3+} probe EPR[8] indicated $E(\Gamma_5^{(2)}) - E(\Gamma_1) = 60°K$. Therefore the resonance originates from $\Gamma_5^{(1)}$ with $x \simeq -0.62$. This identification is consistent with the observation of an extremely weak and broad resonance line corresponding to $g \simeq 0.9$ attributed to $\Gamma_5^{(2)}$ with $x \simeq -0.62$. We have also observed[4] the $\Gamma_5^{(1)}$ EPR in $Tm_{0.5}La_{0.5}As$. The observed g-value indicates a slightly different g value than in pure TmAs, $x \simeq -0.54$.

(c) TmSb. The experimental temperature variation[4] of the g of Tm^{3+} in TmSb closely follows that given by $g \simeq 2.1 - 12/T$. The weak intensity of the signal observed indicates that the resonance comes from an excited state about 250°K above Γ_1. Hence, in analogy to TmAs, we attributed the signal to $\Gamma_5^{(1)}$ with $x = -0.65$.

(d) PrP. For Pr^{3+} in the octahedral, predominantly fourth-order, crystal-field pertinent to PrP, the ground state is a Γ_1 singlet, and the excited states are Γ_3, Γ_4, and Γ_5 with no duplication of symmetry types. The crystal-field-only g-value for the Γ_5 state is 2.00, and this level is expected to be about 400°K above the ground state. Our observations[2] of the g behavior shown in Fig. 6 are in agreement with this expectation, giving a slightly ferromagnetic g-shift with no anomalous behavior observed. However, the small temperature dependence of the observed intensity and g-shift over the temperature range studied makes us attach little significance to the lack of evidence of any anomalous behavior.

(e) $Tb_{0.1}La_{0.9}P$. Pure TbP is an antiferromagnet, but because of the dilution of the Tb^{3+}, $Tb_{0.1}La_{0.9}P$ is paramagnetic down to 0°K. The energy level scheme for Tb^{3+} in a cubic field is identical to that of Tm^{3+} (see inset of Fig. 1). The theoretical variation for the crystal-field-only g-values with respect to x is similar to that given in Fig. 1, but we have to multiply by a factor of 9/7. Figure 7 shows our data[4] of $\Gamma_5^{(2)}$ from which we obtain

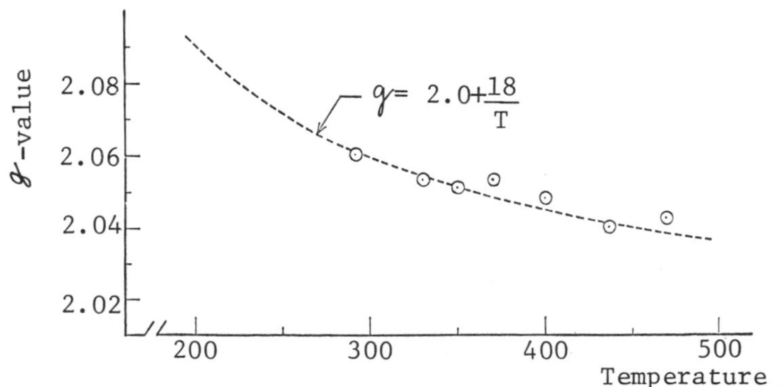

Fig. 6: Temperature dependence of the g-value of the excited $\Gamma_5^{(2)}$ level of Pr^{3+} in PrP.

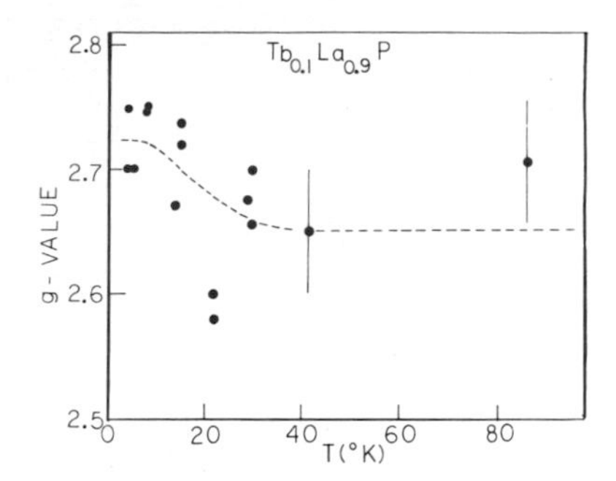

Fig.7: Temperature dependence of the g-values of $Tb_{0.1}La_{0.9}P$.

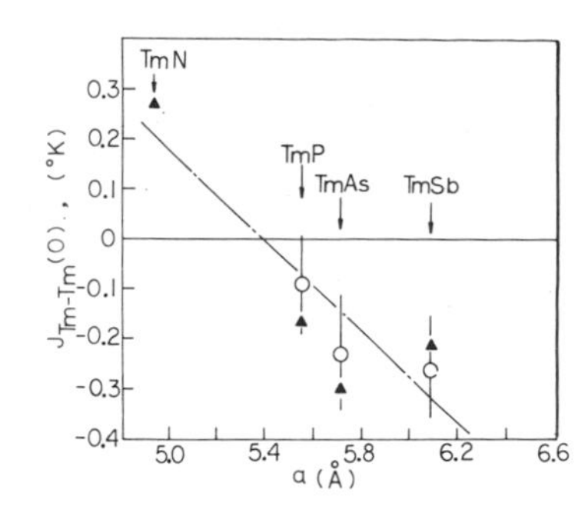

Fig.8: The exchange constant vs the lattice constant. The triangles and circles represent the data obtained by excited state EPR and the Gd^{3+} probe EPR technique, respectively.

$x = -0.96 \pm 0.03$. As shown in this figure, data were taken down to $\sim$ 5°K, even though the signals were weak. This is reasonable since $\Gamma_5(2)$ is only about 35°K above Γ_1 (as discussed below in connection with the behavior of pure TbP).

C. Theory and Discussion

We have considered[4] how the temperature dependence of the observed g-behavior in TmN and TmP can be understood on the basis of Tm-Tm interactions, including the possibility of anisotropic interaction effects. This involves a generalization[4] of the earlier work[2] which treated only isotropic exchange. We have considered exchange small compared to the crystal-field so that the unperturbed Hamiltonian,

$$H_0 = \sum_i (H_{CFi} - \lambda\beta H J_{iz}) \tag{1}$$

includes the crystal-field plus applied magnetic field effects, while the interaction term,

$$H_1 = \sum_{i \neq j} \sum_{ll',mm'} G_{ll'}^{mm'} \tilde{O}_l^m(J_i)\tilde{O}_{l'}^{m'}(J_j) \tag{2}$$

is treated as a perturbation on H_0.

H_1 must have cubic or isotropic symmetry. However, there are many forms of interaction compatible with this. We can divide these into two classes, depending on whether the operator equivalents $\tilde{O}$ contain odd or even powers of angular momentum operators. The simplest odd-odd interaction is the isotropic Heisenberg interaction.

However, the higher degree interactions involving third or fifth powers of angular momentum operators are the ones of interest here. The simplest even-even interaction is the quadrupole interaction.

The resonance frequency is found from a first moment calculation where the transition operator M is truncated to include only transitions between the crystal-field triplet states of interest.

For exchange interaction small compared to crystal-field effects, there are two contributions to the g-shift from the crystal-field-only value. These correspond to the part of the exchange coming from the components of angular momentum parallel and transverse to the applied field respectively. In the case of isotropic exchange, these give a g-shift proportional to the crystal-field-only susceptibility and the small flip-flop type correction, respectively.

For even-even interaction, the parallel component shift vanishes unless the operator equivalent has cubic symmetry. Therefore it vanishes for quadrupole-quadrupole interactions. The flip-flop type term only occurs for odd-odd interactions. Thus we see that quadrupole-quadrupole interactions cannot explain the anomalous g-shift behavior. Since they do not contribute to the flip-flop type term, higher order even-even interactions also do not look promising. Thus, if the explanation of the anomalous g behavior lies in anisotropic Tm-Tm interactions, the pertinent interactions are presumably higher order odd-odd interactions. We have not as yet examined the question of whether there is a specific higher order odd-odd interaction that can yield the observed behavior. We should, however, point out that there is evidence[9] in other systems for the existence in practice of such interactions (Ce^{3+} pairs in $LaCl_3$ and $LaBr_3$, and CeBi and CeSb).

The fact that the higher temperature g-shift in the thulium monopnictides is proportional to the crystal-field-only susceptibility that would pertain in these materials in the absence of all exchange allows us to find the values of $J(0) = \sum_j J_{ij}$, the overall Tm-Tm exchange. (The small exchange values present mean that in fact there is very little difference between the crystal-field-only susceptibility and the experimental susceptibility.) These values[4] are shown by the triangles in Fig. 8, and they agree reasonably with those data, as shown by the circles, which were obtained by the Gd^{3+} probe EPR technique.[8]

ANTIFERROMAGNETIC SYSTEMS

(a) Terbium monopnictides. The EPR of the excited $\Gamma_5^{(2)}$ triplet of Tb^{3+} in TbP and TbAs have been reported in Ref. 3. The intensity data indicate that $\Gamma_5^{(2)}$ is 35°K above Γ_1. When the field at which the derivative signal crosses the baseline is used as the "resonant field," we find that the "g-value" increases to a very large value on approaching T_N (for example, $g \sim 400$ at $T-T_N=1.3°K$). The "g-shifts" can be expressed by $\Delta g = A/(T-T_N)^n$, with A=300, n=1.3 ±0.3, $T_N \simeq 8°K$ for TbP and A=350, n=1.3 ± 0.3, $T_N = 10.5°K$ for TbAs. These T dependences are the same as that of the staggered susceptibility.

(b) Antiferromagnetic CeP. As a complement to our studies of Tb singlet-ground-state antiferromagnets discussed above, we have investigated the Kramers salt CeP at 9.2 GHz above 77°K. In a cubic field, the lowest multiplet (J = 5/2) splits into Γ_7 and Γ_8. The EPR measurements[10] on 5 at.% Ce^{3+} in LaP indicates that the Γ_7 doublet is the ground-state. According to neutron scattering,[11] Γ_8 is $\sim 170°K$ above Γ_7. Our intensity data (Fig. 9) yield that the resonance originates from an excited state at $\sim 200°K$, which is consistent with the neutron result. Based on Bleaney's analysis,[12] even though the transitions in the Γ_8 quartet could give rise to g ranging from 0.3 to 3.2 depending on the direction of applied field and the states involved in the transitions, the transitions between the first and third, and second and fourth, levels of the quartet dominate for a powder sample. These give an average $g \simeq 2.1$. This is borne out by the fact that the linewidth is only ~ 300 Gauss. Our observed g-values are given in Fig. 10. The data can be fitted to $g \simeq 2.1 + 14/(T-\theta')$, with $\theta' = 7°K$. This is reasonable, since we expect the g-shift to be proportional to χ, which is in turn proportional to $1/(T-\theta)$ with[13] $\theta = 5°K$.

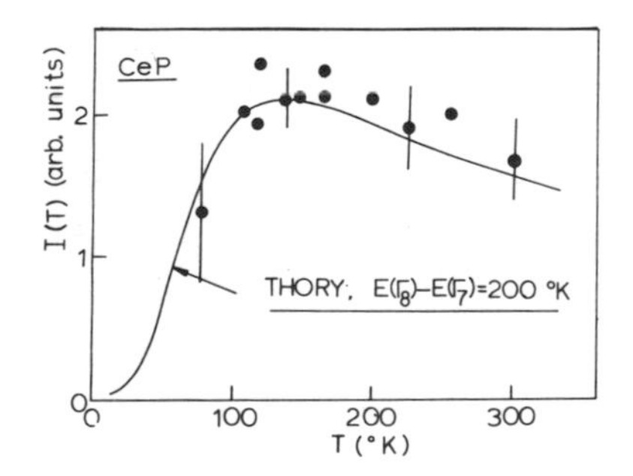

Fig.9: EPR intensity of Ce^{3+} in CeP.

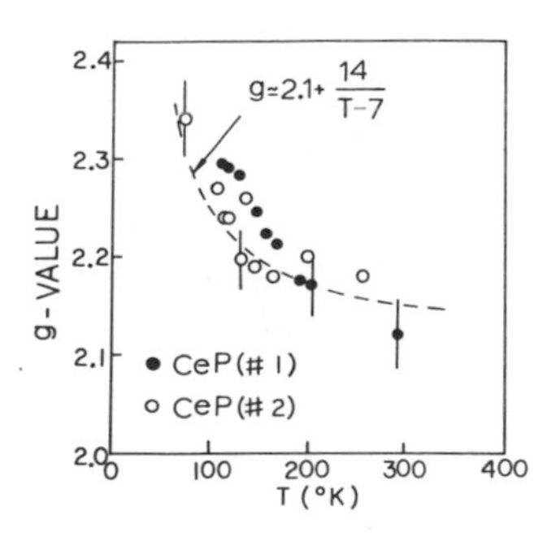

Fig.10: g-values of Γ_8 resonance in CeP.

CONCLUDING REMARKS

It is clear that excited state EPR work is still at an early stage and much more work has to be done in order to understand the physics associated with excited state EPR with confidence. Several problems of interest are: (1) The theory of excited state EPR linewidth and lineshape. (2) Incorporation of magnetic exciton and conduction electron effects into the theory. (3) Improvement of the data for systems already studied and extension to other singlet-ground-state systems, especially alloys diluting the magnetic rare earth species. (4) Completion of the understanding of the low temperature anomalous g behavior. (5) The x values differ quantitatively from those found by neutron scattering. If these differences cannot be explained by uncertainties in the neutron scattering results, they must be understood.

REFERENCES

1. B. R. Cooper, R. C. Fedder, and D. P. Schumacher, Phys. Rev. Lett. 18, 744 (1967); Phys. Rev. 163, 506 (1967)(erratum Phys. Rev. 168, 654 (1968).

2. K. Sugawara, C. Y. Huang, and B. R. Cooper, AIP Conf. Proc. 24, 240 (1975).

3. C. Y. Huang, K. Sugawara, and B. R. Cooper, AIP Conf. Proc. 24, 248 (1975).

4. B. R. Cooper, C. Y. Huang, and K. Sugawara, AIP Conf. Proc. 29, 428 (1976).

5. B. R. Cooper and O. Vogt, J. Phys. (Paris) 32, C1-958 (1971); B. R. Cooper, CRC Crit. Rev. Sol. State Sci. 3, 83 (1972) and references therein.

6. K. Sugawara, C. Y. Huang, and B. R. Cooper, Paper BL-4, Bull. Am. Phys. Soc. 21, 271 (1976).

7. R. J. Birgeneau, E. Bucher, L. Passell, and K. C. Turberfield, Phys. Rev. B4, 714 (1971).

8. K. Sugawara, C. Y. Huang, and B. R. Cooper, Phys. Rev. B11, 4455 (1975).

9. R. J. Birgeneau, M. T. Hutchings, J. M. Baker, and J. D. Riley, J. Appl. Phys. 40, 1070 (1969).

10. K. Sugawara and C. Y. Huang, J. Phys. Soc. Japan 40, 295 (1976).

11. H. Heer, A. Furrer, and W. Halg, J. Magnetism and Magnetic Materials 3, 55 (1976).

12. B. Bleaney, Proc. Phys. Soc. (London) 73, 937 (1959).

13. T. Tsuchida and W. E. Wallace, J. Chem. Phys. 43, 2885 (1965).

ESR OF IMPURITY AS A PROBE FOR THE HOST EXCHANGE INTERACTION IN Tm VAN-VLECK COMPOUNDS

V. Zevin*, D. Davidov*, R. Levin*, D. Shaltiel*
and K. Baberschke**

*Racah Institute of Physics, Hebrew University of Jerusalem, Israel
**Free University of Berlin

ABSTRACT

The relaxation rate of an impurity in Tm Van-Vleck compounds with cubic structure is given for two limiting cases. Analysis of the ESR linewidth of Gd in TmP, TmSb and TmBi in terms of this theory yields information concerning the nature of the host fluctuation spectra and the host exchange parameters.

This paper presents a brief sketch of a theory for impurity relaxation in Van-Vleck compounds with weak exchange coupling. It is shown that the relaxation rate can be expressed in terms of the host exchange interaction, the host crystalline field splitting levels and the impurity-host exchange interaction. Thus, knowledge of the crystalline field levels (say, from an independent neutron scattering measurement) and the impurity-host exchange (from g shift study), enable one to extract the host exchange by fitting our theory to the experimental ESR thermal broadening. The Gd ESR relaxation rates of some Tm Van-Vleck compounds (TmP, TmSb and TmBi) are discussed in view of our theory.

The relaxation of the impurity (Gd) due to fluctuations of the host Van-Vleck ions (Tm) has been calculated in the frame of the Bloch kinetic equations under the following assumptions: (1) The Hamiltonian describing the host Tm ions contains the cubic crystalline field Hamiltonian[1], a bilinear isotropic exchange interaction and the Zeeman Hamiltonian. It is assumed that the crystalline field splitting is larger than the host exchange interaction. This enables one to separate between the high and low frequency components of the host fluctuation spectra. Only the low frequency part of the fluctuation spectra contributes to the impurity (Gd) relaxation rate.[2] (2) We assume that the "width" of the low frequency part of the fluctuation spectra, $\Delta\omega$, is much larger than the impurity relaxation rate, $1/T_2$. (3) The host Tm ions and the conduction electrons are in constant thermal equilibrium with the lattice. It is assumed, also, that the transverse susceptibility associated with the Gd spin at ω_0 is much larger with respect to the other spin species (Tm, conduction electrons) at ω_0. (4) We assume a spin $S = 1/2$ for the impurity or a spin larger than $1/2$(Gd) but in the extreme narrowing regime (in the sense of Zimmerman et al.[3]).

Under all these assumptions the Gd relaxation rate, $1/T_2$, for the case of Tm cubic compounds can be calculated for two limiting cases: (a) $\Delta\omega \gg \omega_0$ and (b) $\Delta\omega \lesssim \omega_0$. In case (a) it is impossible to distinguish between the transverse and the longitudinal components of the fluctuation spectra, and the low frequency part of the fluctuation spectra can be described by a single distribution function, $F(\omega)$. In case (a) the relaxation rate, $1/T_2$, of the impurity can be written as

$$\frac{1}{T_2} = \frac{2\pi}{\hbar}(g_J-1)^2 \frac{\sum_k (J_k')^2 F(0)}{Z} \{c_{44}^2 e^{-E_4/T} + c_{55}^2 e^{-E_5/T} + c_{\bar{5}\bar{5}}^2 e^{-E_{\bar{5}}/T}\} \tag{1}$$

where g_J is the Tm^{+3} Landé g factor, the indices $4,5,\bar{5}$ represent the Γ_4, $\Gamma_5(1)$ and $\Gamma_5(2)$ crystalline field splitting states, respectively, E_α ($\alpha=4,5,\bar{5}$) are the appropriate energy levels, respectively, $c_{\alpha\alpha}$ are non-vanishing matrix elements of J_z acting on the $|\Gamma_\alpha,f_\alpha\rangle$ states ($c_{\alpha\alpha} = \langle\Gamma_\alpha f_\alpha|J_z|\Gamma_\alpha f_\alpha\rangle$), Z is the partition function ($Z = \Sigma e^{-E_\beta/T}$) and J_k' is the exchange interaction parameter between the $^\beta$Gd spin, $\vec{S}(Gd)$, and that of the k-th host Tm ion, $\vec{S}(k)$. This exchange interaction is defined by

$$\mathcal{H}'_{ex} = -(g_J-1)\sum_k J_k' \vec{S}^{(Gd)}\cdot\vec{J}^{(k)} \tag{2}$$

For a Gaussian lineshape of the low frequency part of the fluctuation spectra, the value of $F(0)$ in eq.(1) is related to the second moment M_2 according to $F(0) = (2\pi M_2)^{-1/2}$. The second moment, M_2, of the low frequency part of the fluctuation spectra was calculated using the truncation procedure[4]. Considering only autocorrelation contributions and neglecting pair correlations, one finds for $\Delta\omega > \omega_0$ and for Tm cubic compounds

$$M_2 = \frac{\sum'_{k\neq 1} (g_J-1)^4 J_{1k}^2}{Z[c_{44}^2 e^{-E_4/T} + c_{55}^2 e^{-E_4/T} + c_{\bar{5}\bar{5}}^2 e^{-E_{\bar{5}}/T}]} \tag{3}$$

$$\{2\, c_{44}^2 c_{14}^4 e^{-(E_1+E_4)/T} + 2\, c_{55}^2 c_{25}^4 e^{-(E_2+E_5)/T} +$$

$$+2\, c_{\bar{5}\bar{5}}^2 c_{2\bar{5}}^4 e^{-(E_2+E_{\bar{5}})/T} + 2\, c_{44}^2 c_{34}^4 e^{-(E_3+E_4)/T} +$$

$$+2\, c_{55}^2 c_{35}^4 e^{-(E_3+E_5)/T} + 2\, c_{\bar{5}\bar{5}}^2 c_{3\bar{5}}^4 e^{-(E_3+E_{\bar{5}})/T} +$$

$$+4\, c_{44}^6 e^{-2E_4/T} + 4\, c_{55}^6 e^{-2E_5/T} + 4\, c_{\bar{5}\bar{5}}^6 e^{-2E_{\bar{5}}/T} +$$

$$+4[(c_{44}^2+c_{55}^2)(c_{44}^2 c_{55}^2+c_{45}^4) + (c_{44}+c_{55})^2 c_{45}^4]e^{-(E_4+E_5)/T} +$$

$$+4[(c_{44}^2+c_{\bar{5}\bar{5}}^2)(c_{44}^2 c_{\bar{5}\bar{5}}^2+c_{4\bar{5}}^4) + (c_{44}+c_{\bar{5}\bar{5}})^2 c_{4\bar{5}}^4]e^{-(E_4+E_{\bar{5}})/T} +$$

$$+4[(c_{55}^2+c_{\bar{5}\bar{5}}^2)(c_{55}^2 c_{\bar{5}\bar{5}}^2+c_{5\bar{5}}^4) + (c_{55}-c_{\bar{5}\bar{5}})^2 c_{5\bar{5}}^4]e^{-(E_5+E_{\bar{5}})/T}\}$$

where J_{1k} is the exchange interaction (defined as $-(g_J-1)^2 J_{1k}\vec{J}^{(1)}\vec{J}^{(k)}$) between the spin of the 1-st Tm ion and that of the k-th Tm ion. The prime in (3) excludes the impurity site from the total sum in this equation. Numerical calculations of the second moment and fourth moment indicate that the fluctuation spectra exhibit a Gaussian lineshape. Also pair correlation contributions to M_2 were calculated[5] and their contribution does not exceed 30% of the auto-correlation contribution.

In case (b) ($\Delta\omega < \omega_0$), the low frequency part of the fluctuation spectra can be separated into transverse and longitudinal components. The lineshape of the individual component can be calculated separately using the method of moments. We find for case (b) the following expression

$$\frac{1}{T_2} = \frac{2\pi}{\hbar}(g_J-1)^2 \frac{\Sigma(J_k')^2}{Z} \{c_{44}^2 F_1^* e^{-E_4/T} + c_{55}^2 F_2^* e^{-E_5/T} + c_{\bar{5}\bar{5}}^2 F_3^* e^{-E_{\bar{5}}/T}\}$$

where F_1^*, F_2^* and F_3^* are spectral functions defined elsewhere[5].

The theory above can be used to extract the host exchange of TmP, TmSb and TmBi using the experimental ESR thermal broadening of Gd as measured for TmP:Gd[6], TmSb:Gd[7] and TmBi:Gd[7]. Fig. 1 yields the Gd ESR linewidth as a function of temperature for the system TmSb:Gd[7]. Similar features have been observed also for TmP:Gd[6] and TmBi:Gd[7]. The results in Fig. 1 were observed in ref. (7). These results are different from those of Sugawara et al. (ref. 6) for the same system. New measurements performed on 3% Gd in TmSb are consistent with those of Davidov and Baberschke at high temperatures. The low temperature (i.e. at T=4°K) linewidth, however, is by 150 G larger than that reported by Davidov and Baberschke. This difference is attributed to different residual widths. The discrepancy between these two sets of data and that of Sugawara et al. for TmSb:Gd is not understood at present. Now, using the crystalline field splitting energy levels as measured by others together with the Gd-Tm exchange from g shift study[7,9] (see Table I) the host exchange in the Tm pnictides can be extracted by fitting experiment to theory. The best fit of case a is shown in Fig. 1 for the case of TmSb:Gd. The host exchange $|J|$ as extracted for these compounds are given in Table I.

A difficulty arises in such a fitting procedure because the appropriate limit (case (a) or case (b)) is unknown apriori when the host exchange which determines $\Delta\omega$ (note $\Delta\omega \sim \sqrt{2\pi M_2}$ and M_2 depends on ΣJ_k^2) is not known.

Table I summarizes the various values of the host exchange as determined using case (a) together with the assumption of nearest-neighbor exchange only. Using these values as well as eq.(3) one can estimate the width, $\Delta\omega$, of the low frequency part of the fluctuation spectra. We find that while $\Delta\omega \geq \omega_0$ for TmP:Gd and TmBi:Gd, for TmSb:Gd the condition $\Delta\omega < \omega_0$ is valid. This indicates that the low frequency part of the spectra of fluctuations exhibits structure, case (b) should be used to analyze the TmSb:Gd data. With case (b) we find slightly smaller values of J for TmSb:Gd (see Table I).

Table I

The Tm-Tm exchange interaction, $|J|$, as extracted from the ESR thermal broadening using our theory together with the assumption of nearest-neighbor exchange interaction ($\Sigma J_k^2 = z_0 J^2$ with z_0 as the number of Tm nearest-neighbors). The values of the Gd-Tm exchange parameters, J', are also given.

	$\|J'\|$ (meV)	$\|J\|$ meV
TmP	0.029†	0.06*
TmSb	0.015†	0.006$^{\dagger\dagger *}$, 0.004**
TmBi	0.04†	0.05*

†extracted using the g shift study of Gd in these hosts

*extracted using case (a) of our paper

††extracted using the data in Fig. 1. If however we use the data of Sugawara et al. for TmSb:Gd, a value of $|J|$=0.02 meV is obtained for case (a)

**extracted using case (b)

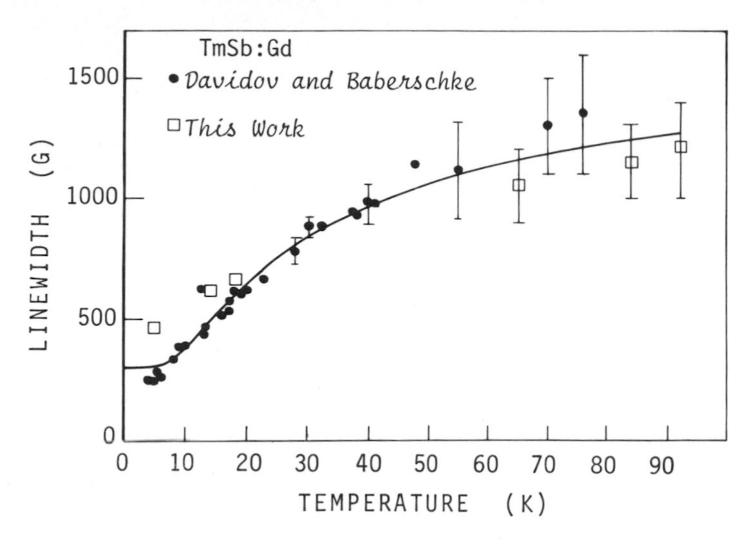

Fig. 1
The ESR linewidth of Gd in TmSb as a function of temperature. The solid line is a theoretical fit of case (a) (equations 1 and 3) which includes, in addition, the residual width as well as the Korringa broadening.

REFERENCES

1. K.R. Lea, M.J.M. Leask and W.P. Wolf, J. Phys. Chem. Solids 23, 1381 (1962).
2. D. Davidov, C. Rettori and V. Zevin, Sol. State Comm. 16, 247 (1975).
3. P.H. Zimmerman, D. Davidov, R. Orbach, L.J. Tao, J. Zitkova, Phys. Rev. B6, 2783 (1972).
4. M. MacMillan and W. Opechowski, Can. J. Phys. 389, 2915 (1963); V. Zevin and B. Shanina, Ukr. Fiz. J. 11, 1089, 1966 (Sov. Phys.); see also D. Davidov, V. Zevin, R. Levin, D. Shaltiel and K. Baberschke, submitted to Phys. Rev.
5. V. Zevin, D. Davidov, R. Levin, D. Shaltiel and K. Baberschke, to be published.
6. K. Sugawara, C.Y. Huang and B.R. Cooper, Phys. Rev. B11, 4455 (1975).
7. D. Davidov and K. Baberschke, Phys. Letters A51, 144 (1975).
8. R.J. Birgeneau, E. Bucher, L. Passell and K.C. Turberfield, Phys. Rev. B4, 718 (1971); R.J. Birgeneau, E. Bucher, J.P. Maita, L. Passell and K.C.Turberfield, Phys. B8, 5345 (1973); H.L. Davis and H.A. Mook, AIP Conf. Proc. 18, 1086 (1974).
9. C. Rettori, D. Davidov, A. Gragevsky and W.M. Walsh, Phys. Rev. B11, 4450 (1975); Sugawara et al., ref. 6.

EPR of Gd^{3+} in Lu and Sc Single Crystals[+]

J. Nagel and K. Baberschke

Inst. f. Atom- u. Festkörperphysik Freie Universität

Berlin, D-1000 Berlin 33, Boltzmannstr. 20, Germany

ABSTRACT

We report spin resonance experiments of Gd^{3+} (300...5800 ppm) doped into Lu and Sc single crystals. The resonance shows an anisotropic single line spectrum, which is explained as the exchange narrowed finestructure of the Gd^{3+} in the hexagonal crystalline electric field. The fit to the theory yields for the axial crystal field parameter $D = 3B_2^o$:

Lu: $D = -63(6)$ G   Sc: $D = -330(30)$ G

and the isothermal g-factors:

Lu: $g = 2.10(1)$ Sc: $g = 2.13(3)$ G

Sc:Gd is strongly bottlenecked, leading to an anormal angular dependence of the resonance field, which is explained in terms of available theories.

INTRODUCTION

Electron-spin resonance of localized moments in metals enables us to determine the dynamical and static interaction with the conduction electrons. If the experiments are performed on single crystals, the interaction with the crystalline electric field (CEF) can be determined in addition. In contrast to other methods (specific heat, magnetization etc.) the ESR method permits the investigation of rather low concentrated alloys and has the further advantage, that the generally very small CEF-splitting of S-state ions can be measured directly.

[+]This work is part of the research program of the "Sonderforschungsbereich 161" der Deutschen Forschungsgemeinschaft

We report in this work ESR-measurements on Gd^{3+} (groundstate $^8S_{7/2}$) in single crystals of hexagonal Lu[7] and Sc. The nonmagnetic metals Sc, Y and Lu are isostructural and their electronic and crystalline structure is very similar to the other (magnetic) rare earth metals so that they form a suitable diluent for them. Magnetization measurements [1] of dilute Tb, Dy and Er in these metals studied the systematic of CEF-parameters so that it was of interest to complete these measurements for the S-state ion Gd^{3+}. The resonance of Gd in Sc was first observed by Salamon[2]. He found a single line whose position changed with crystal orientation. He assumed that the spin-orbit splitting of the conduction electron d-levels is the source of an anisotropic local host susceptibility, leading to an anisotropic g-factor. More recent measurements on Y:Gd[3] and Mg:Gd[4] and the development of the theory of CEF-effects in metals by Plefka[5] and Barnes[6],however,demonstrated the necessity of a new interpretation and further experimental datas. For the analysis of the results we used the theory of Plefka, which was also successful in the explanation of the Y:Gd and Mg:Gd datas. According to this theory the observed line is the exchange narrowed fine structure splitting of the Gd[3] in the hexagonal environment. This narrowing into a completely collapsed single-line spectrum with Lorentzian line shape occurs, when the isothermal Korringa-rate is large compared to the fine structure splitting of adjacent lines. The fine structure splitting in hexagonal environment can be described by the spin Hamiltonion

$$\mathcal{H} = g\mu_B H_o S_z + D(S_z^2 - 1/3S(S+1)) \qquad (1)$$

neglecting CEF-contributions of higher order. For S=7/2 the spectrum contains 7 transitions which are equidistant for $D \ll H_o$. The separation is maximum at $\Theta=0^o$ and minimum approx. at $\Theta \approx 55^o$. (Θ is the angle between the c-axis and the applied field H_o).

If no resonance bottleneck is present, the position of the narrowed line is determined by the first moment of the spectrum defined by (1). If the resonance is bottlenecked, the strong coupling to the conduction electrons results in a detectable deviation of the line position from the center of gravity (see Fig. 2).

EXPERIMENTAL RESULTS AND ANALYSIS

The measurements were performed on a Lu single crystal with 1700 ppm Gd and single crystals of Sc with Gd concentrations in the range 300-5800 ppm. The samples were prepared by a recrystallisation method. The orientation dependence of both systems was measured at X-band frequency at 1.4 K and 4.2 K and the Sc crystals in addition at Q-band frequency. The resonance field H_{res} and the linewidth ΔH changed with respect to the angle Θ between the c-axis of the crystal and the applied field (see Fig. 1). The field for resonance and linewidth were obtained by fitting a superposition of an absorption and dispersion

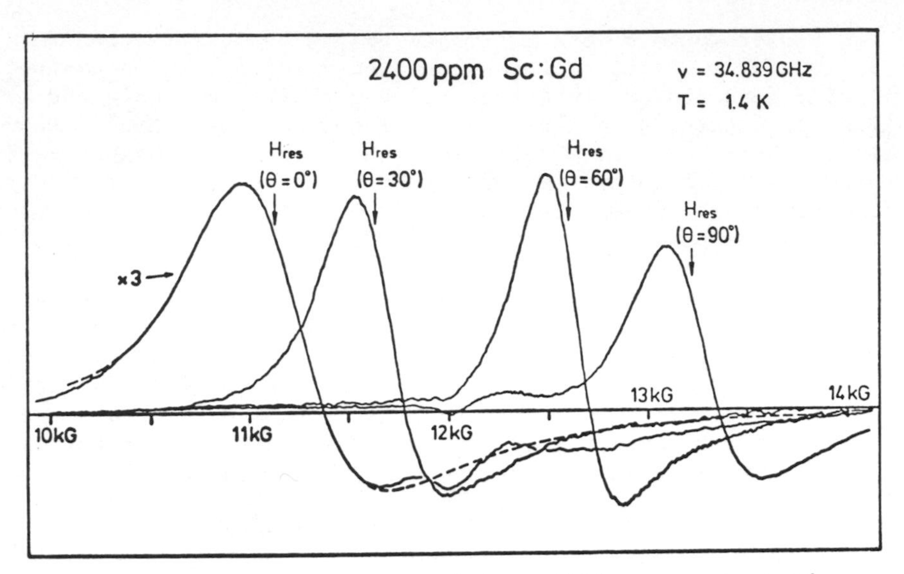

Fig.1: Q-band ESR-spectra of four different orientations.

part of a Lorentzian line to the recorder trace. Deviations from this lineshape occurred only for orientations of maximum splitting.

Lu:Gd Details of this measurement are described in ref. 7. The fit of Plefkas theory to the resonance field datas yields for the CEF-parameter D=-63(6) G and for the g-value g=2.10(1). Measurements of the thermal broadening on polycrystalline samples with different Gd-concentrations show no concentration dependence and a Korringa-broadening of b=80(8) G/K. This independence of impurity concentration and the g-shift lead to the conclusion that Lu:Gd is unbottlenecked.

Sc:Gd The angular variation of the resonance field and the linewidth was much larger than in Lu (Fig.1). The analysis of all measurements yield the same crystalline field parameter D=-330(30) G and g=1.99(1), independent of concentration, temperature and microwave-frequency. The analysis of the linewidth datas, however, showed that an additional narrowing mechanism by Gd-spin-spin interaction is effective at higher Gd concentration.

The g-value of 1.99 and the consistent fitting to the theory shows that Sc:Gd is strongly bottlenecked. The isothermal g-factor g_o and the Korringa-broadening b are parameters in the theory, the fit yields b=200(40) G/K and g_o=2.13(3)[8].

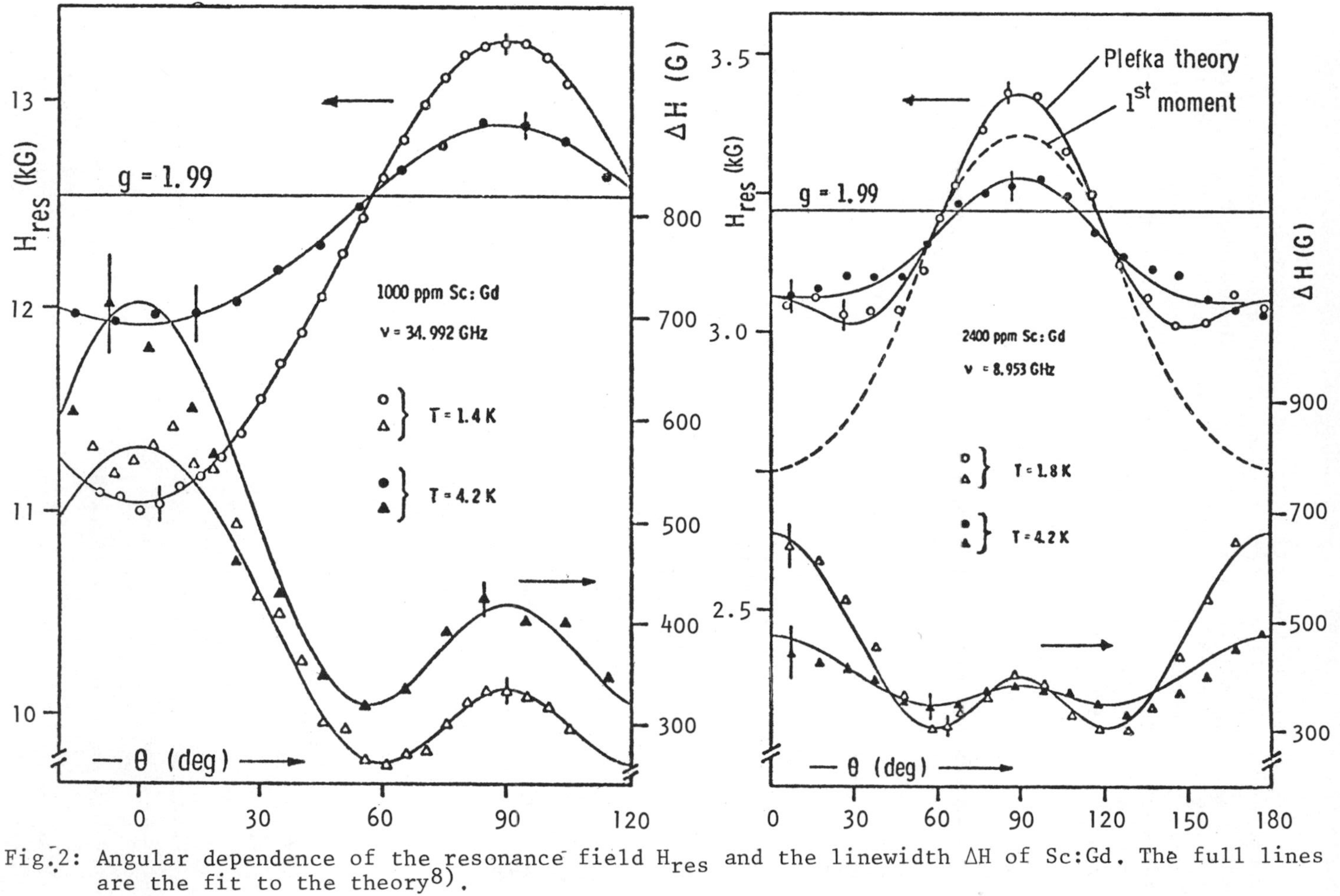

Fig. 2: Angular dependence of the resonance field H_{res} and the linewidth ΔH of Sc:Gd. The full lines are the fit to the theory[8].

DISCUSSION

The dynamical properties of the two systems are extremely different. For Lu:Gd the scattering rate of the conduction electron to the lattice δ_{eL} is large compared to that at the Gd impurities. Therefore the conduction electrons are in thermal equilibrium with the lattice. In the Sc:Gd system δ_{eL} is some orders of magnitude smaller and determined by the Gd-impurity itself, so that decreasing the Gd-concentration is ineffective in breaking the bottleneck. Sc and Lu just represent the lightest and heaviest "rare earth" metal. We attribute this behaviour to different spin-orbit scattering of the conduction electrons at dislocations. The good agreement of the resonance field and linewidth datas with Plefkas theory leads us to the conclusion that it is the correct description. The assumption of an anisotropic g-factor cannot explain the observed temperature dependence of the line position or the angular variation of the linewidth. From our measurements we estimate a possible anisotropy of the g-value to be $|g_{\parallel}-g_{\perp}|<0.02$. In Tab. 1 are summarized the CEF-parameters D for different hexagonal host metals, obtained by ESR-measurements. For comparison also the $A_2^o\langle r^2\rangle$-parameters are shown, which have been determined by magnetization measurements for the non S-state impurity Er^{3+}. The pointcharge model predicts for the A_2^o proportionality to the difference $c/a-\sqrt{8/3}$ (D<0) which is also given in Tab.1. It can be seen that the A_2^o follow at least qualitatively this rule, but that this model does not explain the relative magnitude of the D-parameters. This occurs also in insulators and is due to the complicated mechanisms of the groundstate splitting of Gd. Under these circumstances it seems to be difficult to discuss the different contributions of the conductions electrons to the CEF. Only the positive sign of D for the Mg:Gd system can certainly be attributed to screening effects by the conduction electrons, caused by the charge difference of the host (2+) and the impurity (3+).

Tab. 1: Crystal field parameters for Y, Lu, Sc, and Mg.

Host	D(G)	$c/a-\sqrt{8/3}$	$A_2^o\langle r^2\rangle$ [1] (K) for Er^{3+}
Y[3]	-150(20)	-.062	-122(12)
Lu[7]	- 63(6)	-.048	- 63(5)
Sc[4]	-330(30)	-.040	- 22(2)
Mg	+140	-.010	-------

REFERENCES

1) P.Touborg and J.Høg, Phys. Rev.Lett. 33, 775 (1974)
2) M.B. Salamon, Phys.Rev.Lett. 26, 704 (1971)
3) G.Weimann and B.Elschner, Z. Phys. 261, 85 (1973)
4) P.M.Zimmermann et al., Phys.Rev. B6, 2733 (1972)
5) T.Plefka, Phys. Stat. Sol. A55, 129 (1973); and thesis (THD 73)
6) S.E.Barnes, Phys.Rev. B9, 4789 (1974); and thesis (UCLA, 1972)
7) K.Baberschke and J.Nagel, Phys.Rev. B13, 2793 (1976)
8) The details of the Sc:Gd resonance and the "anomalous" dynamic properties will be published elsewhere.

EPR FINE STRUCTURE OF S STATE IONS

M. Hardiman*, S.E. Barnes+ and J. Pellisson*

*DPMC, Ecole de Physique, 1211 Geneva 4, Switzerland
+Physical Laboratory Trinity College, Dublin 2, Ireland

ABSTRACT

We have recently measured the EPR fine structure spectrum of Gd in Pt and found a b_4 parameter of +86 $\pm$ 3 gauss. The magnitude of this is the largest ever observed for Gd in any host and is some three times greater than in any other pure metal. We discuss the parametrisation of our spectra and then compare the available results for b_4 parameters in metals and intermetallic compounds.

INTRODUCTION

The EPR fine structure of Gd^{3+} in a cubic environment has been observed previously in the pure metals gold[1], palladium[2] and silver[3] and in several of the lanthanum and yttrium pnictides[4,5]. In such cases the spin-Hamiltonian habitually used is

$$H = g\mu_B \underline{H}\cdot\underline{S} + \frac{1}{60} b_4 O_4 + \frac{1}{1260} b_6 O_6$$

where O_4 and O_6 are the fourth and sixth degree cubic operators[6] and b_4 and b_6 are the experimentally determined constants, sometimes known as the spin-Hamiltonian parameters. For a non-S state ion, e.g. Er^{3+}, the Hamiltonian is

$$H = g_J'\mu_B \, H\cdot J + C_4\beta O_4 + C_6\gamma O_6$$

where β and γ are the Stephens multiplicative factors[6] which are constant for a given ion. C_4 and C_6 are the crystal field parameters, which in the point charge model are inversely proportional to the fifth and seventh power of the lattice parameter respectively[6]. Our recent measurements on the PtGd system give values for both b_4 and b_6. In an attempt to understand the large value of b_4 obtained we look first at the systematics displayed by the b_4 parameters in metallic hosts and then compare these with the behaviour of the crystal field parameters.

EXPERIMENTAL RESULTS AND ANALYSIS

The EPR spectra were taken at X band frequency on platinum single crystals containing between 1000 and 1500 ppm of gadolinium. As can be seen from figure 1, in the [001] direction at the lowest temperatures, the seven lines of the Gd fine structure are almost totally resolved. Using the well known formulae for the line positions in strong magnetic field[6], the b_4 parameter may be estimated from this spectrum as 90 ± 8 gauss. The sign of the splitting can be unambiguously determined to be positive from the temperature dependences of the relative line intensities in this direction. In order to measure b_6 and to determine b_4 more accurately we must take into account the line narrowing and shifts caused by the conduction electron induced 'hopping' of the ma-

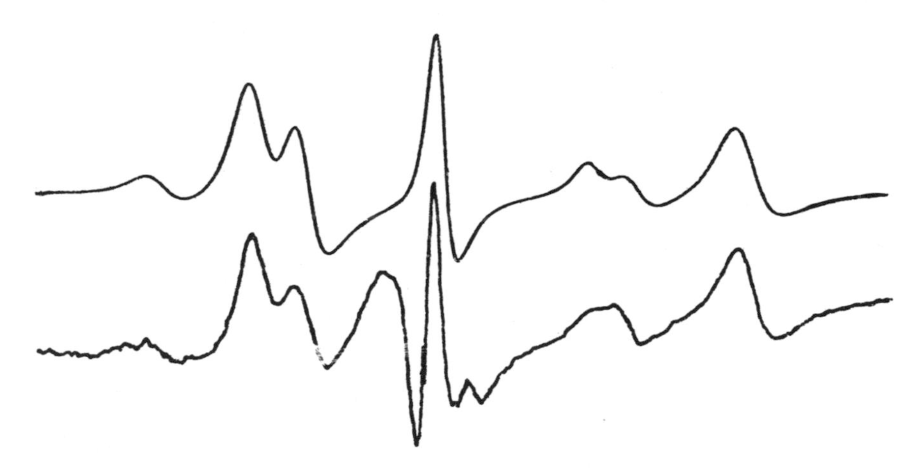

Figure 1 : Experimental (lower) spectrum for a PtGd 1300 ppm single crystal at 1.13 K and 9.162 GHz. The magnetic field is parallel to the [001] direction and the sweep is from 1 to 6 kilogauss. The upper spectrum is that calculated using the values given in the text. The experimental g value is 1.92 ± 0.01.

gnetic excitation between crystal field levels, and the changes in line intensities and positions due to the fact that there is a partial conduction electron bottleneck in this system. These effects are particularly important for the $-1/2 \rightarrow -3/2$, $-3/2 \rightarrow -5/2$ pair, and it is this pair which we use primarily to evaluate b_6. The incorporation of the conduction electron effects has been done here using the theory of Barnes[8]. The resulting theoretical spectrum is also shown in figure 1. The essential fitting parameters used are b^4, b^6, two variables governing the residual linewidths and a variable which gives a measure of the degree of bottleneck. Using the values which give a best fit in the [001] direction, we obtain equally good fits for all other directions in the {110} plane except those around the crossing angle where the structure collapses due to ion-ion exchange narrowing[7]. We obtain finally $b_4 = +86 \pm 3$ gauss and $b_6 = -2.7 \pm 0.5$ gauss. The magnitude of the fourth order term is somewhat surprising, being larger than the maximum value seen in an insulator; for Gd^{3+} in ThO_2, $b_4 = -60.5$ gauss and $b_6 = -0.56$ gauss.

In table 1 we have listed all the values of b_4 measured up until now for Gd^{3+} in metallic systems. To our knowledge no sixth order term has been measured apart from the present work. The four pure metals are all fcc structure with coordination number (CN) 12 and the pnictides have the NaCl structure and CN 6. We note that a re-examination of the experimental data for Au suggests that there may be an ambiguity in the sign of b_4.

DISCUSSION

The origins of the fine structure splitting of an S state ion in an insulator (or a metal) have been, and remain, the subject of much discussion[9]. The argument usually put forward is that there is an admixture of higher states into the ground state by the (Gd) spin-orbit coupling. The ion ground state has then some orbital character and can respond to a crystalline electric field (CEF). If this ground state remains constant in different materials then

Table 1 : b_4 parameters in gauss for Gd^{3+} in metals and intermetallic compounds

		YAs[5] +8 ±1	
Pd[2] +29 ±2	Ag[3] +7.5 ±1	YSb[4] +27 ±3	LaSb[4] +31 ±2
Pt +86 ±3	Au[1] -20 ±1	YBi[4] +47 ±5	LaBi[4] +64 ±3

the S state splittings should reflect the different CEF's.

The most striking feature of table 1 is the increase in $|b_4|$ on descending each column. Any attempts to relate this to a change in the lattice parameter by a $1/a_0^5$ dependence are not successful. It is perhaps worth noting that a parameter which does change dramatically on going down each column (increasing Z) is the host spin-orbit coupling (or nearest neighbour in the case of the pnictides). If such an increase in the host s.o. coupling can enhance the orbital character of the ion ground state, then this will be reflected by a bigger splitting in a given crystal field. It is however difficult to visualize how such an effect could be more important than the self admixture of the ion.

Assuming that both the Gd^{3+} ground state is unchanged from host to host for the pure metals in table 1, and the same total potential V_{CEF} produces the splitting for both S and non-S state ions in the same host, then one might expect that $|C_4|$ should at least increase with increasing $|b_4|$. In table 2 we list the experimental values of $C_4\beta$ and $C_6\gamma$ for Er^{3+} in Pd, Ag and Au. No data for Pt is presently available. We also list again the spin-Hamiltonian parameters for ease of comparison. It is fairly evident that there is no simple correlation between the behaviour of C_4 and b_4 in these hosts.

CONCLUSIONS

The variations in the fine structure splitting of Gd^{3+} in cubic metals cannot be easily understood in terms of the crystal field parameters. It is not clear whether this is due to differences in V_{CEF} for S and non-S state ions[9], or to changes in the S state ion ground state from host to host. To progress with this

Table 2 : Values of $C_4\beta$ and $C_6\gamma$ for Er^{3+}, and of $b_4/60$ and $b_6/1260$ for Gd^{3+} in some cubic metal hosts. All values in μK.

Pd[11] -1340 -6.45 +66.9	Ag[10] -3110 +26.9 +16.7	Host $C_4\beta$ $C_6\gamma$ $b_4/60$ $b_6/1260$
Pt +192 -0.29	Au[10] -1470 +13.5 -44.6	

problem it will be useful to have more experimental information, in particular a clear determination of the sign of b_4 in Au and values for C_4 and C_6 in Pt.

REFERENCES

1. E P Chock R Chui D Davidov R Orbach D Shaltiel and L J Tao Phys Rev Lett 27 582 (1971)

2. R A Devine D Shaltiel J-M Moret J Ortelli W Zingg and M Peter Solid State Commun 11 525 (1972)

3. S Oseroff B Gehmann S Schultz and C Rettori Phys Rev Lett 35 679 (1975)

4. For references see K Baberschke D Davidov and C Rettori Proceedings of LT 14 (Helsinki) Vol 3 p484 (NHP 1975)

5. P Urban private communication

6. A Abragam and B Bleaney EPR of Transition Ions (OUP 1970)

7. M Hardiman J Pellisson S E Barnes P-E Bisson and M Peter Submitted for publication

8. S E Barnes Phys Rev B9 4289 (1974)

9. D J Newman and W Urban Adv in Physics 24 793 (1975)

10. G Williams and L L Hirst Phys Rev 185 407 (1969)

11. A P Murani D Gignoux D Givord Proceedings of LT 14 vol 3 p302 (NHP 1975)

SOFT ELECTRONIC AND VIBRATIONAL MODES NEAR PHASE TRANSITIONS

R. J. Elliott

Department of Theoretical Physics, 12 Parks Road
Oxford, OX1 3PQ, England

ABSTRACT

At a phase transition in an electronic system the appropriate zero frequency response function must diverge. This can occur either through the movement of a finite frequency peak to zero frequency (soft mode) or the divergence of a relaxation type response (central peak). Examples of these types of behaviour in systems with crystal field splitting are discussed. When the coupling to the phonons is important (Jahn-Teller effect), the mixing of the electronic and vibrational modes is significant. This can have a dramatic effect on the elastic constants near a phase transition, and examples of this effect are also discussed.

INTRODUCTION

A large number of crystals containing magnetic ions with crystal field splittings show phase transitions. Basically these are of two types, those which show a co-operative ordering of magnetic moments, and those which order the electric multipoles of the ions. The former (magnetic transitions) are normally caused by exchange interactions, which are much larger than the classical magnetic dipole-dipole interactions. The latter normally arise, not from direct electrostatic interactions between ions, but from larger effective interaction due to the displacement of other ions in the system. Thus this type of phase transition is accompanied by a lattice distortion and can be called a co-operative Jahn-Teller effect (CJTE). Both types of transition can occur in the same material if the J-T interactions giving a transition at temperature T_D are larger than the magnetic ones which give a magnetic transition at T_C. But since the magnetic symmetry is lower it is usual to find only one transition if the magnetic interactions are large. However, the magnetic transition is accompanied by multipole ordering and lattice distortion (magnetostriction) of appropriate symmetry.

Iron group compounds with degenerate crystal field ground states in a cubic or tetrahedral environment often show CJTE at high T, while those with singlet orbital ground states do not.[1,2] In rare earth and actinide compounds the lattice interaction, though smaller than in the iron group, is normally comparable with the exchange. In those systems, magnetic transitions as in UO_2[3] or rare earth metals[4] may show large lattice distortions. Good examples of pure CJTE are given by the rare earth zircons[5] while the pnictides and similar semi-metallic compounds can show both types of transition.[6] There are excellent reviews by Gehring and Gehring[7] and Fulde[8] of these effects.

The appropriate static susceptibility of the system must diverge at the phase transition temperature. Studies of the dynamical sysceptibility at low frequencies show that this may occur in two basic ways. The response may be peaked at a frequency which decreases to zero as T_C is approached from above and from below - a soft mode. There may also be a central peak in the response at w=0 which increases in intensity near T_C. There has been considerable discussion in the recent past as to whether one or both of these phenomena are universal. In this paper we propose to review the position as it relates to systems containing magnetic ions with crystal field effects. The relation of these to other systems, e.g. those showing displacive phase transitions will also be discussed.

THE DYNAMIC SUSCEPTIBILITY

The dynamical sysceptibility of an assembly of non-interacting ions may be written

$$g(w) = \sum_{n,m} \frac{|\langle n|O|m\rangle|^2 (P_n - P_m)}{w + i\varepsilon + w_m - w_n} \qquad (1)$$

where P_n is the occupation probability of crystal field state with energy w_n. This gives the response of the system to a field h, say, so that the interaction energy takes the form $hO(\ell)$ where $O(\ell)$ is the appropriate operator - for the magnetic susceptibility it is the magnetic dipole, in other cases it may be the electric quadrupole or other multipoles. At w=0 this gives the usual Van Vleck formula with two types of contribution. Those coming from pairs where $w_m \neq w_n$ we shall call Van Vleck terms. But other contributions arise from terms when $w_m = w_n$. In this limit

$$(P_n - P_m)/(w_m - w_n) \rightarrow P_n/kT$$

and these give the Curie terms in the susceptibility. The imaginary part of g as $\varepsilon \rightarrow 0$ corresponds to transitions between levels, for the Van Vleck terms these are at finite $w = w_n - w_m$, and for the Curie terms at frequencies $w \rightarrow 0$.

In the case of interacting systems a theory which is related to the well known random phase approximation (RPA) was introduced by Fulde and Peschel[9] and developed by Buyers, Holden and Perreault.[10] It was further generalised by Young[11] and applied by him to the CJTE. For an interaction

$$\sum_{\ell \neq \ell'} J(\ell, \ell') O(\ell) O(\ell') \tag{2}$$

this gives a response function

$$G(\underline{k}, w) = \frac{g(w)}{1 - J(\underline{k}) g(w)} \tag{3}$$

where $J(\underline{k})$ is the fourier transform of $J(\ell, \ell')$. In this approximation the transition temperature is given by the molecular field result when, for the maximum value of $J(\underline{k})$,

$$1 - J(\underline{k}) g(0) = 0. \tag{4}$$

The characteristic frequencies of the excitations correspond to the poles of $G(\underline{k}, w)$. These intersperse the poles of $g(w)$ at each $w_n - w_m$. There is an excitation branch, with dispersion, for each pair of atomic levels. It is straightforward to show that if $g(w)$ contains a Curie-like term, even if it arises from excited degenerate atomic states, then $G(\underline{k}, w)$ has a pole at $w \rightarrow 0$ for all $\underline{k}$. In addition, below the transition in the ordered phase the matrix elements $\langle n|O|n \rangle \neq 0$, and a further static response appears exactly at $w=0$.

The divergence in the real part of $G(\underline{k}, 0)$ as $T \rightarrow T_C$ can arise in two ways. In the presence of a Curie term, a pole remains at $w \sim 0$ and the intensity diverges as $T \rightarrow T_C$. In the absence of such a term, one of the other Van Vleck poles approaches zero. Thus in this approximation a central peak is always present, with diverging intensity, if the single ion contains degenerate states which are populated at T_C. Only if these are absent does a true soft mode appear. Of course, if the ground states are singlets and the doublets have small P_n, the lowest excitation may soften appreciably, but it will not actually tend to zero at T_C.

The absence of a soft mode has been noted in a number of systems, the cases of Pr_3Tl[10] and $TbVO_4$[5] have been studied in particular detail. In both cases the central peak has been observed directly by neutron scattering.[12] (See also papers by Kjems and Holden at this conference).

The approximation discussed so far gives no width to the excitations. For an isolated magnetic system we expect, in analogy with theories of magnetic resonance,[13] that the width of the transitions will be related to a relaxation time by $\Delta w \sim 1/\tau_{mn}$. Moment calculations give a width depending on the root mean square of the exchange,

$$\Delta w \sim \left[\sum_{\ell'} J(\ell, \ell')^2\right]^{1/2}.$$

However near T_C, we may expect critical effects to modify these widths, although no detailed theory is available for complex systems.

OTHER ORIGINS FOR CENTRAL PEAKS

As discussed above, central peaks are not expected in systems where only singlets are populated. The simplest model of this type has only two singlet levels - this can be described by a pseudo-spin $\sigma=1/2$, say, and the Hamiltonian of the Ising model in a transverse field

$$\mathcal{H} = \sum_{\ell\ell'} J(\ell, \ell')\sigma^z(\ell)\sigma^z(\ell') - \sum_{\ell}\Gamma\sigma^x(\ell) \tag{4}$$

In certain circumstances[14] this can be regarded as analagous to the displacive phase transitions which occur because of anharmonicity in, for example, ferroelectric crystals. However, it was in these systems that prominant central peaks were first observed.[15] Subsequent observations have confirmed the effect and have demonstrated that such peaks are very narrow in w - in fact, to date all experiments show them to be narrower than the level of detection employed. The explanation of this phenomenon must lie outside the theory discussed above, and indeed three different types of mechanism have been proposed for it.

(1) The dynamics of the spin system is certainly more complicated in the critical region than that given by the RPA. Improved calculations[16] do give some extra structure at low w as the soft mode becomes overdamped, but it is not a very sharp central peak. Computer simulations of the classical anharmonic model[17] in one and two dimensions show pronounced structure, although the width of the central peak is not easily determined. It has been suggested that, certainly in one dimension, this reflects the motion of

'domains' of parallel spins related to soliton effects.[18] However, it is far from certain that these explanations will be valid in three-dimensional systems or that they predict a sharp enough central peak to explain the experimental results.

(2) Coupling to other degrees of freedom at low w has been suggested by Cowley and Coombs[19] for the anharmonic phonon case. Their mechanism for the present model has been treated by Elliott and Young[20] using terms in the Hamiltonian of the form

$$\mathcal{H}' = \sum_{jj'\underline{k}\underline{q}} V^z(\underline{q})(c_j^+(\underline{k}) + c_j(-\underline{k}))(c_{j'}^+(-\underline{k}+\underline{q}) + c_{j'}(\underline{k}-\underline{q})) \tag{5}$$

in the limit j=j' and $\underline{q}$=0. The spin mode then couples to a response representing fluctuations in the phonon number - in fact, the overall effect is very similar to the addition of a Curie term. However, the simple theory gives the central peak a width related to the phonon relaxation time $1/\tau_{pn}$. Recently it has been suggested[21] that a more detailed treatment of the many-body effects will lead to a critical narrowing of this central peak, but the validity of this approximation is not substantiated.

(3) Finally, there remains the possibility that the narrow central peak is a static impurity effect. In the critical region where the susceptibility is high, an impurity may induce a surrounding region (domain) of order. Recently this idea has been given quantitative treatment which enhances its plausibility.[22]

On the whole the narrowness of the observed central peaks favours the last static explanation, although effect of types (1) and (2) should exist, and give broader features near w=0.

ELASTIC PROPERTIES

In the case of CJTE the co-operative interaction arises through the phonons. The Hamiltonian

$$\mathcal{H} = \sum_{jk} \hbar w_j(\underline{k})(c_j^+(\underline{k})c_j(\underline{k})+\tfrac{1}{2}) + \sum_{jk} \xi_j(\underline{k})(c_j^+(\underline{k})+c(-\underline{k})) \sum_{\ell} O(\ell)e^{i\underline{k}\cdot\underline{\ell}} \tag{6}$$

can be transformed to displaced operators

$$\mathcal{H} = \sum_{jk} \hbar w_j(\underline{k})(\gamma_j^+(\underline{k})\gamma_j(\underline{k})+\tfrac{1}{2}) + \sum_{\ell\ell'} J(\ell, \ell')O(\ell)O(\ell') \tag{7}$$

where

$$J(\underline{k}) = \sum_j K_j(\underline{k}) - \frac{1}{N} \sum_{j\underline{k}} K_j(\underline{k}) \quad \text{and} \quad K_j(\underline{k}) = 2|\xi_j(\underline{k})|^2/\hbar w_j(\underline{k}) \tag{8}$$

In the general case, where the phonons couple to several different operators $O(\ell)$, the commutation rules of the γ are changed. Nevertheless, it is possible to perform the usual RPA decoupling. The resulting susceptibility retains the form (2), but the dynamics of the phonons is important. This may be included(11) by replacing $J(\underline{k})$ in (3) by

$$J(\underline{k}) + w^2 \sum_j K_j(\underline{k})/(w_j^2(\underline{k}) - w^2) \tag{9}$$

$G(\underline{k}, w)$ then has poles arising from both the electronic and phonon frequencies, which are coupled. The mutual repulsion of such pairs is observed(23) when $w_j(\underline{k}) \sim w_m - w_n$.

Such a coincidence of frequencies is certain to happen near T_C between the soft mode and/or central peak and the acoustic phonons. In fact, since $w_a(k) = sk$, say, tends to zero as $k \to 0$, and the width of acoustic mode also tends to zero, while the width of central peak does not, the acoustic mode resonse always lies lower in frequency than the bulk of the electronic resonse when k is small enough. Thus, when this is coupled to the electronic system, it is the acoustic mode which forms the soft mode of the total system. This is compatible with the fact that the crystal is unstable against distortion at a CJTE phase transition.

The softening of elastic constants near T_C has been observed in numerous materials.(8, 24, 25) In view of the other response near $w=0$, there may be variations with frequency. In the case of a central peak with width $1/\tau$, the full softening is only observed at $w \ll 1/\tau$. Thus, neutron or Brillouin experiments made at relatively high w may not show the effect. A good example of this is $TbVO_4$(24) which is expected to have a central peak with a width corresponding to spin-spin relaxation time. In this case full softening is seen by acoustic resonance but not by Brillouin scattering. The difference corresponds to different thermodynamic regimes - the g(w) may be written at low w

$$g(w) = g_I(w) + \frac{g_S - g_I(0)}{1 - iw\tau} + (g_T - g_S)\delta(w) \tag{10}$$

where g_I, g_S, g_T are the isolated, adiabatic and isothermal susceptibilities. If spin-lattice relaxation is much larger than the spin-spin, there will be a further regime when $1/\tau_S > w > 1/\tau_L$.

In conclusion, we may summarise the situation in the following manner. We expect a central peak, rather than a soft mode in the electronic response of these crystals, except in special circumstances. However, the peak will be relatively broad - and not of the very narrow type observed at some displacive phase transitions. This is probably an impurity effect - it may also be found in the CJTE. In appropriate symmetry conditions we do expect a soft acoustic mode at T_C.

REFERENCES

1. J. B. Goodenough, 'Magnetism and the Chemical Bond' (Interscience, N.Y., 1963).

2. R. Englman, 'The Jahn-Teller Effect in Molecules and Crystals' (Wiley, London, 1972).

3. S. J. Allen, Phys. Rev. 166, 530; 167, 492 (1968); J. Faber, G.H. Lander and B.R. Cooper, Phys. Rev. Lett. 35, 1770 (1975).

4. J. J. Rhyne, 'Magnetic Properties of Rare Earth Metals' (Ed. Elliott; Plenum, 1972).

5. R. J. Elliott, R. T. Harley, W. Hayes and S. R. P. Smith, Proc. Roy. Soc. A238, 217 (1972).

6. P. M. Levy, J. Phys. C 6, 3545 (1973); E. Bucher et al., Proc. 1st Conf. on Crystalline Field Effects in Metals, p.221 (Ed. Devine, Unw. Montreal, 1975).

7. G. A. Gehring and K. A. Gehring, Rep. Prog. Phys. 38, 1, (1975).

8. P. Fulde in 'Handbook of Physics and Chemistry of Rare Earths', Ed. Gschneidner and Eysing (to be published).

9. P. Fulde and I. Peschel, Z. Phys. 241, 82, (1971).

10. W. J. L. Buyers, T. M. Holden and A. Perreault, Phys. Rev. B11, 266 (1975); W.J.L. Buyers, AIP Conf. Proc. 24, 27 (1975).

11. A. P. Young, J. Phys. C 8, 3158 (1975).

12. M. T. Hutchings, R. Scherm, S. H. Smith and S. R. P. Smith, J. Phys. C 8, L393 (1975); J. Als-Nielson et al., Proc. Int. Mag. Conf., Amsterdam (1976).

13. R. Blinc and B. Zeks, Adv. Phys. 21, 693 (1972).

14. R. J. Elliott, 'Structural Phase Transitions and Soft Modes', p. 235 (Ed. Samuelson, Anderson and Feder, Oslo Universitetsforlaget, 1971); R. B. Stinchcombe, J. Phys. C 6, 2459 (1973).

15. G. Shirane and J. D. Axe, Phys. Rev. B8, 1965 (1973); R. Comes et al., Phys. Rev. B8, 571 (1973).

16. M. A. Moore and H. C. W. L. Williams, J. Phys. C 5, 3168, 3185, 3222 (1972); T. N. Tommett and D. L. Huber, Phys. Rev. B11, 1971 (1975).

17. T. Schneider and E. Stoll, Phys. Rev. Lett. 31, 1254 (1973)

18. A. R. Bishop, E. Domany and J. A. Krumhansl,'Ferroelectrics' (to be published).

19. R. A. Cowley and G. Coombs, J. Phys. C 6, 121, 143 (1973).

20. R. J. Elliott and A. P. Young, J. Phys. C 7, 2721 (1974).

21. J. Ruvalds, E. N. Economou and N. L. Ngai, 'Light Scattering in Solids', p.770 (Ed. Balkanski, Leite and Porto, Flammarian, 1975).

22. C. M. Varma and P. Hohenberg (to be published).

23. R. J. Birgeneau, L. C. van Uitert, J. K. Kjems and G. Shirane, Phys. Rev. B10, 2512 (1974); J. K. Kjems, W. Hayes and S. H. Smith, Phys. Rev. Lett. 35, 1089 (1975).

24. J. B. Saundercock et al., J. Phys. C 5, 3126 (1972).

25. E. Pytte, Phys. Rev. B8, 3954 (1973).

INDUCED JAHN TELLER TRANSITION IN $PrCu_2$

H.R. Ott[+,§], K. Andres*, P.S. Wang[§], Y.H. Wong[§] and B. Lüthi[§]

[+]Laboratorium für Festkörperphysik, ETH-Hönggerberg, 8093 Zürich, Switzerland
*Bell Telephone Laboratories, Murray Hill, N.J. 07974, USA
[§]Department of Physics, Rutgers University, New Brunswick, N.J. 08903, USA

ABSTRACT

We have investigated the low temperature phase transition of $PrCu_2$ by means of several experimental methods on single crystal specimens. We have measured the magnetic susceptibility, the electrical resistivity, the thermal expansion and the magnetostriction, several elastic constants and the thermal conductivity. We conclude that $PrCu_2$ undergoes an induced type Jahn-Teller effect at 7.5 K.

Magnetic ordering in singlet ground state systems is a well known effect and has recently been investigated in many rare earth intermetallic compounds. Apparently less frequently studied is its analogue for structural phase transitions, the cooperative Jahn-Teller effect. Whereas in the case of magnetic ordering a supercritical exchange interaction can lead to long range order among singlet ground state ions, the structural order occurs if the total quadrupole coupling between the ions exceeds a critical value. This situation was found before in insulating materials[1)] but never met in metallic systems. The first such example appears now to be the intermetallic compound $PrCu_2$ where the crystal electric field J=4 ground state is completely split into singlets due to the low symmetry of the crystal lattice. TmCd is another intermetallic compound exhibiting a Jahn-Teller effect but in this case the ground state of the Tm^{3+}-ion is a non-magnetic doublet Γ_3 and hence a transition may be expected for any finite value of the magnetic ion-lattice coupling constants. In addition it should be mentioned that singlet ground state systems usually exhibit a parastructural effect which is due to the strain- or volume dependence of the crystal electric field energy levels[2].

The first indication of a low temperature phase transition in $PrCu_2$ was found from specific heat measurements[3] where an anomaly was found around 7 K. More recently, Wun and Phillips[4] have re-examined this anomaly and they suggested a structural transition to be the cause for this behaviour. We have now investigated the transition in more detail by studying various physical properties on single crystals of $PrCu_2$.

$PrCu_2$ crystallizes in an orthorhombic structure and we have investigated the magnetic susceptibility, the electrical resistivity, the thermal expansion and the magnetostriction along the three principal axes of the crystal lattice. Moreover we studied certain elastic constants and the thermal conductivity in an arbitrary direction.

In fig. 1 we show the temperature dependence of the magnetic susceptibility along the three principal axes. A pronounced anisotropy

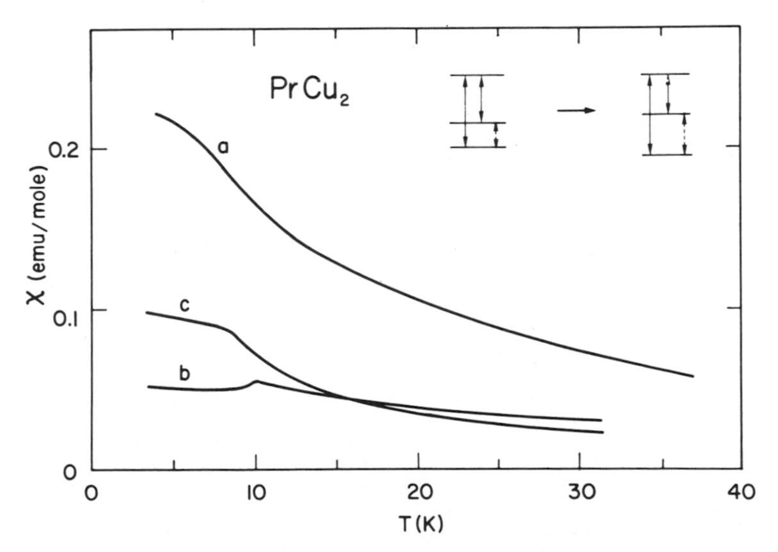

Fig. 1: Magnetic susceptibilities in the three principal directions measured in a field of 1 kOe. Solid arrow=J_z-interaction, broken arrow=J_z^2-interaction.

is observed and small, but distinct features may be seen below 10 K. The general behaviour can be understood by considering a three singlet level scheme where the lowest two levels are coupled via J_z^2. The Van Vleck susceptibility is then arising from matrix elements of J_x, J_y and J_z with higher lying levels. Without the knowledge of the appropriate wavefunctions, however, a quantitative analysis is not possible.

Fig. 2 shows the temperature dependence of the velocity of shear sound waves propagating along the a-direction of the crystal lattice. A strong softening may be seen for the polarisation vector $\vec{R}$ parallel to the b-direction while no temperature dependence is observed for $\vec{R}$ parallel to $\vec{c}$. The solid line is calculated using a three singlet level system with the excited states at 6 K and 18 K respectively and only the lowest two levels are quadrupole coupled. From this fit we deduce a rather strong magnetoelastic coupling constant which is about four times larger than the quadrupole-quadrupole coupling constant due to other sources[1,2].

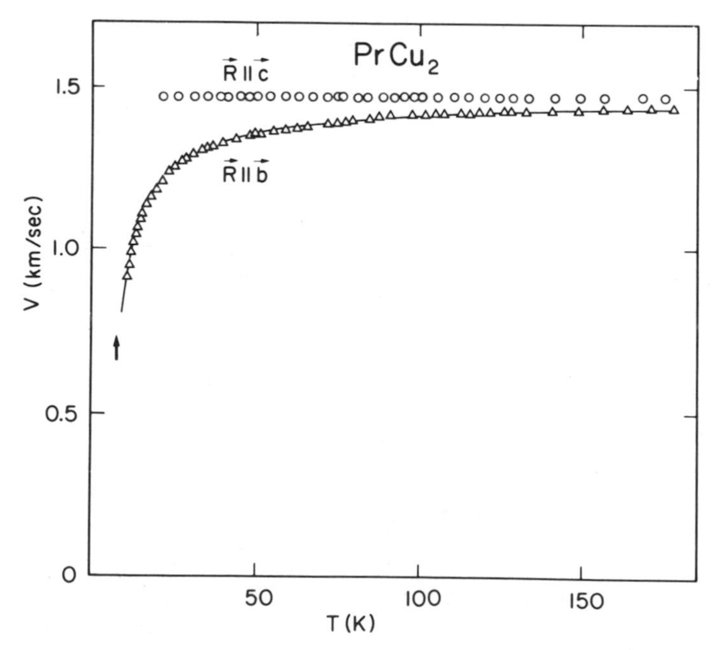

Fig. 2: Temperature dependence of shear sound waves propagating along the a-axis. $\vec{R}$ denotes the polarisation vector of the shear waves. Solid line is calculated (see text).

The temperature dependence of the linear thermal expansion coefficients α_i along the three principal axes is shown in fig. 3. Distinct extremal values of α_i around 7.5 K are observed giving evidence for an actual distortion of the crystal lattice. The ion-lattice coupling is thus very strong in $PrCu_2$ and we have indeed a so called induced Jahn-Teller effect in this compound in the manner mentioned above. From our measurements a net volume change is observed around the transition.

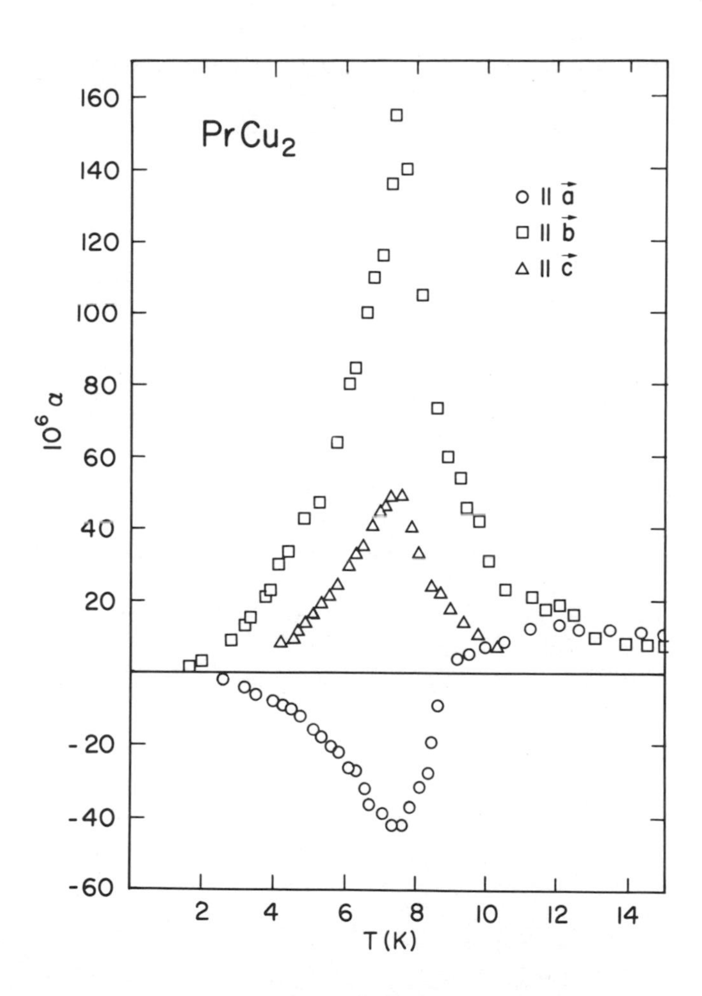

Fig. 3: Linear thermal expansion coefficients α_i along the three principal crystal axes.

From all these measurements and from additional experimental results which we shall not discuss in detail here, we conclude that $PrCu_2$ exhibits a cooperative Jahn-Teller effect at 7.5 K. In comparison with the earlier reported Jahn-Teller transition in TmCd[5] we observe a smaller strain occurring at the transition. This is, however, not surprising considering that the transition is only of induced type. Moreover we find that a magnetic field of 30 kOe does not shift the transition in temperature. In TmCd the transition temperature was raised appreciably even in small magnetic fields.

More experimental data are needed for a complete description of this induced Jahn-Teller effect in $PrCu_2$. Certain strains and additional elastic constants have to be investigated and a determination of the crystal electric field energy levels has to be made. This will then facilitate a comparison with theoretical calculations. Such work is now in progress.

REFERENCES

1. see e.g. G.A. Gehring and K.A. Gehring, Rep. Progr. Phys. 38, 1 (1975)

2. M.E. Mullen, B. Lüthi, P.S. Wang, E. Bucher, L.D. Longinotti, J.P. Maita and H.R. Ott, Phys. Rev. B10, 186 (1974)

3. K. Andres, E. Bucher, J.P. Maita and A.S. Cooper, Phys. Rev. Letters 28, 1652 (1972)

4. M. Wun and N.E. Phillips, Phys. Letters 50A, 195 (1974)

5. B. Lüthi, M.E. Mullen, K. Andres, E. Bucher and J.P. Maita, Phys. Rev. B8, 2639 (1973)
   H.R. Ott and K. Andres, Solid State Comm. 15, 1341 (1974)

CRYSTAL FIELDS AND DYNAMIC JAHN TELLER COUPLING OF AN ELECTRONIC QUARTET TO Γ_5 LATTICE MODES IN THE FINE STRUCTURE OF Er^{3+} IN Pd-Er

J.M. Dixon

Physics Department, University of Warwick

Coventry, CV4 7AL, ENGLAND

ABSTRACT

The angular variation of the field for resonance H, for Er^{3+} in Pd-Er, has been investigated theoretically without introducing static lattice distortions from cubic symmetry and couplings to excited electronic states. In the limit of strong anharmonicity, using a dynamic lattice model, it is possible to reproduce the splittings of one of the E.P.R. lines in the regime of both strong and weak inversion splitting.

INTRODUCTION

The E.P.R. spectrum of Er^{3+} in single crystals of Pd observed by Devine et al [1,2] was interpreted in terms of a simple Γ_8 but there were deviations from this simple picture. To understand these we label the eigenstates of the Γ_8 in a magnetic field as $|i\rangle$ where i=1,2,3,4 and such that $|i\rangle$ has the energy level E_i with $E_1<E_2<E_3<E_4$. The angular variation of the field for resonance H was not the same for the $1\leftrightarrow2$ and $3\leftrightarrow4$ transitions as would be expected for a simple Γ_8. In one of the samples, with the field along two different [1,1,1] directions, the resolved lines were similar (Zevin and Shaltiel [3]).

The difference between the $1\leftrightarrow2$ and $3\leftrightarrow4$ transitions was explained by Zingg et al [4] by the presence of the low lying electronic Γ_6 level and other authors [3] have suggested static lattice distortions but this latter cannot explain the above.

This paper discusses the possibility that similar resolved spectra (with the transitions $1\leftrightarrow 2$ and $3\leftrightarrow 4$ separated) could be obtained with the field along <u>any</u> $\langle 1,1,1\rangle$ direction, the system being always cubic and using only a Γ_8 electronic state.

THE MODEL

We use a cluster model for the Pd ions neighbouring the Er^{3+} ion. The Pd cluster will be described by the hamiltonian H_{modes} and its interaction with the rare earth by H_{int}. Our effective Hamiltonian is thus

$$H = H_{ion} + H_{int} + H_{modes}$$

We form collective co-ordinates thus:-

$$Q_o = \tfrac{1}{2}(q_1+q_2+q_3+q_4) \qquad (\Gamma_1)$$

$$\left.\begin{aligned} Q_1 &= \tfrac{1}{2}(q_1-q_2+q_3-q_4) \\ Q_2 &= \tfrac{1}{2}(q_1+q_2-q_3-q_4) \\ Q_3 &= \tfrac{1}{2}(q_1-q_2-q_3+q_4) \end{aligned}\right\} \qquad (\Gamma_5)$$

where q_i (i=i,2,3,4) are defined as body diagonal deformations as shown in figure 1.

The collective displacement Q_o corresponds to a breathing mode of the cluster which we henceforth neglect. The kinetic and potential energy of vibration of the Γ_5 modes gives rise to the hamiltonian

$$H_{modes} = (1/2M)(P_1^2+P_2^2+P_3^2)+(M\omega^2/2)(Q_1^2+Q_2^2+Q_3^2)+BQ_1Q_2Q_3$$

P_1, P_2 and P_3 are momenta conjugate to Q_1, Q_2 and Q_3 respectively and ω is the angular frequency of the modes. M is the mass of a Pd ion and the last term describes anharmonic elastic vibrational energies (White and Pawlowicz[5]).

We describe the electronic Γ_8 with an effective spin $\tilde{S}=3/2$ and H_{ion} by

$$H_{ion} = g\beta(H_x\tilde{S}_x+H_y\tilde{S}_y+H_z\tilde{S}_z) + f\beta(H_x\tilde{S}_x^3+H_y\tilde{S}_y^3+H_z\tilde{S}_z^3)$$

Bleaney[6] where we have taken account of the exchange coupling J_{fs} between the local moment and the conduction electrons by adding

$$\{ (g_J-1)J_{fs}g_eN(E_F)/2(1-N(E_F))\bar{V} \} \beta\, \underline{H}.\underline{J}$$

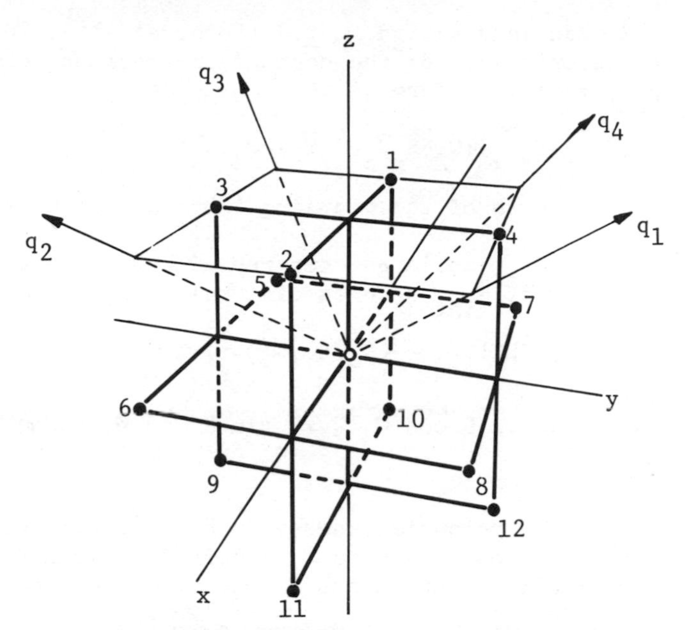

Fig. 1. Body Diagonal Deformations of a f.c.c. Structure.

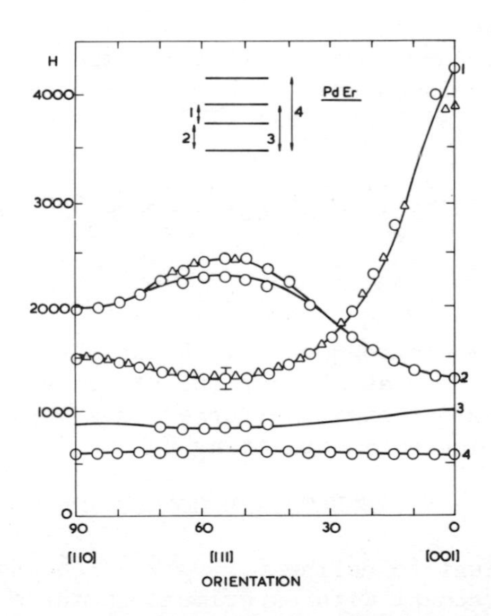

Fig. 2. The angular variation of the field for resonance for the transitions within the vibronic Γ_8 quartet. Solid line, theory; circles and triangles, expt.

to the ionic Zeeman interaction $g_J\ \underline{H}.\underline{J}$ (Peter et al[7]). Thus g and f above will henceforth include the conduction electron local moment coupling. H_{int} takes the form

$$H_{int}=\epsilon\{Q_1(\tilde{S}_x\tilde{S}_y+\tilde{S}_y\tilde{S}_x)+Q_2(\tilde{S}_x\tilde{S}_z+\tilde{S}_z\tilde{S}_z)+Q_3(\tilde{S}_y\tilde{S}_z+\tilde{S}_z\tilde{S}_y)\}$$

where ϵ is the strength of the coupling (Abragam and Bleaney[8]).

We assume that H_{int} is large compared with H_{ion} so we perform a transformation so that H_{int} becomes diagonal using the states

$$|\alpha^{\pm}\rangle=\mp\frac{i}{2}|\tfrac{3}{2}\rangle+\frac{1}{2}(C\mp SE^*)|+\tfrac{1}{2}\rangle\mp\frac{1}{2}(C\pm SE)|-\tfrac{1}{2}\rangle-\frac{i}{2}|-\tfrac{3}{2}\rangle$$

where $Q_1=R\cos\theta, Q_2=R\sin\theta\cos\varphi$, $Q_3=R\sin\theta\sin\varphi$, $c=\cos\theta, s=\sin\theta$ and $E=\exp i\varphi$

In general V, the potential energy in H_{modes}, will be much larger than H_{int} or H_{ion} so now we assume B is large and look for minima in V. If $B>0$ four minima occur at $\cos\theta=-\frac{1}{\sqrt{3}}$, $\varphi=\pi/4$ (A); $\cos\theta=1/\sqrt{3}$, $\varphi=\frac{3\pi}{4}$(B); $\cos\theta=-\frac{1}{\sqrt{3}}$; $\varphi=\frac{5\pi}{4}$ (C); $\cos\theta=+\frac{1}{\sqrt{3}}$, $\varphi=\frac{7\pi}{4}$ (D), and $R=R_o$

Similarly there are four minima at A', B', C', D' when $B<0$ but at different angular positions.

A secular determinant is now set up with eight Born-Oppenheimer product states of $|\alpha^{\pm}\rangle$ at each potential minimum together with approximate vibrational eigenstates of H_{modes}. We then add W as a perturbation which will have matrix elements within and between these states where

$$W=\frac{1}{2M}\{R_o^{-2}P_\theta^2+R_o^{-2}S^{-2}P_\varphi^2+\frac{1}{4}\hbar^2R_o^{-2}\cot^2\theta-\frac{1}{2}\hbar^2R_o^{-2}S^{-2}\}+BR_o^3CS^2\cos\varphi\sin\varphi$$

$$+\ H_{ion}$$

and P_θ, P_φ are momenta conjugate to θ and φ respectively. The lowest eight states transform as $2\Gamma_8$ for the $B>0$ case and as $\Gamma_8+\Gamma_6+\Gamma_7$ when $B<0$. We will consider only the former case when the two Γ_8's are separated by the inversion splitting 2δ.

RESULTS AND CONCLUSION

It is convenient to define $E_1=(4g+7f)/2\sqrt{6}$ and $E_2=(4g+13f)/8\sqrt{3}$. We can obtain agreement with experiment in three ways, the best of which gives the fit to the field for resonance data shown in fig. 2. In this case $x=0.28$[9]. $E_1=3.2$, $E_2=-0.96$, $\delta=19.62\text{cm}^{-1}$ $J_{fs}=-0.023\text{eV}$ and where the transitions here are solely <u>within</u> the lowest vibronic Γ_8 state and are identified in the same way as ref. 1and 2.

We obtain a value of x approximately a factor of two smaller than Devine et al and of opposite sign. Our model explains how one can obtain identical spectra with the field along any 1,1,1, direction and in each case "line 2" will be split. The observation of an excited Γ_6 could be explained by the presence of an excited vibronic Γ_6 state and not a purely electronic crystal field Γ_6 state.

We conclude that if a dynamic ion-lattice coupling is assumed, a splitting of "line 2" seen by Devine et al can be explained without resort to static lattice distortions from cubic symmetry and without using excited electronic states.

REFERENCES

1. Devine, R.A.B., Zingg, W. and Moret, J.M., Solid State Communic. 11, p.233-237, (1972a).

2. Devine, R.A.B., and Moret, J.M., Phys. Letts. 41A, 11 (1972b).

3. Zevin, V and Shaltiel, D., (private communication).

4. Zingg, W., Bill, H., Buttet, J. and Peter, M., Phys. Rev. Lett., 32. 1221. (1974).

5.

5. White, G.K. and Pawlowicz, A.T., J. of Low Temp. Phys., Vol. 2, no. 5/6, 1970, P.631-639.

6. Bleaney, B., Proc. Phys. Soc., 73, 939 (1959).

7. Peter, M., Dupraz, J. and Cottet, H., Helv. Phys. Acta, 40, 301 (1967).

8. Abragam, A. and Bleaney, B., "Electron Paramagnetic Resonance of Transition Ions", Clarendon Press, Oxford (1970).

9. Lea., K.R., Leask, M.J. and Wolf, W.P., J. Phys. Chem. Solids, 23, 1381, (1962).

SPIN DYNAMICS OF THE S = 2 MAGNET IN A CUBIC CRYSTAL FIELD

T. Egami[+] and M.S.S. Brooks[++]

[+]Department of Metallurgy and Materials Science,
University of Pennsylvania, Philadelphia Pa 19174 USA
[++]European Institute for Transuranium Elements, Euratom
P.O. Box 2266, D-75 Karlsruhe

ABSTRACT

The S = 2 magnet in a cubic field is studied when the ground state is a non-magnetic doublet. The system exhibits induced moment magnetism with both Van Vleck and Langevin contributions to the susceptibility. In the paramagnetic phase, as the temperature approaches T_c from above, the longitudinal spectrum contains a softening mode the frequency of which extrapolates to zero at a temperature below T_c. A central peak due to excitations within the degenerate triplet is also present. The structure of equations for the Green functions is similar to that obtained by Young and Elliott for a ferroelectric; consequently the two modes are strongly mixed at long wavelengths and the low frequency response may be expressed in Debye form with a relaxation time that diverges at T_c. Results are in agreement with dynamic susceptibility theories of Buyers and Lines.

INTRODUCTION

In the strong crystal fields of lanthanide and actinide systems the magnetic moment contains, in general, both Van Vleck and Langevin contributions[1,2]. The magnetic phase transition is then driven by a central peak[2] in contrast to closely studied cases of pseudo-spin systems of the Van Vleck type[4] where the frequency of a soft mode falls to zero at T_c. The S = 2 magnet in a cubic field is the simplest system of mixed susceptibility for which a full pseudo-spin theory[4] may be employed. The total longitudinal moment is a sum of Van Vleck and Langevin parts and the dynamics of the two parts are strongly coupled through the exchange field: this is shown to produce a typical structure in the equations for the Green functions. The frequency of the softening mode does not vanish at T_c but a central peak narrows and diverges, as found by Buyers[2] and Lines[3] for other cubic systems.

MOLECULAR FIELD AND DYNAMIC SUSCEPTIBILITY THEORY

The system is characterized by the Hamiltonian (with spin equal to 2)

$$H = -\sum_{n \neq m} J(\underline{n}-\underline{m})\underline{S}_n \cdot \underline{S}_m + H\sum_n \tilde{O}_1^0(\underline{n}) + B_4^0 \sum_n \left[\hat{O}_4^0(\underline{n}) + \left(\tfrac{10}{7}\right)^{1/2} \hat{O}_4^4(\underline{n})\right] \quad (1)$$

The $\tilde{O}_n^m$ are spherical tensor operators. H is the applied magnetic field, J the exchange interaction and B_4^0 the only non-zero crystal field parameter. $\tilde{O}_4^4$ is the symmetric combination $\frac{1}{2}(\tilde{O}_4^4 + \tilde{O}_4^{-4})$. The eigenstates of the CEF are classified as a Γ_3 doublet and a Γ_5 triplet. When $B_4^0 < 0$ the doublet lies lowest; this corresponds to an easy direction along one of the cubic axes in the ordered phase.

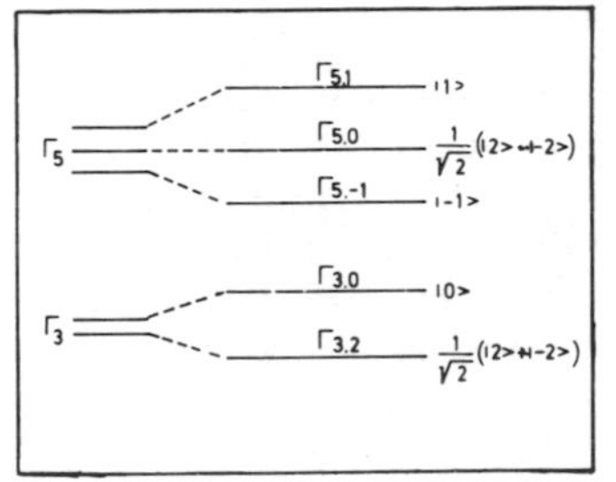

The multiplets are separated by a gap $\Delta = -15\ B_4^0$. The level scheme for the system is shown in fig.1. The ionic susceptibility is obtained by setting J equal to zero and following the usual perturbation expansion due to Van-Vleck. The susceptibility contains both Van Vleck (from $\Gamma_{3,2};\Gamma_{5,0}$) and Langevin (from $\Gamma_{5,1};\Gamma_{5,-1}$) contributions:

Fig.1. Energy level scheme

$$\chi_0 = \chi_0^{VV} + \chi_0^{L} = \frac{2[\beta + \frac{4}{\Delta}(e^{\beta\Delta} - 1)]}{3 + 2e^{\beta\Delta}} \quad (2)$$

The enhanced susceptibility is recovered when the total field is assumed to contain a molecular field (MF) contribution (obtained by replacing the exchange interaction by $\phi\sum_n \tilde{O}_1^0(n)$ with $\phi = -2J(0)\langle\tilde{O}_1^0\rangle$) in addition to H

$$\chi = (\chi_0^{VV} + \chi_0^{L}) / [1 - (\chi_0^{VV} + \chi_0^{L})\phi] \quad (2)$$

Since $\Delta \neq 0$ a critical value of exchange, $\Delta/8$, is required for ordering at zero temperature. A graph of Curie temperature against ratio of ϕ to Δ is shown in fig. 2.

The single-ion and the exchange enhanced dynamic susceptibilities (Green functions) of the Buyers[2] and Lines[3] theories are readily deduced

$$G(q,\omega) = 1/[1/g_0(\omega) - 2J(q)] \quad (3)$$

where

$$g_0(\omega) = -8\Delta f_1/(\omega^2 - \Delta^2) + 2\beta f_2 \quad (4)$$

and $f_1 = (1 - e^{-\beta\Delta})/z$; $f_2 = e^{-\beta\Delta}/z$; $z = 2 + 3e^{-\beta\Delta}$

The values of the Curie temperature from (4) are identical with the MF values of fig.2.

PSEUDO-SPIN THEORY

We employ the method of a unitary transformation to pseudo-spin space described in previous papers[4]. To the spin Hamiltonian (1) is applied the transformation A, thus the Hamiltonian is $\tilde{\mathcal{H}} = A\mathcal{H}A^{-1}$. The form of A required to diagonalize (1) is

$$A = \exp[i\theta\hat{O}_4^4/3(\tfrac{35}{2})^{1/2}] \quad \text{where } \hat{O}_4^4 = \tfrac{1}{2}(\tilde{O}_4^4 - \tilde{O}_4^{-4})$$

This transformation takes $\Gamma_{5,2} \to |-2\rangle$, $\Gamma_{3,0} \to |2\rangle$ and leaves other states unchanged. θ is chosen, so that $\tilde{\mathcal{H}}$ is as close to diagonal form as possible, in some given approximation. After extremely lengthy but straigthforward algebra one finds the transformed longitudinal Hamiltonian as

$$\tilde{\mathcal{H}} = B_4^0 \sum_n [\tilde{O}_4^0(n) - \tfrac{15}{4}\sin\theta L_3(n) + \tfrac{15}{4}L_1(n)] - \sum_{n,m} J(n-m)[\cos\theta L_3(m) + L_2(m) + \sin\theta L_1(m)][\cos\theta L_3(n) + L_2(n) + \sin\theta L_1(n)] \tag{5}$$

$$\text{where } L_3 = \tfrac{2}{15}(6\tilde{O}_1^0 + \tilde{O}_3^0), \quad L_2 = \tfrac{1}{5}(\tilde{O}_1^0 - \tfrac{2}{3}\tilde{O}_3^0), \quad L_1 = \tfrac{4}{15}\sqrt{\tfrac{10}{7}}\,\bar{O}_4^4 \tag{6}$$

Under the transformation the Z-component of spin becomes

$$\tilde{O}_1^0 = \sin\theta L_1 + \cos\theta L_3 + L_2 \quad [\text{If } J = \tilde{O}_1^0;\ J = L_1 + L_2 \text{ when } \theta = \tfrac{\pi}{2}] \tag{7}$$

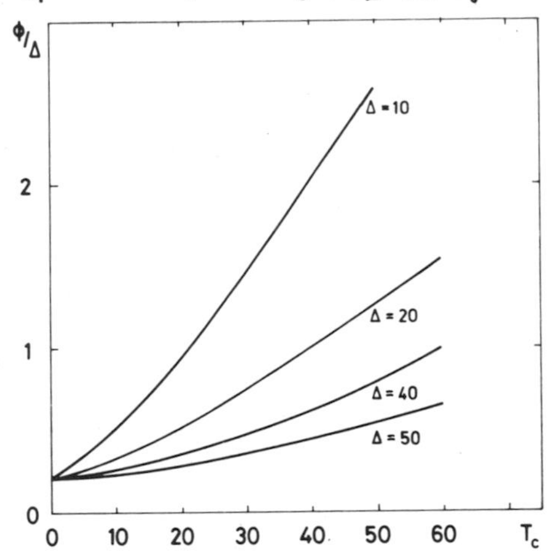

Fig.2. Curie temperature as a function of the ratio of molecular field to energy gap.

The angle of the transformation, is determined variationally[4]. This introduces the possibility of improving upon MFA or RPA. However in this paper we restrict ourselves (a) to simplest approximation (RPA) (b) to the paramagnetic phase where, in RPA, $\theta = \pi/2$. Equation (7) then clearly shows that the longitudinal fluctuation has two components in pseudo-spin space: 1) L_1 which generates the softening mode excitation, 2) L_2 which is diagonal in pseudo-spin space and is responsible for the Langevin susceptibility.

From standard linear response theory (e.g. Lines[3]) one finds

$$\mathcal{G}^{JJ}(q,\omega) = g_0^{JJ}(\omega) + g_0^{JJ}(\omega) V_q \mathcal{G}^{JJ}(\omega) \tag{8}$$

$$\text{where } \mathcal{G}^{A_1A_2}(q,\omega) = \langle\langle A_1(q); A_2(-q)\rangle\rangle_\omega \text{ and } V_q = -2J(q) \tag{9}$$

and g_0 is the Green function when J is set equal to zero (ionic Green function). Also, from the equation of motion for L_1

$$\mathcal{G}^{L_1A}(q,\omega) = g_0^{L_1A}(\omega) + g_0^{L_1A}(\omega) V_q \mathcal{G}^{JA}(q,\omega)$$

From (8) and (10) it is possible to deduce the equations for the Green functions of the softening mode and the intra-triplet mode.

If one defines

$$\mathcal{G}_0^{AA}(q,\omega) = g_0^{AA}(\omega)/[1 - g_0^{AA}(\omega)\,2J(q)] \qquad (A = L_1, L_2) \tag{11}$$

as the enhanced, non-interacting, Green functions for each of the modes one finds

$$\mathcal{G}^{L_1L_1} = \mathcal{G}_0^{L_1L_1} + \mathcal{G}_0^{L_1L_1} V_q\, \mathcal{G}^{L_1L_2} \;;\; \mathcal{G}^{L_1L_2} = \mathcal{G}_0^{L_1L_1} V_q\, \mathcal{G}^{L_2L_2} \tag{12}$$

$$\mathcal{G}^{L_2L_2} = \mathcal{G}_0^{L_2L_2}(1 + V_q\, \mathcal{G}^{L_1L_2})$$

This set of equations is identical in structure to those obtained by Young and Elliott[5] for ferroelectric response in a pseudo-spin model. Equations (12) are in closed form with solutions

$$\mathcal{G}^{L_1L_1} = \mathcal{G}_0^{L_1L_1}/[1 - \mathcal{G}_0^{L_1L_1}\Sigma_1] \;;\; \mathcal{G}^{L_2L_2} = \mathcal{G}_0^{L_2L_2}/[1 - \mathcal{G}_0^{L_2L_2}\Sigma_2] \tag{13}$$

where the self-energies are

$$\Sigma_1(q,\omega) = V_q^2\, \mathcal{G}_0^{L_2L_2}(q,\omega) \;;\; \Sigma_2(q,\omega) = V_q^2\, \mathcal{G}_0^{L_1L_1}(q,\omega) \tag{14}$$

It is easy to reduce equations (13) and add them to recover the Buyers' result (4) if one uses equation (11). However equations (13) allow us to exploit the analogy with ferroelectrics to show the problem as one of two interacting modes. One cannot obtain the Green function $\mathcal{G}_0^{L_2L_2}$ from its equation of motion but, since it is a diffusive mode[5,6] a relaxation time, τ, may be used to show that

$$\mathcal{G}_0^{L_2L_2}(q,\omega) = \chi_0^{L_2}(q)/[1 - i\omega\tau\, \chi_0^{L_2}(q)] \tag{15}$$

where

$$\chi_0^{L_2}(q) = \chi_0^{L}/[1 - \chi_0^{L}\, 2J(q)] \tag{16}$$

is the static, enhanced Langevin susceptibility which diverges at a temperature, T_2, given by $\chi_0^L = 1/2J(0)$

Using (13) and (16) we may write down the Green function for L_1 in an identical form to the ferroelectric case[5,7]

$$\mathcal{G}_{L_1L_1}(q,\omega) = \frac{8\Delta f_1}{\bar{\omega}_0^2 - \omega^2 - L - \dfrac{i\omega\tau_1}{1 - i\omega\tau_1} + 2i\omega\gamma_s} \tag{17}$$

where

$$\bar{\omega}_0^2 = \Delta^2 - 16\Delta f_1 J(q)$$

$$L = 32\Delta f_1\, \chi_0^{L_2}(q) J(q)^2 \qquad \tau_1 = \chi_0^{L_2}(q)\tau \tag{18}$$

and a relaxation time, $1/\gamma_s$, for the L_1 mode has been added. The behaviour of $\mathcal{G}^{L_1L_1}$ is as for a ferroelectric[5,7]: (a) for $\omega\tau \gg 1$ there are poles at $\pm\bar{\omega}_0$ and the softening mode frequency extrapolates to zero at a temperature, T, given by $\bar{\omega}_0 = 0$ which is below T_C. Whether or not the mode is underdamped depends upon γ_s . (b) For $\omega\tau \ll 1$ and small γ_s (17) may be expressed in Debye form; at $q = 0$

$$\frac{\mathrm{Im}\, \mathcal{G}^{L_1L_1}(\omega)}{\mathcal{G}^{L_1L_1}(0)} = \frac{\omega\tau^* L}{(1 + \omega^2\tau^{*2})\bar{\omega}_0^2} \tag{19}$$

where

$$\tau^* = \frac{\bar{\omega}_0^2\, \tau\, \chi_0^{L_2}(0)}{\bar{\omega}_0^2 - L} \tag{20}$$

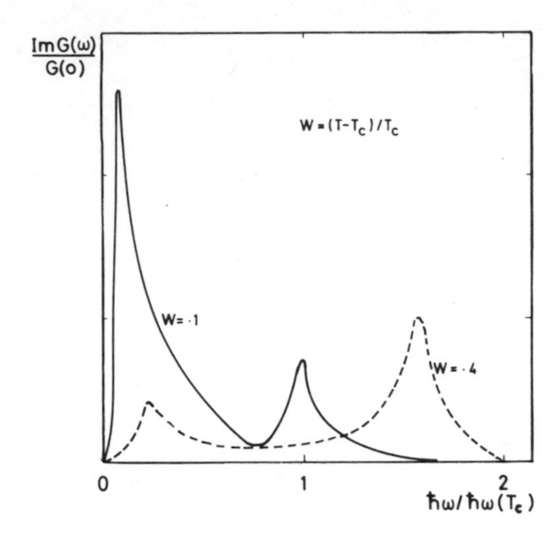

Fig.3. Normalized imaginary part of the Green function as a function of the normalized frequency.

and the relaxation τ diverges at $\bar{\omega}_0^2 - L = 0$; this is now the condition determining T_c and it agrees with Buyers'[2] theory and equation (3). The solution for $G^{L_2L_2}$ shows zero response at $\omega = \bar{\omega}_0$ and for $\omega\tau \ll 1$ $q = 0$ one again finds a Debye response

$$G^{L_2L_2}(\omega) = \frac{\chi_0^{L_2}(0)\,\bar{\omega}_0^2}{(\bar{\omega}_0^2 - L)}\left(\frac{1}{1 - i\omega\tau_2^*}\right)$$

The total neutron cross section diverges at T_c and the weight of the scattering is predominantly from $G^{L_1L_1}$ when T_c is near to T_1, but shifts to $G^{L_2L_2}$ as T_c nears T_2. The narrowing of the central peak is given by (20). The results are exhibited graphically in fig.3.

DISCUSSION

The longitudinal response of the S = 2, doublet ground state, magnet has been analysed in the paramagnetic phase by a pseudo-spin technique in RPA. The results are similar to those obtained by Buyers[2] and Lines[3] for other cubic magnets. If the response is regarded as being derived from two interacting modes there is a close analogy with studies[5,7] of a ferroelectric transition. The low frequency response near T_c is dominated by a central peak which arises from mixing between the off-diagonal and diagonal contributions to the magnetic moment.

REFERENCES

1. J.H. Van Vleck, The Theory of Electric and Magnetic Susceptibilities, Oxford University Press, 1932
2. W.J.L. Buyers, AIP Conf. Proc. 24, 27 (1975)
3. M.E. Lines, Phys. Rev. B11, 1134 (1975)
4. T.Egami and M.S.S.Brooks, Phys.Rev. 12B, 1021 (1975)
   Y.L.Wang and B.R. Cooper, Phys.Rev. 185, 696 (1969)
5. A.P. Young and R.J. Elliott, J.Phys. C7, 2721 (1974)
6. A.I. Akhiezer, V.G. Bar'Yakhtar, S.V. Peletminskii, "Spin Waves" North Holland, 1968.
   L.D. Landau and I.M. Khalatnikov, Doklady Akad. Nauk SSSR 996, 469 (1954)
7. G.J. Coombs and R.A. Cowley, J.Phys. C6, 121 (1973)

CRITICAL BEHAVIOUR OF THE MAGNETIZATION IN TbSb and $Tb_{0.6}Y_{0.4}Sb$

K. Carneiro*, N. Hessel Andersen*, J.K. Kjems+, and

O. Vogt[x]

* University of Copenhagen, Universitetsparken 5
DK-2100 Copenhagen Ø, Denmark
+ Research Establishment Risø, DK-4000 Roskilde, Denmark
x Laboratorium für Festkörperphysik, ETH, Zürich, Switzerland

ABSTRACT

Neutron elastic scattering has been used to study the temperature dependence of the sublattice magnetization in TbSb and $Tb_{0.6}Y_{0.4}Sb$. The transition from type II antiferromagnet into a paramagnet is found to be continuous in both cases. This does not support the recent theoretical prediction that the transition in TbSb should be of first order, since the Hamiltonian shows no stable fixed point within the ε-expansion. In pure TbSb we find the critical exponent $\beta = 0.20 \pm 0.01$ in the region $4 \times 10^{-3} < (T_N - T)/T_N < 2 \times 10^{-1}$. In $Tb_{0.6}Y_{0.4}Sb$ the exponent is significantly different. We find a sharp transition with $\beta = 0.37 \pm 0.05$.

INTRODUCTION

The alloy system $Tb_{1-x}Y_xSb$ is an example of a singlet ground state system in which the magnetic properties[1] reflect the competition between the concentration dependent effective exchange interaction and the concentration independent crystalline electric field. TbSb crystallizes in the cubic NaCl-structure and for impurity concentrations less than the critical value of $x < 0.6$ the system orders from a van Vleck paramagnet into a type II antiferromagnet[2] with $T_N \approx 15$ K for $x = 0$. Here we report the results of neutron scattering experiments at temperatures close to T_N for x = 0 and x = 0.4 which may help to elucidate the critical behaviour of this system.

Several theoretical predictions relate to this class of transitions. For pure TbSb Bak et al.[3] have suggested that the transition is of first order. Their argument is based on symmetry and not on the type of interactions involved. Since Mukamel and Krinsky[4)] found that the Hamiltonian for a type II antiferromagnet with magnetization parallel to the propagation vector, $\vec{m}||\vec{k}$, does not possess a stable fixed point within the ε-expansion and they suggest the new rule that this leads to a first order transition.

The effect of randomness introduced by dilution was considered by Harris[5)]. He concluded that in systems with a specific heat exponent $\alpha<0$, randomness does not change the critical behaviour, whereas for $\alpha>0$ the transition gets smeared over a temperature range $x^{1/\alpha}T_N$. Recent resistance measurements[6)] have been interpreted to give $\alpha=0.31$ indicating an appreciable smearing. It should, however, be kept in mind that Harris only considered the limit of $x<<1$, whereas the case of larger x to our knowledge is unexplored. In particular, the effect of the crystal field on the critical behaviour has not been considered theoretically; but as it appears from our results, such an effect cannot be excluded.

EXPERIMENTAL

The samples belong to a well characterized series of crystals which were used in the resistivity measurements by Hessel-Andersen et al.[6)]. The volumes of these crystals are typically a few mm^3, which make them well suited for a study of the sublattice magnetisation via Bragg intensities since extinction can be neglected.

The samples were mounted in a variable temperature cryostat, and the sublattice magnetization was measured at the DR-3 reactor at Risø, using a triple axis spectrometer with the energy transfer set to zero. This was done to discriminate against the unwanted inelastic part of the critical scattering close to T_N. The neutron energy was chosen to be 14.7 meV and two 5 cm pyrolytic graphite filters were used to suppress neutrons from higher order reflections in the monochromator crystal.

Scans were performed through $(1\frac{1}{2},\frac{1}{2},\frac{1}{2})$ and $(\frac{1}{2},1\frac{1}{2},1\frac{1}{2})$ lattice points of the chemical cell and the integrated intensities were identified with the square of the ordered moment. At each temperature we also measured the intensity slightly away from the magnetic Bragg reflection. This scattering is due to the critical fluctuations and it peaks at T_N.

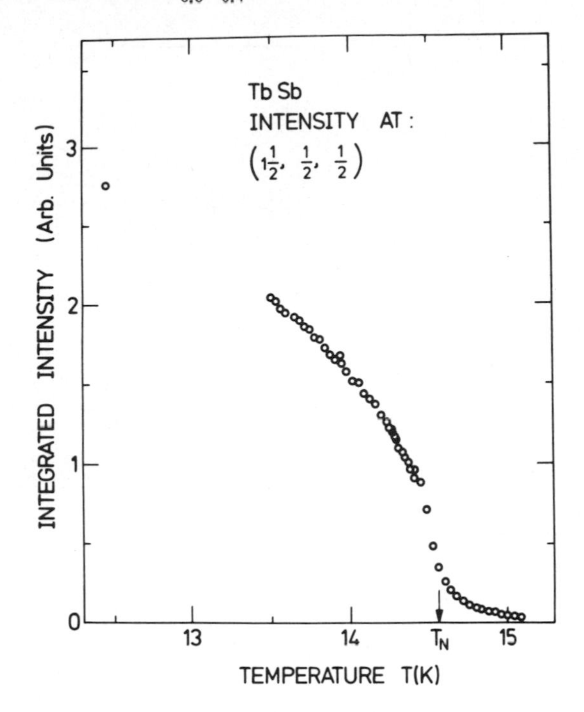

Fig. 1: Temperature dependence of the integrated intensities measured at the $(1\frac{1}{2},\frac{1}{2},\frac{1}{2})$ lattice point of the chemical unit cell in TbSb.

RESULTS AND DISCUSSION

The temperature dependent intensities of the $(1\frac{1}{2},\frac{1}{2},\frac{1}{2})$ magnetic reflection for pure TbSb is shown in Fig. 1. It shows a continuous transition at T=14.56 K, which agrees well with an observed maximum in the critical scattering. We performed the experiments both on heating and cooling and found no evidence of hysteresis. Hence it appears that we see no evidence for a first order transition as predicted by Bak et al.[3]; but since their argument is a qualitative one, we have not shown that it is necessarily wrong, since the jump in the intensity might be very small. In this respect it is interesting to note that in Cr[7], a case where their theory appears to be verified, the observed discontinuity is about 25% of the low temperature intensity, compared to an upper limit of approximately 5% inferred from Fig. 1.

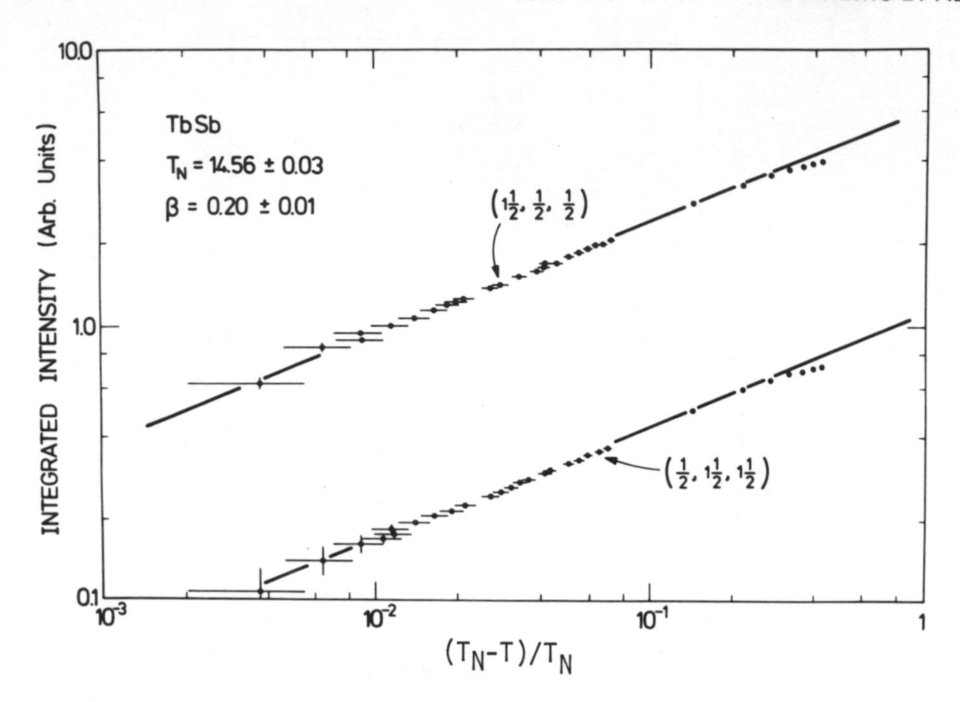

Fig. 2: Integrated intensities vs reduced temperature for the sublattice magnetization in TbSb for two magnetic reflections. The errorbars include counting statistics as well as uncertainties in T_N and correction for critical scattering. The full lines correspond to the power law $I \propto (1-T/T_N)^{2\beta}$.

With no evidence for a first order transition we have analyzed our results in terms of a second order phase transition. The tail at higher temperatures in Fig. 1 is due to critical scattering and intensities below T_N must be corrected for this. To this end the critical scattering below T_N has been estimated by scaling from the observed intensities above T_N using measurements away from the superlattice point to determine the scalefactor. We have employed this correction procedure to the data shown in Fig. 1 and plotted the results on a log-log scale in Fig. 2 with $T_N = 14.56$ K. The error bars take into account the counting statistics, the uncertainty in the critical scattering corrections as well as the uncertainty in the determination of $T_N = 14.56 \pm 0.03$ K. From Fig. 2 we deduce that the reduced spontaneous sublattice magnetization, M, follows a power law in the temperature range $4\times10^{-3} < 1-T/T_N < 2\times10^{-1}$

$M \propto (1-T/T_N)^\beta$ with β = 0.20 ± 0.01 corresponding to the full lines drawn in the figure. The fact that both ($1\frac{1}{2},\frac{1}{2},\frac{1}{2}$) and ($\frac{1}{2},1\frac{1}{2},1\frac{1}{2}$) reflections give identical critical exponents although the absolute intensities differ by a factor of 5 justifies the neglect of extinction in the analysis.

Measurements on a sample of $Tb_{0.6}Y_{0.4}Sb$ with T_N = 8.55 ± 0.08 K give qualitatively similar results in the sense that the transition is continuous with no obvious sign af smearing. However, the critical exponent β = 0.37 ± 0.05 is changed significantly from pure TbSb value.

Our results show that the critical behaviour of $Tb_{1-x}Y_xSb$ depends strongly on concentration, but that the transition both in the pure case (x=0) and for finite x can be characterized as being of second order. It is at the moment not clear how to relate our observations to the theoretical predictions mentioned above, since we do not see evidence of a first order transition in TbSb, nor do we observe any obvious signs of smearing because of the randomness in $Tb_{0.6}Y_{0.4}Sb$. Further experiments, where we measure the critical exponents γ and ν, are in progress. Later we shall consider the complete series, used in the resistivity measurements.

It is a pleasure to acknowledge discussions with Jens Als-Nielsen relating to this work.

REFERENCES

1. B.R. Cooper and O. Vogt, Phys. Rev. B1, 1218 (1970)

2. H.R. Child, M.K. Wilkinson, J.W. Cable, W.C. Koehler, E.D. Wollan, Phys. Rev., 131, 922 (1963)

3. P. Bak, S. Krinsky, D. Mukamel, Phys. Rev. Lett. 36 52 (1976)

4. D. Mukamel and S. Krinsky, J. Phys. C: Solid St. Phys. 8 L 496 (1975)

5. A.B. Harris, J. Phys. C: Solid St. Phys. 7, 1671 (1974)

6. N. Hessel Andersen, P.E. Lindelof, H. Smith, O. Splittorff, . Phys. Rev. Lett. 37, 46 (1976). The value for α is given in N. Hessel Andersen. Thesis. (1976, unpublished)

7. A. Arrott, S.A. Werner, H. Kendrick, Phys. Rev. 153, 624 (1967)

CRYSTAL ELECTRIC FIELD EFFECTS ON THERMAL AND TRANSPORT PROPERTIES

B. Lüthi*, P.S. Wang*, Y. H. Wong*, H.R. Ott[+] and E. Bucher[§].

*Department of Physics, Rutgers University, New Brunswick, N.J. 08903, US
[+]Laboratorium of Festkörperphysik, ETH, CH-8049 Zürich, Switzerland
[§]Fachbereich Physik, Universität Konstanz, D-755 Konstanz, Germany

ABSTRACT

A review is given for crystal field effects on thermal and transport properties in intermetallic rare earth compounds. In thermal properties we emphasize the recently discovered effects in thermal expansion, elastic constants and magnetostriction. In the transport properties we survey electrical and thermal conductivities in concentrated compounds.

INTRODUCTION

It is convenient to divide crystal electric field (CEF) effects into 3 categories[1]: Resonance effects, thermal effects and transport properties. Here we would like to review thermal and transport effects. In the field of thermal properties, several new effects (elastic constants, thermal expansion, magnetostriction) have been investigated recently. They are somewhat complementary to the other effects (inelastic neutron scattering, magnetic susceptibility) inasmuch as they probe electric quadrupole transitions instead of magnetic dipole transitions. Because one probes with these effects the coupling of the magnetic ion to the lattice, one can elucidate more the structural properties of a solid rather than the magnetic ones. Some aspects of thermal CEF-effects have been reviewed recently[1].

In the transport properties we briefly survey typical transport coefficients (electrical conductivity, thermal conductivity, thermoelectric power, magnetoresistance for concentrated compounds, in comparison to dilute rare earth compounds.

THERMAL EFFECTS

The CEF effects on the specific heat (Schottky anomaly) and the magnetic susceptibility are well known and have been used extensively in the study of CEF levels[2]. During the last few years similar effects have been investigated for thermodynamic quantities such as elastic constants[3], thermal expansion[4] and magnetostriction[4]. Here we would like to discuss some aspects of these effects, which have been less emphasized before[1]. All these quantities, as derivatives of the free energy, can be expressed as thermodynamic averages over the energy levels $E(\Gamma_i)$ and their derivatives with respect to H, V, ε_Γ where ε_Γ is the strain of symmetry Γ. This has been shown before[1,3,4].

Although there are interesting new developments in intermetallic rare earth compounds using specific heat and magnetic susceptibility measurements (confirming a theoretical conjecture of the order of a magnetic phase transition[5], spin arrangements in a magnetic field[6]) space does not permit to review them here.

1) Thermal Expansion

A CEF effect in thermal expansion was first found in TmSb[4], which has a Γ_1 groundstate followed by Γ_4(25K) and Γ_5(56K)[7]. The analysis gave the Grüneisen constants $\gamma_i=-\partial \ln E_i/\partial \ln V$ of $\gamma(\Gamma_4)=-1.3$, $\gamma(\Gamma_5)=+1.05$ whereas a point charge model (PCM) gives $\gamma \simeq 5/3$. Using a magnetoelastic Hamiltonian[1,3,8] for the volume strain ε_v

$$H(c_B) = -G_1 N\varepsilon_v - G_{11}\sum_i \varepsilon_v O_4 - G_{12}\sum_i \varepsilon_v O_6 \qquad (1)$$

with the CEF operators, O_4, O_6 we obtain from the γ's $G_{11}=-36.3$mK, $G_{12}=-0.54$mK. From PCM, we get $G_{11}=+\frac{5}{3}B_4=+22$mK, $G_{12}=\frac{7}{3}B_6=-0.07$mK. For these estimates we used the experimentally determined B_4, B_6 (Ref. 7). Next we consider

TmTe. Elastic constant measurements gave an approximate level splitting of $\Gamma_6-\Gamma_8(10K)-\Gamma_7(17K)$ for the Tm^{2+} ion[1]. The thermal expansion can be analyzed as for TmSb, giving $\gamma(\Gamma_8)=0.3$, $\gamma(\Gamma_7)=-0.65$. PrSb has a $\Gamma_1-\Gamma_4(73K)-\Gamma_3(129K)-\Gamma_5(229K)$[7]. Therefore, only $\gamma(\Gamma_4)$ can be observed; $\gamma(\Gamma_4)\simeq 5$, which leads to $G_{11}\cong 0.43K$ compared to a PCM estimate of -0.12K.

The hydrostatic pressure dependence (ε_v dependence) of other physical properties such as the magnetic susceptibility and the Knight shift shows similar departures from PCM estimates as the ones from thermal expansion[9].

In analogy to specific heat measurements, one observes strong anomalies of the thermal expansion near phase transitions. This was shown for the LnSb series[10], as well as for $PrCu_2$ and TmCd[11]. The same comment, regarding the order of the transition, can be made here as for the specific heat.

2) Elastic Constants

A rich field of study has opened up with the investigation of elastic effects of CEF systems. Previously[1] the similarity to a susceptibility was emphasized, the symmetry strain ε_Γ playing the role of the external field and the elastic constants corresponding to the zero field susceptibility. Since elastic effects have been reviewed recently[1], we shall keep the discussion short.

Single Ion Effects: In the case of noninteracting ions one can determine magnetoelastic coupling constants G_i^2. Evaluating strain susceptibilities using also octupole contributions does not influence the pure quadrupole susceptibility very much. This can be seen using point charge expressions for the higher magnetoelastic coupling constants G_{21} in[1,8]:

$$H\,(c_{11}-c_{12})=-G_2\sum(\varepsilon_2 O_{2_i}^2+\varepsilon_3 O_{2_i}^o)-G_{21}\sum(\varepsilon_2 O_{4_i}^2+\varepsilon_3 O_{4_i}^{2,4}) \quad (2)$$

where ε_2, ε_3 are Γ_3-strains, O_ℓ the quadrupole ($\ell=2$) and octupole ($\ell=4$) operators. Eq. (2) is the analogous H_{me} for $c_{11}-c_{12}$ as Eq. (1) is for c_B. This was done in SmSb and also for the other pnictides[12]. In Fig. 1 we show a result for $c_{11}-c_{12}$ for PrSb where the full curve corresponds to the quadrupole susceptibility $\chi(O_2^2)$ whereas the dotted one gives the total strain susceptibility using

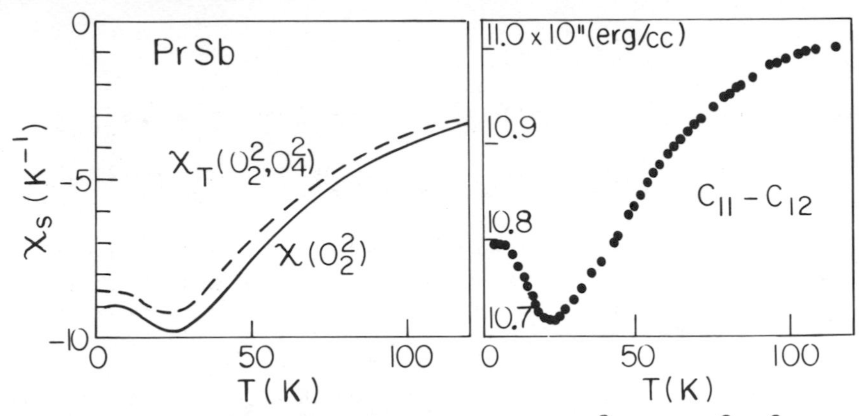

Fig. 1: Strain Susceptibilities $\chi(O_2^2)$, $\chi(O_2^2,O_4^2)$ and elastic constant $c_{11}-c_{12}$ for PrSb.

$\ell=2,4$ contributions (Eq. (2)). Also plotted is the corresponding experimental elastic constant result. One notices that the inclusion of higher order ($\ell>2$) terms gives small changes and can usually be neglected for the study of the temperature dependence of the elastic constants in this rare earth compound. Note also that in the case of thermal expansion $|G_{12}|<<|G_{11}|$. We shall discuss the magnetoelastic coupling constants in the section on magnetostriction.

Cooperative Jahn-Teller Effect: If the strain susceptibility diverges, the corresponding elastic constant softens and the magnetic-lattice interaction (e.g. Eq. (2)) can induce a structural transition, which can be viewed as a quadrupolar ordering. In intermetallic rare earth compounds conduction electrons can also contribute to such an ordering[1]. Many cases have been studied[1,10,11] and we refer to recent reviews of this effect[1,13].

Rotational Strain Interaction: An interesting application of the magnetic ion lattice interaction is the inclusion of the asymmetric part of the strain tensor for the study of sound velocities in a magnetic field. Melcher[14] first carried out experiments in MnF_2 to test this type of theory. Some of our cubic rare earth compounds (PrSb, TmSb) are ideally suited to test such theories, because they are cubic, have neither magnetic nor structural phase transitions and the magnetoelastic coupling constants and CEF parameters are known. Using the principle of rotational invariance, the full magnetoelastic Hamiltonian for a c_{44} reads[15]:

$$H_{str} = G_3(J_xJ_z + J_zJ_x)\,\varepsilon_{xz}$$

$$H_{rot} = -20B_4\omega_{xz}[(J_xJ_z+J_zJ_x)(J_x^2-J_z^2)+(J_x^2-J_z^2)(J_xJ_z+J_zJ_x)]$$
$$+2G_3\varepsilon_{xz}\omega_{xz}(J_x^2-J_z^2)+40B_4[(J_xJ_z+J_zJ_x)^2-(J_x^2-J_z^2)^2]\omega_{xz}^2 \quad (3)$$

Here ε_{xz}, ω_{xz} are the symmetric and antisymmetric tensor components. These terms can also be obtained from a PCM expansion[1,16]. They lead to different sound velocities in a field H_z for the 2 modes (k_zR_x, k_xR_z). Fig. 2 shows a typical result for TmSb at T=2K. The full line is calculated with the parameters B_4=0.013K G_3=20K determined from neutron scattering[7] and $c_{44}(T)$[10]. The agreement of theory and experiment is superb. Other modes and a full discussion can be found in Ref. 16.

3) Magnetostriction

Magnetostriction experiments in the paramagnetic region allow us in principle to determine the magnitude <u>and</u> sign of the magnetoelastic coupling constants G_2, G_3. Elastic constant measurements on the other hand give only G_2^2, G_3^2. For pure quadrupole contributions the magnetostriction $\varepsilon(H)=(\frac{\delta\ell}{\ell})_H-(\frac{\delta\ell}{\ell})_0$ in the direction of magnetic field applied along the [100] axis of cubic crystals is given by:

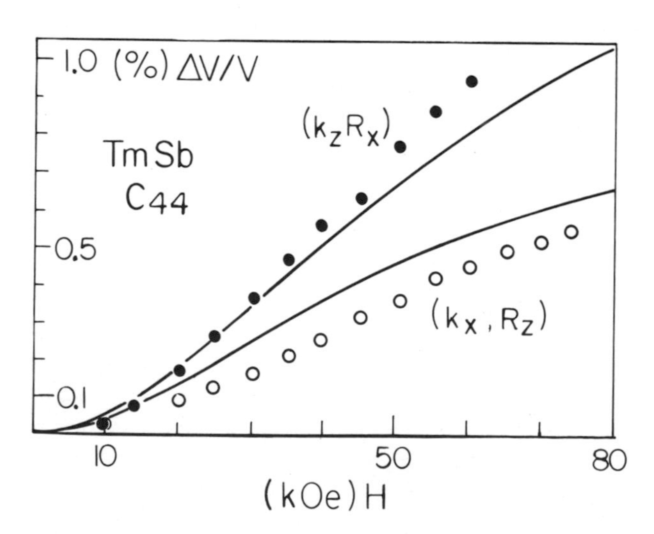

Fig. 2: TmSb, Relative velocity change $\Delta v/v$ vs H for 2 modes (k_zR_x), (k_xR_z) with theoretical curves.

$$\varepsilon(H)=\sqrt{\frac{2}{3c^{o}_{\Gamma}}}\left[\left(\frac{\partial F}{\partial \varepsilon_{\Gamma}}\right)_{H}-\left(\frac{\partial F}{\partial \varepsilon_{\Gamma}}\right)_{o}\right]=\sqrt{\frac{2}{3c^{o}_{\Gamma}}}\left(\frac{\partial F}{\partial \varepsilon_{3}}\right)_{H}=\sqrt{\frac{2}{3}}G_2\langle O^o_2\rangle_H \frac{N}{c^o_{\Gamma}} \qquad (4)$$

because $\langle O^o_2\rangle_o=0$ for cubic materials in the paramagnetic phase. In Eq. (4) we neglected similar octupole contributions from both higher ε_3 and ε_v coupling, because, as in thermal expansion and elastic constants, these contributions are in general small. Another contribution, not included in Eq. (4), is exchange striction. The appropriate Hamiltonian $H=\sum(\frac{\partial\lambda}{\partial\varepsilon}\varepsilon_v J_i J_j+\frac{\partial K}{\partial\varepsilon}\varepsilon_v O_i O_j)$, where λ, K are the exchange and quadrupole constants respectively, can give a noticeable effect on magnetostriction, as discussed below. In Fig. 3 we show experimental results for various cubic substances (PrSb, HoSb, ErSb, TmSb, TmCd). One finds magnetostriction values typically of the order of 10^{-5}-10^{-4} for H~15kOe at low temperatures. The sign of the magnetostriction, and therefore the sign of the dominant contribution G_2, follows the sign of the Stevens factor α_J and is in agreement with the sign from PCM. The full lines are based on Eq. (4) with G_2 taken from elastic constant measurements[10,21].

We give now a discussion for each substance:
PrSb: The magnetostriction is rather small as expected for a singlet ground state system without phase transition. However, it is still a factor of 6 larger than what one expects from elastic measurements. Therefore, exchange

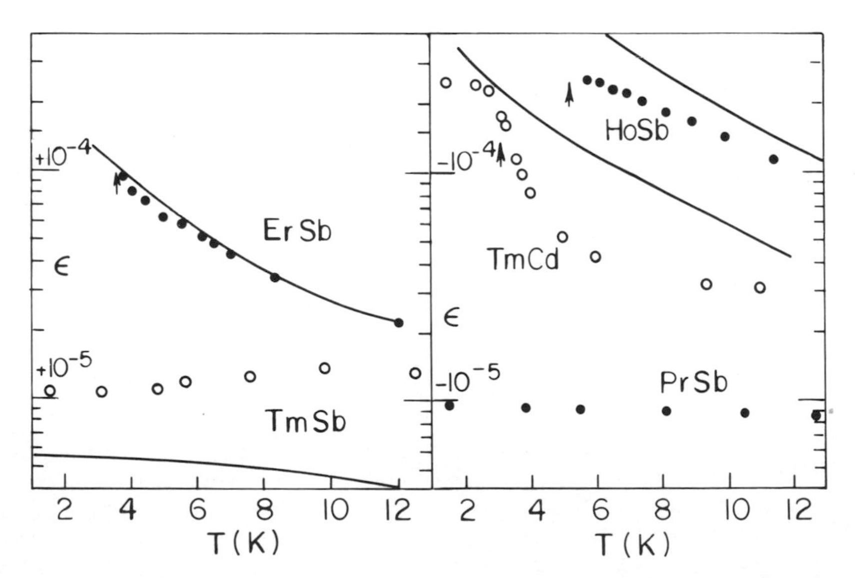

Fig. 3: Semilogarithmic plot of the magnetostriction along (100) axis in 15kOe (TmCd 10kOe). Full lines fit using Eq. (4) with G_2 from Ref. 10, 21.

striction contributions are important[17].
HoSb: With Γ_3 ground state one expects a large magnetostriction as is observed indeed. The quadrupolar contribution is ca 20% larger than what is observed experimentally. The difference can again be attributed to exchange striction. HoSb orders magnetically at T_N=5.25K.
ErSb: The lowest CEF state is Γ_8 and again one observes a large magnetostriction. Note the difference in sign to HoSb, in agreement with signs of α_J for Er^{3+}, Ho^{3+}. G_2 determined from Eq. (4) agrees with the upper limit derived from elastic constant measurements[10].
TmSb: Experimentally we observe a shallow maximum around 10K which cannot be explained with Eq. (4), giving only 50% of the effect. In this case exchange contributions are negligible, which, however, does not exclude necessarily large $\partial\lambda/\partial\varepsilon$. At the moment we do not understand this discrepancy.
TmCd: Magnetostriction results in this compound were analyzed before in a different way[11]. Our analysis gives a factor of 2 larger effects than observed. TmCd has a coop. J-T transition at 3.16K. The discrepancy might be explained by a quadrupole striction from $\partial K/\partial\varepsilon$. Close to T_a one notices a strong temperature dependence of $\varepsilon(H)$.

The simple approach, explaining the magnetostriction solely as a quadrupole-strain effect, gives in some cases (ErSb, HoSb) quantitatively correct results as shown above. In other cases contributions from exchange striction must however be present. A more detailed account of this aspect will be given elsewhere[17].

The sign of the quadrupole coupling constant G_2 is given by the sign of α_J and agrees with PCM as mentioned above. This fact means, that for many cubic rare earth intermetallic compounds an effective PCM gives approximately correct CEF parameters[7] B_4 and magnetoelastic coupling constants[1] G_2. But the magnetoelastic coupling G_{11} connected with volume strain (hydrostatic effects) turns out to be quite different[4,9].

TRANSPORT PROPERTIES

From a study of transport properties one can get information about electron scattering mechanisms. In the past, these experiments were performed mostly on dilute rare earth alloys[18]. Here we draw our attention to the concentrated case of rare earth intermetallic compounds. There are significant differences between the two cases: in the dilute case an isolated impurity acts as a scattering center via potential scattering, exchange scattering or aspherical Coulomb charge scattering. In concentrated systems only deviations from the periodic potential, i.e. temperature induced excited spin and quadrupole states act as scattering mechanisms. For these we use the exchange and aspherical Coulomb charge processes, which we write in a simplified form as[1,19]:

$$V = \sum_{kk'} j_{ex}(\underline{sJ}) + \sum_{kk'\ell\ell'} V_Q a^+_{k'\ell'} a_{k\ell} \qquad (5)$$

Eq. (5) is the microscopic analogue of the ion ion interaction Hamiltonian. It has been used for many calculation of transport properties[19].

As long as one is only interested in the case of non-interacting ions the calculation for the concentrated case proceeds very similarly to the dilute case[20]: With V_1 denoting any deviations in the lattice periodicity, one gets for the thermal conductivity due to electrons

$$\frac{\kappa(T)}{T} = \int_0^\infty \frac{x^2}{1+\cosh x} F(x,t)dx \qquad (6)$$

with $F(x,t)=\{V_1^2 + \sum_{\alpha\beta} |\langle\alpha|V|\beta\rangle|^2 (n_\alpha + n_\beta) \dfrac{\frac{1}{2}(1+\cosh x)}{\cosh\frac{\Delta\alpha\beta}{T}+\cosh x}\}$ and similar expressions for the electric resistivity.

Experimentally the electrical resistivity was measured for the following rare earth compounds[21]: TmCd, TmSb, $PrCu_2$, $PrCu_5$ but for the thermal conductivity there exists only one brief report[22]. Both transport coefficients display similar features: Electrical and thermal conductivities show characteristic deviations from normal behavior ($\sigma(T)$=const., κ/T=const. at low temperature) due to CEF scattering. In Fig. 4 we show a compilation of thermal conductivity results for various compounds (TmSb, ErSb, $PrPt_5$, $PrCu_5$, $PrCu_2$) together with a calculated fit using Eq. (6). Values of κ_o/T_o are listed in the figure caption. Instead of κ/T=constant, one notices an initial strong decrease of κ/T (T), which becomes weaker towards higher T. One notices a fair overall agreement between theory and experiment except for those cases where the assumption of noninteracting ions is no longer valid (ErSbT_N=3.53K, $PrCu_2 T_a$=7.5K). For the calculation we took the level scheme quoted in the literature. Note that the cross terms in V due to exchange and quadrupole scattering do not give any contribution in these cases. For further details concerning the thermal conductivity, possible phonon contributions and experiments on LaSb, see Ref. 22.

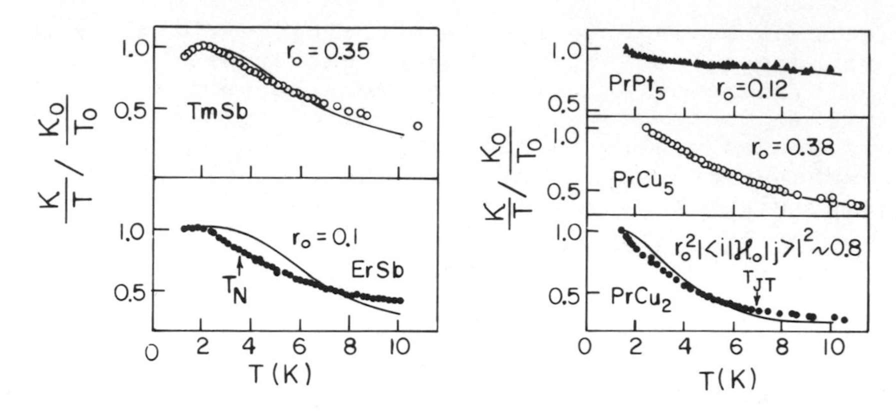

Fig. 4: Thermal conductivities, comparison of experiments with theory (full lines). Fitting parameters r indicated in figure. Values for κ_o/T_o, are 0.83(TmSb), 0.71(ErSb), 7.9($PrPt_5$), 19.75 ($PrCu_5$), 15($PrCu_2$) mW/K^2cm, $r_o = V/V_1$.

Thermal and electrical conductivities cannot distinguish very well between exchange and aspherical Coulomb charge scattering. Magnetic dipole and electrical quadrupole matrixelements contribute in a similar way to these transport coefficients. This is no longer the case for the thermoelectric power, where the two mechanisms contribute quite differently[23]. Another example is superconductivity, where the exchange interaction is a pair breaking mechanism, but aspherical Coulomb charge scattering is a pair enhancing mechanism. This fact has been explored in dilute rare earth alloys[24]. Superconductivity is of course no longer observed in concentrated compounds with magnetic ions.

Finally one should mention magnetoresistance, where in the dilute case the two mechanisms can be differentiated (Friederich and Fert Ref. (18)). Such experiments have not yet been performed in concentrated compounds.

REFERENCES

1. B. Lüthi, Joint MMM-Intermag Conference Pittsburgh 1976, AIP Conference Proc. in press.
2. J. H. Van Vleck, The Theory of Electric and Magnetic Susceptibilities, Oxford University Press 1932; W. E. Wallace, Rare Earth Intermetallics, Acad. Press 1973.
3. B. Lüthi, M. E. Mullen and E. Bucher, Phys. Rev. Lett. 31, 95 (1973).

4. H. R. Ott and B. Lüthi, Phys. Rev. Lett. 36, 600 (1976), AIP Conference Proc. 29 333 (1975).
5. D. Mukamel, S. Krinsky and P. Bak, AIP Conference Proc. 29, 474 (1975).
6. T. O. Brun, G. H. Lander, F. W. Korty and J. S. Kouvel, AIP Conference Proc. 24, 244 (1974).
7. R. J. Birgeneau, E. Bucher, J. P. Maita, L. Passell and K. C. Turberfield, Phys. Rev. B8, 5345 (1973).
8. E.R. Callen and H.B. Callen, Phys. Rev. 129, 578 (1963).
9. R. P. Guertin, J. E. Crow, L. D. Longinotti, E. Bucher, L. Kupferberg and S. Foner, Phys. Rev. B12, 1005 (1975); H. T. Weaver and J. E. Schirber, AIP Conference Proc. 24, 49 (1974).
10. M. E. Mullen, B. Lüthi, P. S. Wang, E. Bucher, L. D. Longinotti, J. P. Maita and H. R. Ott, Phys. Rev. B10, 186 (1974).
11. H. R. Ott and K. Andres, Solid State Comm. 15, 1341 (1974). K. Andres et. al. AIP Conference Proc. in press.
12. V. Dohm, Z. Phys. B23, 153 (1976); P. A. Lindgård private communication.
13. G. A. Gehring and K. A. Gehring, Rep. Progr. Phys. 38, 1 (1975); R. L. Melcher in Physical Acoustics Vol. 12, W. P. Mason and R. N. Thurston editors, Acad. Press.
14. R. L. Melcher in Proc. Internat. School of Physics, E. Fermi, Course LII E. Burstein editor, London, 1972.
15. V. Dohm and P. Fulde, Z. Phys. B22, 359 (1975).
16. P. S. Wang and B. Lüthi to be published; P. S. Wang, thesis Rutgers University 1976.
17. H. R. Ott, to be published.
18. A. Friederich and A. Fert, Phys. Rev. Lett. 33, 1214 (1974); N. Anderson, P.E. Gregers-Hansen, E. Holm, H. Smith and O. Vogt, Phys. Rev. Lett. 32, 1321 (1974); B. R. Cooper, et. al. to be published.
19. R. J. Elliott, Phys. Rev. 94, 564 (1954); L.L. Hirst, Solid State Comm. 5, 751 (1967); P. Fulde and I. Peschel, Adv. in Phys. 21, 1 (1972).
20. P. Thalmeier and P. Fulde, to be published.
21. B. Lüthi, M. E. Mullen, K. Andres, E. Bucher, J.P. Maita, Phys. Rev. B8, 2639 (1973); K. Andres, P. S. Wang, Y. H. Wong, B. Lüthi and H. R. Ott, AIP Conference Proc. in press (unpublished).
22. Y. H. Wong and B. Lüthi, AIP Conference 29, 534 (1975). Y. H. Wong (to be published).
23. J. Sierro, E. Bucher, L. D. Longinotti, H. Takayama and P. Fulde, Solid State Comm. 17, 79 (1975); H. Takayama and P. Fulde, Z. Physik B20, 81 (1975).
24. R. W. McCallum et. al. Phys. Rev. Lett. 34, 1620 (1975).

MAGNETORESISTANCE OF PRASEODYMIUM

J.A.G. Temple and K.A. McEwen

Department of Pure and Applied Physics

University of Salford, Salford M5 4WT, UK

ABSTRACT

The effect of the crystal field levels on the magnetoresistance of Pr is calculated numerically on a simple model and agreement is shown with experimental measurements at 4.2K up to 20 tesla. Results at higher temperatures are discussed.

INTRODUCTION

The crystalline electric field in Pr removes the 9-fold degeneracy of the ground state. The Boltzmann population of these different energy levels at a finite but low temperature gives a contribution to the resistivity which has been calculated numerically employing the full level system of the ground multiplet on both crystallographic sites together with the magnetization calculated in a molecular field approximation. An applied field alters both the energy of a level and also the wavefunction, thereby altering the scattering matrix elements and hence the resistivity.

CALCULATIONS

In a rigid lattice model of Pr the ions and electrons are governed by a Hamiltonian

$$\mathcal{H} = \sum_i \left\{\frac{\underset{\sim}{p}_i^{\,2}}{2m} + U(\underset{\sim}{r}_i) + 2\,\mu_B \underset{\sim}{\sigma}_i \cdot \underset{\sim}{H}\right\} - \sum_{j=1}^{N} g\mu_B \underset{\sim}{J} \cdot \underset{\sim}{H} + \mathcal{H}_{cf} + \mathcal{H}_1 \qquad (1)$$

where $\underset{\sim}{p}_i$ is the momentum of the ith conduction electron spin $\underset{\sim}{\sigma}_i$

in a periodic potential $\mathcal{U}(\underset{\sim}{r}_i)$, μ_B is the Bohr magneton and $\underset{\sim}{J}$ the total angular momentum of the j^{-th} ion. The crystal structure of Pr is dhcp; the two types of lattice site are classified by their nearest neighbour environments as cubic or hexagonal and $\mathcal{H}_{cf}$ represents the crystal field at each site as :

$$\mathcal{H}_{cf} = B_2\, O_2^{\,o} + B_4\, O_4^{\,o} + B_6(O_6^{\,o} + \frac{77}{8}\, O_6^{\,6}) \quad \text{hexagonal sites}$$

$$\mathcal{H}_{cf} = B_2\, O_2^{\,o} + B_4\, (O_4^{\,o} + 20\sqrt{2}\, O_4^{\,3}) + B_6(O_6^{\,o} - \frac{35}{\sqrt{8}}\, O_6^{\,3} + \frac{77}{8}\, O_6^{\,6}) \tag{2}$$

for the cubic sites, where $O_\ell^{\,m}$ are the Stevens operator equivalents. The internal field H is

$$H_\alpha = H_{applied} + \frac{1}{(g\mu_B)^2}(\lambda M_\alpha + \mu M_\beta) \tag{3}$$

for α,β = cubic, or hexagonal and λ , μ are Weiss molecular field parameters.
The interaction between ions and conduction electrons is

$$\mathcal{H}_1 = -\sum_{ij} \Gamma\, \delta(\underset{\sim}{r}_i - \underset{\sim}{R}_j)\, \underset{\sim}{\sigma}_i \cdot \underset{\sim}{J}_j\, (g-1) \tag{4}$$

The Boltzmann equation for the electronic conductivity may be solved assuming isotropic scattering at the fermi level, and with transitions between states evaluated in the first Born approximation, neglecting the effect of the magnetic field on the conduction electrons to yield for the relaxation time[1]

$$\frac{1}{\tau} = \frac{m^* k_f}{\pi\hbar^3} \sum_{fi} N_i\, |M_{if}|^2\, \frac{2}{1+\exp(-\varepsilon_{if}/k_B T)} \tag{5}$$

where N_i is the average occupancy of the initial state and M_{if} represents the transition probability to each of the allowed final states with energy change ε_{if}. The properties of electron-conduction in metals ensures that the states involved are localized near the fermi energy and hence this is the only property of the electrons needed; the part of the Hamiltonian in braces in (1) being redundant. We take N_i to be the Boltzmann population of the ith crystal field level with $|M_{if}|^2$ given by Fermi's golden rule using $\mathcal{H}_1$ then the final form may be written[2]

$$\rho = \rho_o\, \mathrm{Tr}\,(PQ) \tag{6}$$

where the matrices P, Q are given by

$$P_{ij} = \frac{e^{-\varepsilon_i/k_B T}}{z} \cdot \frac{2}{1+\exp(-\varepsilon_{if}/k_B T)} \tag{7}$$

$$Q_{ij} = |\langle i|Jz|j\rangle|^2 + \tfrac{1}{2}\,\{|\langle i|J+|j\rangle|^2 + |\langle i|J-|j\rangle|^2\} \qquad (8)$$

with $|i\rangle$ a linear combination of $|J, Jz\rangle$ states produced as the ith crystal field level, ε_i its energy. We have used the relation $\rho = m^*/(ne^2\tau)$ between relaxation time and resistivity so that

$$\rho_o = \frac{N(g-1)^2|\Gamma|^2 m^*}{e^2\hbar\, E_f}\,\frac{3\pi}{8} \qquad (9)$$

and depends on N the number of atoms per unit volume, the strength of the electron-ion coupling and mass and fermi energy of conduction electrons.

RESULTS

If the phonon contribution at a given temperature is independent of the internal field then the quantity $\rho(H,T) - \rho(0,T)$ describes the magnetic field enhancement of the resistivity. The experimental results of McEwen and Roeland(3) on polycrystalline Pr at 4.2K were least squares fitted to the polycrystalline average of the calculations yielding the two parameters ρ_{oz}, ρ_{ox} shown in table 1. The fit which is good up to high fields is shown in figure 1. The difference in magnitude of ρ_{ox} and ρ_{oz} is attributed to the different character of the trunk compared with the roots-and-branches effective masses of the fermi surface.

As the temperature is raised the model predicts a smaller initial change in resistivity with applied field but preliminary experiments(4) so far indicate that $\rho(H,T) - \rho(0,T)$ is initially negative exhibiting a minimum which becomes broader and deeper with increasing temperature. If these observations are confirmed on single crystals grown from much purer stock it may be that the coupled electron-exciton system is important and calculations of this are now being performed.

Table 1.

(Kelvin)	B_2	B_4	B_6		(Kelvin)
cubic(5)	0.00	1.390	1.240	$\lambda^{(5)}$ =	− 1.020
hexagonal(5)	4.29	1.390	1.240	$\mu^{(5)}$ =	4.140

ρ_{ox} = 0.18 μΩ cm $\qquad$ ρ_{oz} = 0.96 μΩ cm

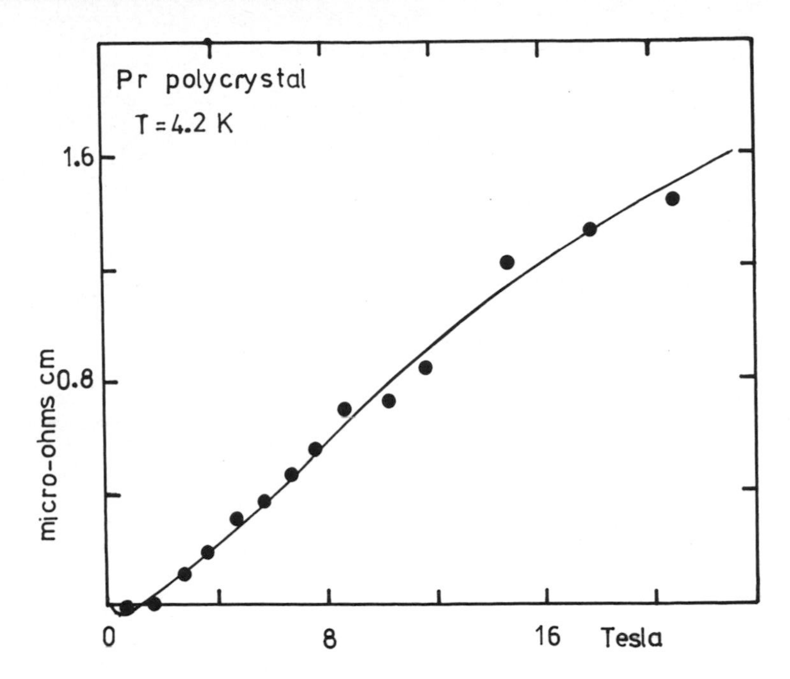

Figure 1

Results of $\rho(H,T) - \rho(0,T)$ as calculated in the text (solid curve) compared with experimental results of McEwen and Roeland[3] (circles) for polycrystalline praseodymium.

References

1. Van Peski-Tinbergen T, Dekker A J. Physica (29) 917 (1963)
2. Hessel-Anderson N, Gregers-Hansen P E, Holm E, Smith H. Phys. Rev. Letters (32) 1321 (1974)
3. McEwen K A, Roeland L W (to be published)
4. McEwen K A, Webber G D (private communication)
5. Rainford B D, Proc AIP Conf. 5 "Magnetism and Magnetic Materials" (1971)

RESISTIVITY MEASUREMENTS ON THE SINGLET GROUND STATE SYSTEM

$Tb_cY_{1-c}Sb$. EFFECTS OF ELECTRON-EXCITON SCATTERING

N.Hessel Andersen[x], P.E.Lindelof[x], H.Smith[x], O.Splittorff[x], and O.Vogt[+]

[x]Physics Laboratory I, H.C.Ørsted Institute, University of Copenhagen, Denmark

[+]Laboratorium für Festkörperphysik, ETH Zürich, Switzerland

ABSTRACT

The electrical resistivity of $Tb_cY_{1-c}Sb$ has been measured as a function of temperature and of Tb-concentration. The resistivity contribution from scattering of conduction electrons by magnetic excitons in the Tb 4f-electron system has been calculated in the paramagnetic regime in order to examine the competing effects of the crystal field and the exchange interaction. Excellent agreement with experiment is obtained for the entire range of c with one value of each of the two adjustable parameters, the Fermi momentum and the electron-ion exchange constant.

For rare earth crystals with non-Kramers ions it commonly occurs that the crystal field only ground state of the 4f-electrons is a singlet. If the magnetic interaction exceeds a certain critical value determined by the crystal field splitting such a system will order at the lowest temperatures. The metallic[1] compounds $Tb_cY_{1-c}Sb$ have proven to be an ideal system for studying this competing effect of crystal field and indirect (RKKY-type) exchange interaction[2]. Whereas the crystal field stays virtually constant the exchange interaction is proportional to the concentration c. For all concentrations $c > 0.4$ a transition from Van Vleck paramagnet to a type-II antiferromagnet is observed[2]. It is thus possible to study the singlet ground state behaviour in the most interesting range around the critical value for ordering at zero temperature.

We have investigated the coexistence of crystal field and magnetic exchange in this system by measuring the electrical resistivity of 15 2-6mm long single crystals of $Tb_cY_{1-c}Sb$ with various values of the Tb-concentration in the temperature range from 1.5 to 300 K. The resistivity was determined by a 4-terminal d.c.-method and the temperature measured with a Au(+0.03 at % Fe) versus Chromel thermocouple thermometer. For the paramagnetic regimes we have calculated the resistivity taking into account the crystal field as well as the magnetic exchange and compared the results with the measured data. Excerpts from our investigations are shown in fig. 1 for the temperature range up to 30 K. The residual resistivity has been subtracted from the experimental data and each resistivity versus temperature curve has an arbitrary origin. Since the absolute value of the measured resistivity is rather poorly defined for some of the very small samples we have adjusted the curves to have equal high temperature slopes. The Tb-concentrations indicated on fig. 1 were determined by neutron activation analysis given an uncertainty of 0.02 in c. We emphasize that although these values of c differ significantly (up to a difference in c of 0.14) from the nominal ones no concentration gradients were observed from electron microprobe analysis.

For the diluted systems ($c \ll 1$) we have previously explained the measured resistivity anomaly by a single ion model[3]. The conduction electrons, interacting with the 4f-electrons via an exchange interaction, are scattered independently from each crystal field splitted Tb-ion (purely fourth order crystal field), giving the total resistivity simply as the contribution from one ion times the number of ions. The electron-ion exchange interaction (in the form of a localized Heisenberg model) makes possible a variety of elastic and inelastic scattering processes. These are evaluated in the first Born-approximation and the resistivity is calculated by solving the linearized Boltzmann equation. Since the Fermi surface is essentially unknown, we take it to be spherical. Due to the singlet nature of the ground state for a Tb-ion in a cubic crystal field the contribution of these processes to the resistivity vanishes at zero temperature. At temperatures much greater than a characteristic energy in the 4f-multiplet it approaches a constant, the spin disorder resistivity limit. As seen in fig. 1 the agreement with the experiments is excellent for the lower concentrations. All the theoretical curves have been fitted to the experiments at 10 K (for TbSb just above the transition temperature). The deviation observed for the curves with $c \ll 1$, increases with temperature similar to the experimental observations in YSb. This we attribute to the electron-phonon scattering. Although a small increase in the phonon contribution may be expected in going from YSb to TbSb, it is most likely that the deviations observed at the high concentrations are incompatible with the single ion model. This is not surprising since the exchange interaction for $c > 0.4$ is strong enough to induce an antiferromagnetic transition. The observed

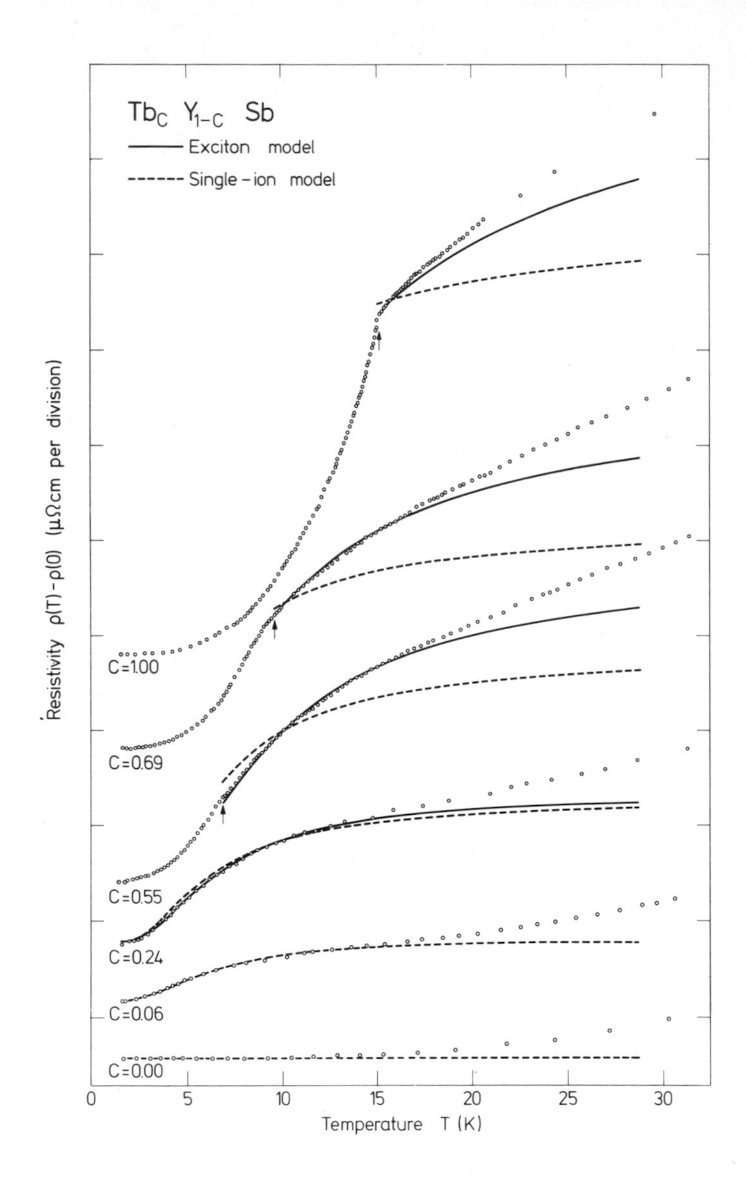

Fig. 1: Electrical resistivity measurements on $Tb_cY_{1-c}Sb$. The zero point of each set of data has been arbitrarily chosen. Solid lines represent the results from the exciton-model, dotted lines from the single-ion model. For c=0.06 the dotted line is also representative for the exciton-model.

transition temperatures are in good agreement with the results of molecular field calculations (indicated with an arrow).

In the presence of an indirect exchange interaction between the ions the crystal field levels broaden into bands of magnetic excitons[2]. Quite recently we have presented an extended experimental effort and a model calculation for the paramagnetic phases taking into account this broadening of the crystal field levels[4]. With the same localized form of the electron-ion exchange interaction, as used in the single-ion model, the scattering cross-sections may, in the first Born-approximation, be expressed in terms of the imaginary part of the dynamical ion-ion susceptibility function $\chi(\vec{q},\omega)$[5]. Knowing the scattering cross-sections the resistivity may be calculated in a variational approach to the Boltzmann equation. The value of the Fermi momentum k_F is an important parameter in the calculations because it determines which excitons affect the resistivity. In particular, if k_F is very small, only the low momentum excitons, whose energies are relatively higher than in the single ion model can scatter the conduction electrons. Consequently, the resistivity increases less rapidly towards its saturation value than predicted by the single ion model. In a RPA approach similar to the one used by Holden and Buyers[6] we have calculated $\chi(\vec{q},\omega)$ with a pseudo-fermion representation of the crystal field states. In order to obtain an explicit expression for $\chi(\vec{q},\omega)$ only the three lower crystal field levels have been taken into account. As the neglected states have energies larger than 90 K this truncation is without importance for $T<30K$. For the ion-ion exchange interaction we have in the actual calculation taken only couplings to the six second nearest neighbours into account, and the strength, increasing linearly with c, has been chosen such that the result of the molecular field calculation of the transition temperature agrees with the experimentally observed value of 15.1 K for TbSb. For the alloys with $c<1$, in which the Tb-ions do not form a spatially periodic lattice we assume that the excitonic modes are still reasonably well-defined and that their dispersion may be calculated approximately by letting the Tb-ions be periodically distributed on a hypothetical lattice with a spacing proportional to $c^{-1/3}$. Furthermore, we take for the Fermi momentum the rather small value $k_F = 0.3\pi/a$ (a is the lattice constant of TbSb), for the entire range of concentrations. In fig. 1 we have shown the results of this calculation. In the fitting procedure (similar to that used for the single-ion model) we have, for the alloys that order magnetically, taken into account contributions from the disorder in the spatial configuration of the magnetic ions, as calculated within MFA. It is clearly seen that the electron-exciton scattering picture gives excellent agreement with the experimental results for the abovementioned choice of the Fermi momentum. As seen on fig. 1 the exciton model approaches as it should the result of the single-ion model for small Tb concentrations. For high temperatures the dispersion in the exciton branches disappears and the exciton-model approaches the single-ion result. This means that the saturation values taken by the exciton model in our fitting procedure should scale with concentration. This is indeed the case.

REFERENCES

1. N.Hessel Andersen, P.E.Gregers-Hansen, E.Holm, F.B.Rasmussen, and O.Vogt, in proc.Int.Conf.Magnetism, Moscow, U.S.S.R. 22-28 August 1973. Vol. VI, p.234 (1974).

2. B.R.Cooper and O.Vogt, J. de Physique, 32, C1-958 (1971).

3. N.Hessel Andersen, P.E.Gregers-Hansen, E.Holm, H.Smith, and O.Vogt, Phys.Rev.Letters 32, 1321 (1974).

4. N.Hessel Andersen, P.E.Lindelof, H.Smith, O.Splittorff, and O.Vogt, Phys.Rev.Letters 37, 46 (1976).

5. P.Fulde and I.Peschel, Adv.Phys. 21, 1 (1972).

6. T.M.Holden and W.J.L.Buyers, Phys.Rev. B9, 3797 (1974).

CRYSTAL FIELD EFFECTS IN THE TRANSPORT PROPERTIES OF RARE EARTH ALLOYS*

B.V.B. Sarkissian

Physics Department, Imperial College, London SW7, England

ABSTRACT

The influence of crystal field splittings on the electrical resistivity of dilute Sc-RE and Y-RE alloys (RE = Gd, Tb, Dy) is analyzed and accounted for by theory. A comparison of the magnetoresistivity of these alloys with the preliminary results of calculations involving the Zeeman splitting of solute energy levels suggests that crystal field effects can be seen in the magnetoresistance also.

INTRODUCTION

Crystal field effects for non S-state heavy rare earth solutes dissolved in Y and Sc have been studied with bulk magnetization measurements by Touborg and Høg (1). For very dilute single crystals they established the ground state of the solute and obtained the (hexagonal) crystal field parameters. Continuing our own studies (2) of the properties of such alloys we have extended our measurements to the dilute regime ($\leqslant$ 1%) in the hope of observing crystal field effects in the resistivity and magnetoresistivity of these alloys. Such effects for the resistivity of RE solutes in gold were first observed by Murani (3). In the present work we show that resistivity data for the non S-state solutes can be fitted by calculations of the scattering from solutes whose splittings are described by the parameters of ref. (1) when the theoretical expression for the resistivity derived by Hirst (4) is used. An extension of his approach to include Zeeman splitting of solute levels can explain the sign and field dependence of the magnetoresistivity. In Gd containing alloys significant crystal field effects are not to be expected; temperature-dependence of the magnetic scattering can arise however either from Kondo-like effects or solute-solute interactions.

EXPERIMENTAL METHODS AND RESULTS

The samples were chill-cast rods of arc-melted alloys. Standard 4 probe DC resistivity measurements were made in a He^4 cryostat, and in some cases in a He^3 cryostat. The magnetic field was provided by a 50 kilo-Oersted superconducting solenoid.

The magnetic scattering is shown for some alloys in Fig. 1 as $\rho_m(T)/\rho_{sd}$, where $\rho_m(T)$ is $\rho_{alloy}(T) - \rho_{host}(T)$ and ρ_{sd} (the spin disorder resistivity) is the magnetic scattering in the high temperature limit, together with calculated curves. The higher temperature deviations ($>$ 8K for Tb and $>$ 15K for Dy) are attributed to deviations from Matthiessen's rule as phonon scattering increases. In the dilute Gd alloys a temperature dependence of the magnetic scattering is observed. Although this goes approximately as log T it does not scale with the composition, and it seems more likely to be due to solute-solute interactions than to a Kondo effect with positive J_{sf}.

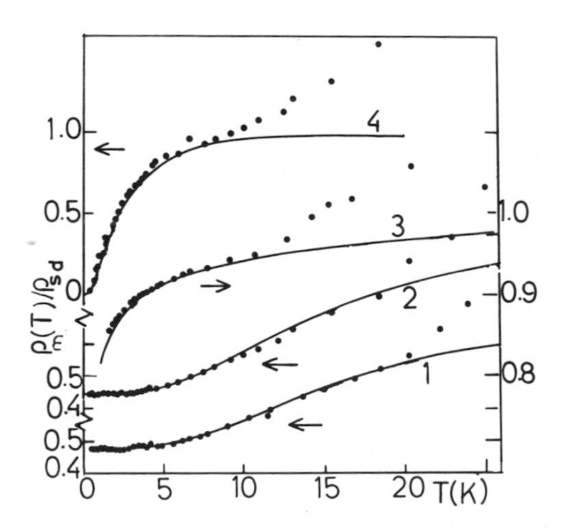

Fig. 1 : Plots of $\rho_m(T)/\rho_{sd}$ versus the temperature (see the text) for different alloys. The solid lines are calculated curves. From the bottom to the top the curves shown are (1) Y 0.2% Dy, (2) Sc 0.1% Dy, (3) Y 0.1% Tb, (4) Sc 0.1% Tb.

Fig. 2 shows the longitudinal (magnetic field parallel to current) and transverse magnetoresistivity of a ScTb alloy, together with a calculated curve that takes into account the Zeeman splitting of the crystal field levels. Preliminary calculations, of which further details will be given elsewhere, show that these effects dominate in both the size and the anisotropy of the magnetoresistance. Similar measurements have been made on the other alloys. A rather complicated field dependence for the Gd alloys suggests the presence of asymmetric or "skew" scattering of the type discussed for such alloys by Fert (5).

ANALYSIS

Hirst (4) has described the scattering of conduction electrons of spin s by rare-earth impurities of total angular momentum J by the Hamiltonian

$$H_{sf} = V_c - 2(g_J - 1)\, J_{exch}\; J \cdot s \tag{1}$$

where V_c is the spin independent term and J_{exch} the sf exchange coupling.

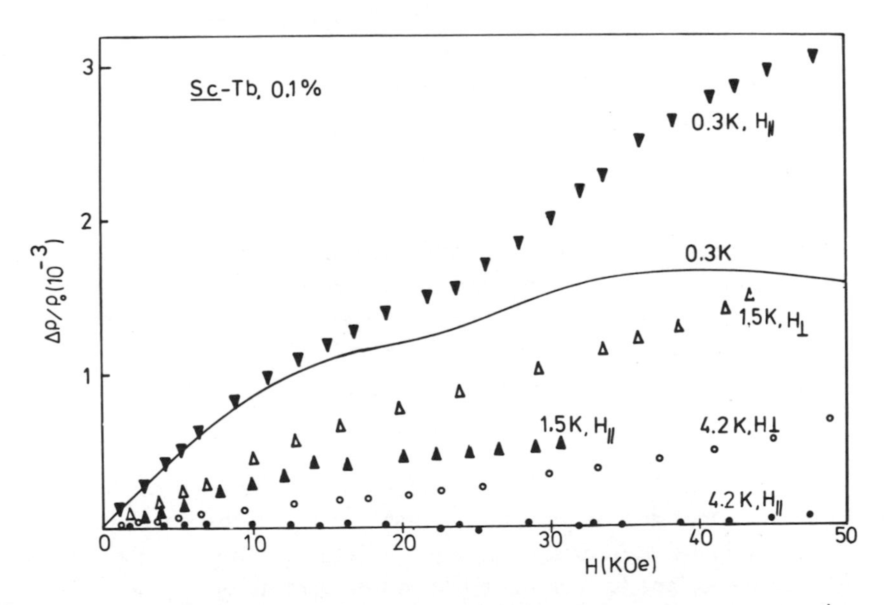

Fig. 2 : Longitudinal (II) and transverse (⊥) magnetoresistivity $\Delta\rho/\rho_0$ $(= (\rho(H,T) - \rho(0,T))/\rho(0,T))$ for Sc 0.1% Tb. The solid theoretical curve is an average $(\frac{1}{3}\rho_{||} + \frac{2}{3}\rho_{\perp})$ of values calculated using equation (2).

The resultant magnetic resistivity is given by

$$\rho_m = C \sum_{I,I'} P(I) \frac{(\Delta_{II'}/kT)}{\exp(\Delta_{I,I'}/kT)-1} \times \frac{1}{2} \sum_{s,s'} |\langle I's'|\underline{J}.\underline{s}|I s\rangle|^2 \tag{2}$$

where I labels the levels of the solute state J in the given crystal field and applied field which are occupied with the Boltzmann probability $P(I)$. The constant C contains features of the conduction electrons and J^2_{exch}, as well as the solute concentration. The eigenstates I and their energies are calculated by diagonalizing the crystal field Hamiltonian for the parameters given in ref. (1), and the matrix elements computed.

At high temperatures ($kT \gg \Delta$) the above expression reduces to

$$\rho_m = \rho_{sd} = CJ(J+1)/4$$

and this quantity and the residual resistivity of non-magnetic origin are used as adjustable parameters in fitting calculated curves to the experimental data, with the assumption that the lattice scattering is equal to that of the pure host. Minor deviations at low temperatures are found for the more concentrated alloys and attributed to solute-solute interactions of the type known (2) to lead to spin-glass freezing at higher concentrations. From ρ_{sd} we have deduced values of J_{exch} of 0.07 eV, 0.14 eV, 0.11 eV and 0.17 eV for $\underline{Sc}$Tb, $\underline{Sc}$Dy, $\underline{Y}$Tb and $\underline{Y}$Dy respectively. (The last figure applies to a 0.2% alloy; the others to 0.1% alloys.) These values may be compared with the NMR value of 0.06 eV for $\underline{Sc}$ Gd (6).

DISCUSSION

The total magnetic scattering at arbitrary temperature will contain both elastic and inelastic contributions, and the shapes of the $\rho_m(T)$ curves will depend on both the level separations and the character of the ground state. For a singlet ground state $\rho_m(T)$ tends to zero as T tends to zero, but for a Kramers ion an elastic scattering term will remain. Such effects can be seen in Fig. 1.

In discussing the variation of ρ_m with solute for rare earths in Lu Hirst pointed out that deviations from proportionality to the de Gennes factor ($(g_J - 1)^2 J(J+1)$) should arise from crystal field splittings. We have now observed such variations for the Y and Sc-based alloys, but we also find variations from the constant J_{exch} implicit in his discussion.

The ability of the magnetoresistance to test crystal field parametrizations (through its sensitivity to both level positions and level character) has been shown and will be discussed more fully elsewhere.

REFERENCES

*This work was supported by the Science Research Council

(1) P. Touborg and J. Høg, Phys. Rev. Lett. 33, 775 (1974)

(2) B.V.B. Sarkissian and B.R. Coles, Comm. on Phys. 1, 17 (1976)

(3) A.P. Murani, J. Phys. C, Metal Phys. 3, 153 (1970)

(4) L.L. Hirst, Solid State Comm. 5, 751 (1967)

(5) A. Fert, J. Phys. F, Metal Phys. 3, 2126 (1973)

(6) F.Y. Fradin, J.W. Ross, L.L. Isaacs and D.J. Lam, Phys. Lett. 28A 276 (1968)

ACKNOWLEDGEMENTS

The author is extremely grateful to Professor B.R. Coles for his encouragement and interest in this work and for many helpful discussions. He also wishes to express his gratitude to Drs. B. Southern, B. Rainford, N. Rivier and P.J. Grout for their help with the calculations and computational work.

SHORT RANGE VAN VLECK ANTIFERROMAGNETISM IN SUPERCONDUCTING $La_{1-y}Tb_ySn_3$ DETECTED BY NEUTRON SCATTERING *

H. E. Hoenig [+] and P. v. Blanckenhagen [x]

[+] Physikal.Institut d. Universität, D-6000 Frankfurt
[x] Kernforschungszentrum, IAK, D-7500 Karlsruhe, Germany

ABSTRACT

Short range antiferromagnetic order with a correlation length of about 25 Å has been detected in superconducting polycrystalline $La_{0.85}Tb_{0.15}Sn_3$ by diffuse neutron scattering. SQUID magnetometry has been applied to obtain crystal field (CF) states and -splittings of the Tb^{3+} ions and the superconducting transition temperatures in the $La_{1-y}Tb_ySn_3$ series up to y = 0.35.

The problem of coexistence between superconductivity and magnetic order in dilute magnetic alloys [1] has been revived recently by experiments which investigate the microscopic nature of the magnetic order. [2] Short range ferromagnetic order of a correlation length of about 15 Å has been shown to coexist with superconductivity in the system $Ce_{.8}Tb_{.2}Ru_2$. The present experimental work on $La_{1-y}Tb_ySn_3$ is intended to give a different example for coexistence where again the type of the magnetic order has been detected by neutron scattering.

Polycrystalline $La_{1-y}Tb_ySn_3$ samples were prepared from 4N La, 6N Sn and 3N Tb metal by levitation melting in a cold crucible for the neutron experiments and in an argon arc furnace for all the other experiments. Most of the samples were analyzed

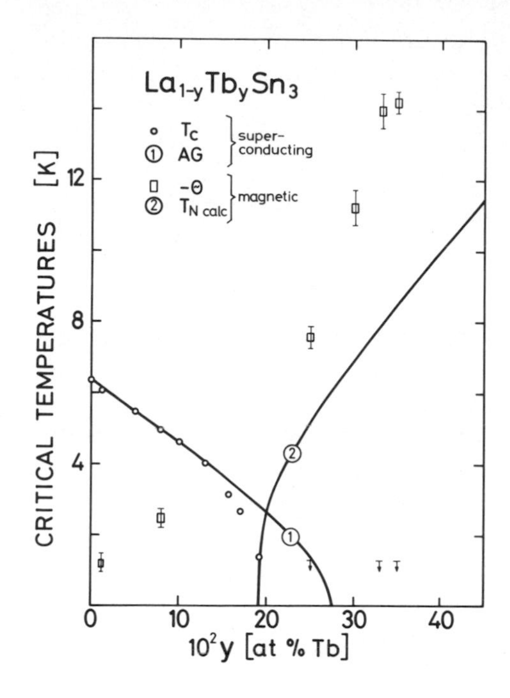

Fig. 1 Superconducting transition temperature T_c, Curie-Weiss temperature Θ, the Abrikosov-Gorkov (AG) dependence and the calculated Néel-temperature T_N as function of Tb concentration for $La_{1-y}Tb_ySn_3$. The arrows indicate upper limits for T_c.

with an electron microprobe and with elastic neutron scattering. All analyzed samples were of Cu_3Au structure. Sn-segregation which has been reported in Ref. 3 turned out to be an effect of corrosion attack limited to the sample surface.

The superconducting transition temperatures T_c (Fig. 1) were derived from measurements of the magnetization and the (upper) critical field with a SQUID-magnetometer [4] using small samples which were crushed under oil just before cool down. For Tb concentrations larger than y = .2 no superconductivity was detected above 1.4 K. The ac mutual inductance technique had given erroneous results for $T_c < T_c$ (Sn).

The magnetic susceptibility was also measured with the SQUID-magnetometer in a field of 0.14 T (in order to suppress superconductivity) in the temperature range from 1.5 to 25 K. The "single ion" susceptibility of Tb in $La_{1-y}Tb_ySn_3$ was determined with a sample of Tb concentration y = .0124 up to which coupling effects are unimportant. The paramagnetism due to the $LaSn_3$ host was taken into account. For increasing Tb concentration antiferromagnetic coupling was observed. The Curie Weiss temperatures Θ as derived from the susceptibility data are plotted in Fig. 1. For concentrations y where the Curie-Weiss temperatures are about equal to or larger than the corresponding superconducting transition temperatures T_c we observed a strong deviation of T_c (y) from the Abrikosov-Gorkov (AG) dependence.

The specific heat was measured with a heat pulse method. Schottky anomalies centered at 5 K were observed [5] with an entropy contribution per mole Tb at 10 K of Rln (4.5 ± .2) for Tb concentrations up to y=.35.

The thermodynamical data indicate that the CF groundstate of Tb^{3+} in $LaSn_3$ is a nonmagnetic singlet, separated by $\delta = 10 \pm 1K$ from a next higher triplet state in the CF-parameter [6] range $J = 6$, $W < 0$, $x = -0.9 \pm 0.2$. The antiferromagnetic coupling is therefore of Van Vleck type. A simple molecular field treatment as discussed in Ref. 7 gives a concentration dependence of the (MF) Néeltemperature as plotted in Fig. 1 as curve 2 suggesting an influence of the magnetic order on the suppression of superconductivity. At the calculated MF-Néeltemperatures the susceptibility only shows a gradual departure from the MF-behaviour and no abrupt change as characteristic for a bulk transition.

The magnetic correlation in $La_{1-y}Tb_ySn_3$ was studied with diffuse elastic neutron scattering at the FR 2 reactor at Karlsruhe using a spectrometer with double monochromator and multidetector. Fig. 2 gives the results for the y = .35 sample at small scattering angles for three different temperatures. We observe diffuse scattering at the $(00\frac{1}{2})$ and $(01\frac{1}{2})$ positions which strongly depends on temperature and magnetic field. For a $La_{.846}Tb_{.154}Sn_3$ sample at 1.7 K a magnetic field of 4.1 T vertical to the scattering plane reduces the integrated peak intensity $I(00\frac{1}{2})$ (which is 1.6 % of I(111) in this case) by 45 %. In table 1 we plotted the integrated diffuse scattering intensity $I(00\frac{1}{2})$ reduced to the integrated nuclear scattering intensity I(111) corrected for solid angle together with the linewidth $\Delta q = (4\pi/\lambda)\sin\Delta\Theta$ as function of Tb concentration y and of temperature T. Here $\Delta\Theta$ denotes half the angular width of the $(00\frac{1}{2})$ peak. Because of the discussed dependence on temperature and magnetic field the contribution from chemical short range order to the diffuse scattering at the $(00\frac{1}{2})$ and $(01\frac{1}{2})$ scattering angles seems to be negligible. The observed diffuse scattering is very similar to the one found for MnO [8] above the Néel-temperature and may be analyzed accordingly with respect to magnetic short range parameters. Since we could not detect any change with temperature or field of the fundamental reflections we can attribute

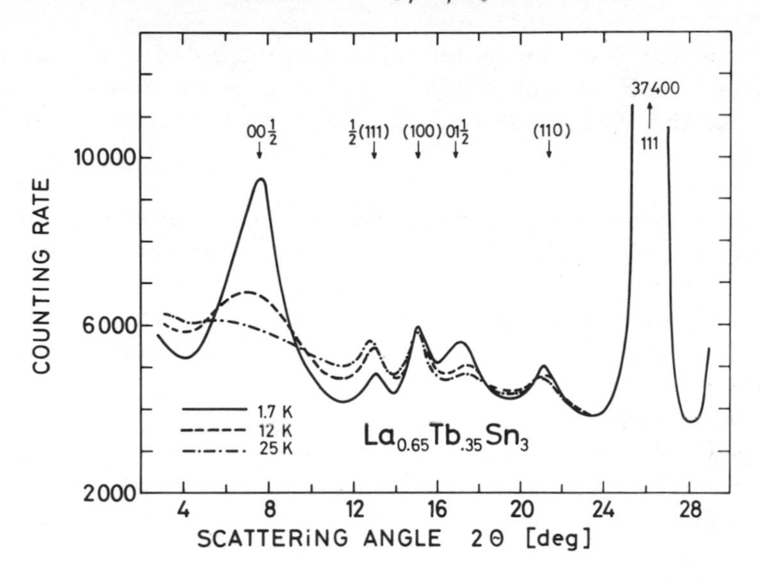

Fig. 2 Neutron (λ = 1.25 Å) counting rate versus scattering angle for $La_{.65}Tb_{.35}$ at 3 different temperatures

Table 1 Ratio of scattering intensities at the $(00\frac{1}{2})$ and (111) scattering angles and the linewidth $\Delta q = (4\pi/\lambda)\sin\Delta\Theta$ as function of temperature T and concentration y for $La_{1-y}Tb_ySn_3$

y	T[K]	$10^2 * I(00\frac{1}{2})/I(111)$	Δq [$Å^{-1}$]	$2\pi/\Delta q$ [Å]
.083	1.7	.6 ± .2	-	-
.154	1.7	1.6 ± .15	.27 ± .03	23. ± 3
	2.8	1.5	.30 ± .04	20.6 ± 4
	4.2	1.3	.29 ± .06	21.5 ± 5
	10.	.6	-	-
.25	1.6	5. ± 1.	.24 ± .05	26. ± 5
	4.6	4.	.24 ± .06	26. ± 6
	6.	3.	.26 ± .09	24. ± 8
.35	1.7	6.3 ± .6	.15 ± .01	43. ± 3
	4.2	6.5	.21 ± .02	30. ± 3
	12.	3.8	.30 ± .05	21. ± 3
	17.	1.5	.23 ± .09	28. ± 11

the $(00\frac{1}{2})$ and $(01\frac{1}{2})$ peaks to antiferromagnetic order of short range. The magnetic correlation length which can be estimated from the width Δq of the $(00\frac{1}{2})$ peak as $2\pi/\Delta q$ is plotted in the last column of table 1.

For the sample $La_{.846}Tb_{.154}Sn_3$ we have observed coexistence of short range Van Vleck antiferromagnetism and of superconductivity, where the magnetic correlation length $2\pi/\Delta q = 23$ Å is about one order of magnitude smaller than the superconducting coherence length which we estimated to be 200 Å on the basis of our resistivity and specific heat data.

Our results for the concentration dependence of the superconducting transition temperature now are qualitatively understood within the framework of the theory due to Klose and Entel [9] for the short range order regime. The CF-theory for superconductors [10] for the given parameters does not explain the observed enhanced pairbreaking but taking into account dispersion of magnetic excitons and assuming a reduction of the effective CF level separation from 10 to 3.5 K we obtain a good fit of theory and experiment.

REFERENCES

1. Y.A.Izumov, Y.N.Skryabin, phys.stat.sol.(b) 61, 9 (1974)
2. S.Roth, K.Ibel, W.Just, J.Phys. C 6, 3465 (1973)
3. R.P.Guertin, J.E.Crow, A.R.Sweedler, S.Foner, Solid State Commun. 13, 25 (1973)
4. J.W.Dawson, H.B.Gray, H.E.Hoenig, R.G.Rossmann, J.M.Schredder, R.H.Wang, Biochemistry 11, 461 (1972)
5. H.E.Hoenig, H.Happel, H.K.Njoo, H.Seim in R.A.B.Devine (Ed.) Proc. 1st Conf. on Crystalline Electric Field Effects in Metals and Alloys (Montreal 1974) p. 298
6. K.R.Lea, M.J.M.Leask, W.P.Wolf, J.Phys.Chem.Solids 23, 1381 (1962)
7. B.R.Cooper, O.Vogt, Phys. Rev. B 1, 1218 (1970)
8. I.A.Blech, B.L.Averbach, Physics 1, 31 (1964)
9. P.Entel, W.Klose, Journal Low Temp. Phys. 17, 529 (1974)
10. J.Keller, P.Fulde, J.Low Temp. Phys. 4, 289 (1971)

* This work was supported by part by the Deutsche Forschungsgemeinschaft via Sonderforschungsbereich 65.

CRYSTAL FIELD EFFECTS IN LAVES AND OTHER RARE EARTH IRON AND RARE EARTH COBALT COMPOUNDS

Uzi Atzmony

Nuclear Research Centre Negev
P.O.Box 9001
Beer Sheva, Israel

ABSTRACT

Single ion crystalline electric field (CEF) calculations are shown to acount for the spin orientations, spontaneus R^{3+} moments and hyperfine interaction parameters in cubic RFe_2, $R^1_xR^2_{1-x}Fe_2$ and RCo_2 Laves phases. The occurence of non-major axes of easy magnetization is discussed in both a phenomenological and a single rare earth ion approach. Ranges of CEF parameters applicable to each of these compounds are given. The magnetocrystalline anisotropy of non-cubic R-Co alloys are treated in a similar way.

INTRODUTION

The present work reviews some of the aspects of the interaction between the well shielded 4f electrons of the rare earth and the crystalline electric field (CEF) in rare earth (R)-transition metal (M) intermetallic compounds. This interaction dominates the magnetic anisotropy properties of the R-M compounds. The work deals mainly with the cubic Laves phase compounds RFe_2 and RCo_2. The RFe_2 have been observed to exhibit large magnetic anisotropy[1], extremely large magnetostrictive strains[2] and high Curie temperatures. The Fe-Fe interaction is the dominating exchange in these compounds. The Curie temperatures of the RCo_2 compounds are much lower indicating weaker exchange that is attributed mainly to the R-Co interaction. This is reflected by the ferromagnetism of YFe_2 (T_C=500K) and by the strongly enhanced Pauli paramagnetism of YCo_2. Both the single ion CEF theory and the phenomenological treatment are outlined. Then, the relevant experimental results are discussed. The application of the CEF theory to RCo_5 and R_2Co_{17} is briefly outlined.

THEORY

Crystal Field Interaction

The magnetocrystalline anisotropy Hamiltonian is

$$\mathcal{H}_{anis} = \Sigma_1 (E_J I + \mathcal{H}_{exch} + \mathcal{H}_{crys}) = N(E_J I + \mathcal{H}_{exch} + \mathcal{H}_{crys}) \quad (1)$$

where N is the number of rare earth ions per unit volume. E_J is the energy of the spin J level. $\mathcal{H}_{exch}$ is the exchange Hamiltonian:

$$\mathcal{H}_{exch} = 2(g_J - 1)\mu_B H_{exch} \vec{J}.\vec{n} \quad (2)$$

$\vec{n}$ is a unit vector parallel to the exchange field, H_{exch}. $\mathcal{H}_{crys}$ is the CEF Hamiltonian and is given by

$$\mathcal{H}_{crys} = \sum_{k,q} A_k^q \ \langle r^k \rangle \ U_k^q \quad (3)$$

where U^q_k are the Racah operators and A^q_k are related to the strength of the CEF. When mixing of J states can be disregarded the Racah operators are replaced by Stevens operators equivalents, O^q_k .
For a 4f ion in a cubic symmetry only the fourth and the sixth terms are retained i.e.

$$\mathcal{H}_{crys} = A_4(1-\sigma_4) \langle r^4 \rangle \ [\ U_4^0 + (5/14)^{1/2}(U_4^{+4} + U_4^{-4})] + A_6 \langle r^6 \rangle \ [\ U_6^0 + (7/2\ )^{1/2}(U_6^{+4} + U_6^{-4})] \quad (4)$$

The z axis is parallel to the [001] direction. The values of $\langle r^n \rangle$ of the 4f ions and of σ_4 the shieldings parameters have been calculated by Freeman and Watson. $\mathcal{H}_{anis}$ can be calculated and diagonalized as a function of four parameters: H_{exch} , A_4 , A_6 and the direction of $\vec{n}$. With the eigenvalues E_i, and the wave functions $|\psi_i\rangle$ the following can be done.
1. The free energy per ion can be calculated:

$$F_R(\vec{n}_j,T) = -kT \ln Z(\vec{n}_j,T) = -kT \ln \Sigma_1^m \exp\{-E_i/kT\} \quad (5)$$

m is the number of levels.The easy direction of magnetization at a given temperature is the direction of $\vec{n}_j$ for which F_R has its lowest value.
2. Spin orientation diagrams (SOD)[3] of ternary combinations $R^1_x R^2_{1-x} Fe_2$ can be constructed: A SOD presents the direction of easy magnetization for various values of x and T. The avarage free energy per ion in a ternary compound is

$$F(x,\vec{n}_j,T) = xF_{R^1}(\vec{n}_j,T) + (1-x)F_{R^2}(\vec{n}_j,T) \quad (6)$$

3. The spontaneous magnetic moment of the rare earth sublattice, $M_R(\vec{n}_j,T)$, the contributions of the 4f electrons to the magnetic

hyperfine field, $\vec{H}_{4f}(\vec{n}_j,T)$, and to the electric field gradient, $q_{eff}(\vec{n}_j,T)$, acting on the rare earth nucleus are calculated:

$$\vec{M}_R(\vec{n}_j,T) = M_R(0)\ \{\Sigma[1/J<\psi_i|\vec{J}|\psi_i>\ \exp(-E_i/kT)]\}/Z(\vec{n}_j,T) \quad (7)$$

$$\vec{H}_{4f}(\vec{n}_j,T) = \Sigma\{<\psi_i|\vec{N}|\psi_i>\ \exp(-E_i/kT)\}/Z(\vec{n}_j,T) \quad (8)$$

$$q_{eff}(\vec{n}_j,T) = q_{fi}\{\Sigma[<\psi_i|3J_{z'}^2-J(J+1)|\psi_i>\exp(-E_i/kT)/[J(2J-1)]]\}/Z \quad (9)$$

Where N is the magnetic hyperfine operator and q_{eff} is assumed to be axially symmetric along z'.

Bulk Anisotropy Constants.

The bulk magnetocrystalline free energy of a cubic system is expanded into power series of the direction of cosines, α_i, of $\vec{n}$ with respect to the cubic axes:

$$E(\vec{n}_j,T) = K_0 + K_1(\alpha_1^2\alpha_2^2+\alpha_2^2\alpha_3^2+\alpha_3^2\alpha_1^2) + K_2\alpha_1^2\alpha_2^2\alpha_3^2 + K_3(\alpha_1^4\alpha_2^4+\alpha_2^4\alpha_3^4+\alpha_3^4\alpha_1^4) + \dots \quad (10)$$

The K_i-s are the temperature dependent bulk magnetic anisotropy constants. The direction of $\vec{n}$ is the direction for which $E(\vec{n}_j,T)$ has it lowest value. The inclusion of the eighth power cosine term in Eq. 10 has been shown to yield also non-major axes of $\vec{n}$ in a cubic crystal[4]. Figure 1 represents in the $K_1' = K_1/K_3$, $K_2' = K_2/K_3$ plane regions with different possible axes of magnetization. The spin may rotate either continuously or discontinuously (for example on crossing lines AB and AC in Fig. 1 respectively). The bulk magnetic anisotropy constants corresponding to the rare earth, K_i^R, can be calculated by substituting $E(\vec{n}_j,T)$ by $F_R(\vec{n}_j,T)$ (Eq.5). For each value of T,F_R is calculated for several directions of $\vec{n}_j$ and the $K_i^R(T)$-s are derived by a least squares procedure. The difference $F_{add}(\vec{n}_j,T) = E(\vec{n}_j,T) - F_R(\vec{n}_j,T)$ is attributed to other sources of anisotropy energy.

EXPERIMENTAL RESULTS

Direction of Easy Magnetization and SOD

The ^{57}Fe Mössbauer effect, (ME), is a highly effective tool for the study of the direction of easy magnetization in polycrystalline cubic Laves phase ^{57}Fe containing RM_2 compounds[5,6]. Pending on the angle between $\vec{n}$ and the local $\bar{3}m$ axis the iron ions form one to four magnetically inequivalent sites.

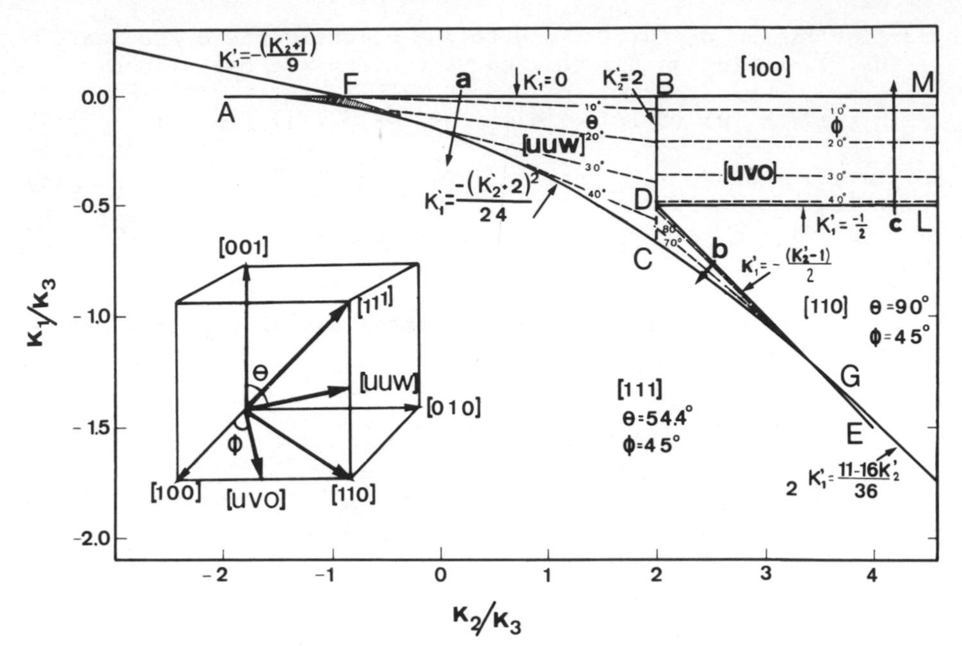

Fig. 1: Boundaries of regions corresponding to different easy axes in the $K_1' = K_1/K_3$ and $K_2' = K_2/K_3$ plane. For details see Ref. 4.

The directions of $\vec{n}$ were determined in the RFe_2 compounds (R= Ce, Sm, Gd, Tb, Dy, Ho, Er, Tm, Lu)[7-9] and in ^{57}Fe doped RCo_2 compounds (R= Pr, Nd, Gd, Tb, Dy, Ho, Er)[10-12] at various temperatures (4.2-300K). Experimental SOD of ternary systems, $R^1_xR^2_{1-x}Fe_2$, with R^1= Tm, Er, Tb, and R^2= Ho, Dy , for 4.2<T<300K were constructed[3,8]. As an example the SOD of $Ho_xEr_{1-x}Fe_2$ is shown in Fig. 2.

The following results should be emphasized:
1. In most of the binary compounds $\vec{n}$ is parallel either to the cubic [111] ($TbFe_2$, $TbCo_2$, $ErFe_2$, $ErCo_2$, $TmFe_2$, YFe_2) or to the [100] ($GdFe_2$, $DyFe_2$, $DyCo_2$, $PrCo_2$) axis at all temperatures up to 300K or T_C.
2. In some binary compounds spin-rotation-like transitions are found (Table 1). The reorientations take place over finite ranges of temperatures at which $\vec{n}$ rotates continuously from one major axis of cubic symmetry to another.
3. In the SOD-s the (x,T) plane is devided into two ($Dy_xR_{1-x}Fe_2$) or three ($Ho_xR_{1-x}Fe_2$) domaines[3]. In each of these the magnetization is along one of the major cubic axes. Transitions regions between domains of major axis spin alignement have been observed. Within these regions $\vec{n}$ was found to assume non-major directions.

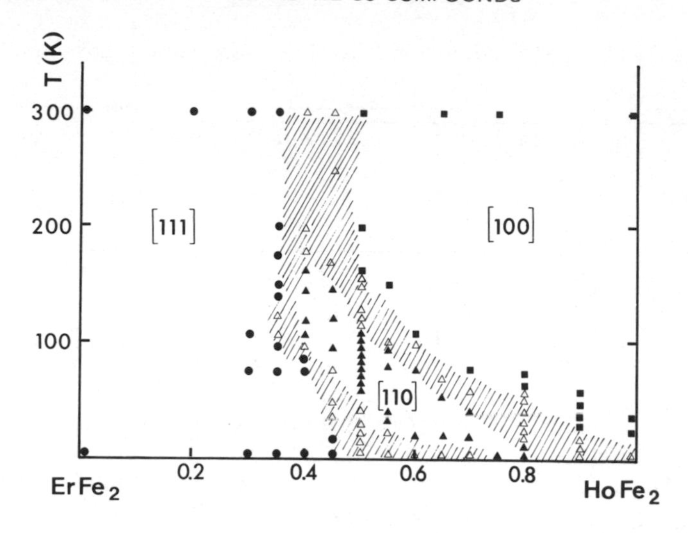

Fig 2: Experimental SOD of the $Ho_xEr_{1-x}Fe_2$ system[3,4,8]. The shaded region corresponds to non-major symmetry axes of easy magnetization. (See detailes in Ref. 4).

The SOD boundaries were also dtermined by sound velocity measurements as anomalies appear in the temperature dependence of the elastic modouli at points corresponding to the upper part of the transition region[8]. The spin rotation process was investigated by neutron diffraction measurements of an aligned powder sample[14] of $Ho_{0.6}Tb_{0.4}Fe_2$, and the angle between $\vec{n}$ and [100] was directly determined as a function of temperature. Further, the direction of $\vec{n}$ was determined by magnetization measurements of single crystals. Gignoux et al.[12] investigated $GdCo_2$ and $HoCo_2$(Table 1). In $HoCo_2$ they found that $\vec{n}$ is parallel to [110] at low temperatures. This disagreement with the ME findings can be overcome by the assumption that in $HoCo_2$, as in $HoFe_2$, $\vec{n}$ is parallel to a non-major axis of cubic symmetry, probably in the vicinity of the [110] axis.

Application to Theory

The above results can be reproduced by both single ion CEF calculations and by bulk anisotropy phenomenological treatment(Eq.10). From the observed directions of easy magnetization in the binary compounds, ranges of K_1' and K_2' and of $\mu_B H_{exch}(0)$, A_4 and A_6 were determined. The former may be done by using Fig. 1 and the latter from the single ion CEF calculations. Values of $\mu_B H_{exch}(0)$ can be estimated from other measurements. It was found to be around -150K in RFe_2 compounds and -100K in RCo_2. The predicted direction of $\vec{n}$ is not very sensitive to the value of $\mu_B H_{exch}(0)$ in the range (-100 -

Table 1: Spin rotations in RFe_2 and RCo_2 compounds.

	First range		Second range		Third range		
Compound	$\vec{n}$	range (K)	$\vec{n}$	range (K)	$\vec{n}$	range (K)	
$CeFe_2$	[001]	4.2-150	[uuw]	150-230*			ME[9]
$NdCo_2$	**	4.2- 45	[001]	45-110			ME[6]
$SmFe_2$	[110]	4.2-140	[uuw]	140-240	[111]	240-	ME[13]
$GdCo_2$	[001]	4.2-200	[uuw]	200-290	[110]	290-	ME[11]
$HoFe_2$	[uv0]	4.2- 14	[001]	14-			ME[8]
$HoCo_2$	†[uuw]	4.2- 16	[001]	16- 95			ME[6]
$HoCo_2$	[110]	4.2- 14	[001]	14- 78			mag[12]
$Ho_{0.6}Tb_{0.4}Fe_2$	[110]	4.2- 90	[uuw]	90-150	[111]	150-	ME[3],ND[14]
$Tm_{0.2}Ho_{0.8}Fe_2$	[110]	4.2- 30	[uvo]	30- 40	[001]	40-	ME[3,15]

ND= neutron diffraction. mag= magnetization. ME=Mössbauer effect.
* $\vec{n}$ jumps within 5K to 15° from [001] in $(1\bar{1}0)$ plane and than slowly rotates towards [111] and terminate at T_C 30° away from [001].
** Unlike other compounds the results below 45K could not be fitted assuming that $\vec{n}$ is confined to a major cubic plane.
† [111] at 4.2K.

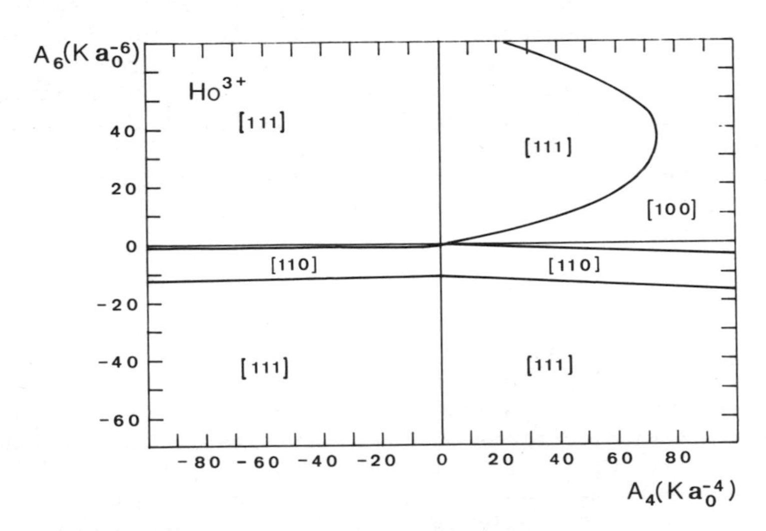

Fig 3: Easy axes diagram for Ho^{3+} at 4.2K. Direction of easy axis as a function of crystal electric field parameters A_4 and A_6 for an exchange field $\mu_B H_{exch}(0)= -100K$.

-150K). For a given $\mu_B H_{exch}(0)$ an easy axis diagram[16], namely boundaries between regions of different spin alignments in the A_4, A_6 plane, can be plotted at each T. As an example the easy axis diagram for Ho^{3+} with $\mu_B H_{exch}(0)$=-100K at 4.2K is shown in Fig. 3. In drawing this diagram only directions parallel to major cubic axes were taken into acount. The temperature dependence of $\vec{n}$ enable reduction of the applicable ranges of A_4 and A_6. Further reduction of these ranges is done by assuming that A_4 and A_6 are independent of the trivalent rare earth involved. This was found to be possible for the heavy rare earth ions and not for the light ones. As $\vec{n}$ is the same in both the heavy RFe_2 and RCo_2 similar ranges of A_4 and A_6 are obtained. Compounds containing light R elements are discussed below.

It was recently shown that the existence of non-major cubic axes of $\vec{n}$ and its continuous rotation could also be predicted by single ion CEF calculations[4]. Hamiltonian (1) was calculated for thirty directions of $\vec{n}$ which were confined to the $(1\bar{1}0)$and (001) planes including the three major cubic directions. Values of A_4 and A_6 were found for which $\vec{n}$ is parallel to non-major cubic directions. For example the calculations for Ho^{3+} with $A_4=36Ka_0^{-4}$, $A_6/A_4=-0.045a_0^{-2}$ and $\mu_B H_{exch}$=-150K predicted direction of easy magnetization that rotates continuously with T in the (001) plane from [110] at 0 to [100] above 28K. This is in good agreement with the experimental results in $HoFe_2$ and $HoCo_2$. Predicted regions of spin rotation are very sensitive to the value of A_6/A_4 and less to that of A_4. By using eq. 6 SOD for the heavy rare earths ternary compounds were constructed[3,4]. The SOD of $Ho_xEr_{1-x}Fe_2$ with $\mu_B H_{exch}$=-150K, $A_4=36Ka_0^{-4}$ and $A_6/A_4=-0.043a_0^{-2}$is shown in Fig. 4. The calculated SOD using these values are in good agreement with the experiment[8].

By calculating the free energy difference $F_{Ho}([110],14K)$ - $F_{Ho}([001],14K)$ (Eq. 5) and comparing it with the difference observed from magnetization curves of $HoCo_2$, Gignoux et al.[12] found A_4= $25.1Ka_0^{-4}$ $A_6/A_4=36a_0^{-4}$ and $\mu_B H_{exch}$=60K. These are within the calculate ranges of A_4 and A_6, above. K_1 and K_2 measured by Clark et al.[1] at room temperature enable determination of $A_4=36Ka_0^{-4}$ in $ErFe_2$ (Eq.10). This value was used in most of the calculations. In particular, the temperature dependences of the K_i-s were calculated[4,17].

Some discrapencies between the theoretical and experimental SOD subsist. This was attributed to the anisotropy of interactions other than the 4f CEF. The relative strength of these interactions could be estimated by calculating the additional free energy terms, $F_{add}(\vec{n}_j,T)$. This was done for the SOD of $Ho_xR_{1-x}Fe_2$, neglecting the existence of transition regions and assuming that $F_{add}(\vec{n}_j,T)$ does not depend on the rare earth involved[3]. It was found that these are less than 10% of the calculated single ion contributions.

The compounds containing light R elements that were ivestigated are: $PrCo_2$, $NdCo_2$and $SmFe_2$. To predict the spin reorintation around

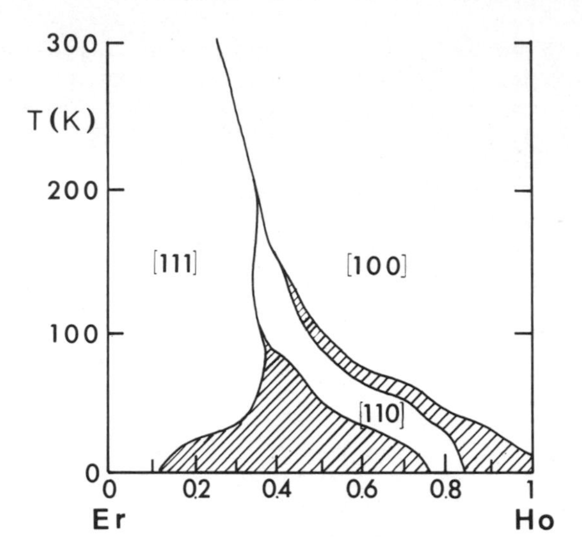

Fig 4: SOD of the $Ho_xEr_{1-x}Fe_2$ system. The shaded region corresponds to non-major symmetry axes of easy magnetization. The parameters used in the calculations are $\mu_B H_{exch}$= -150 K, A_4=36Ka_0^{-4} and A_6/A_4=-0.043 a_0^{-2}.

180 K the lowest $-A_4/A_6$ is 0.18a_0^{-2} and A_4 in exess of 150K$a_0^{\ 4}$. (Ref. 13, 16). The spin reorientation in $NdCo_2$ from [110] to [001] at 43±2 K can be predicted for 25<A_4<40 Ka_0^{-4}, -0.088<A_6/A_4<-0.068 a_0^{-2}, on a single A_4 vs. A_6/A_4 curve[6,18]. Non-major cubic axes of magnetization that exist in both compounds could not be predicted by the single ion model. Thus, the presence of additional contributions to the magnetic anisotropy free energy is indicated. Powder samples of both materials were investigated by magnetization measurements and the results could be explained by using Eq.7 for the R^{3+} sub-lattice magnetization[18,19].

The predicted hyperfine interactions of the rare earth nuclei in the Laves phase compounds depend on the direction of easy magnetization (Eqs. 8,9). Yanovsky et al.[15] studied the ME of ^{170}Yb with sources $Tm_xHo_{1-x}Fe_2$ between 4.2 and 80K. The experimental values for $H_{eff}(T)$ and eqQ(T) in $TmFe_2$ were fitted to a [111] direction of $\vec{n}$ by adjusting A_4, A_6 and $\mu_B H_{exch}$ together with $eq_{fi}Q$, H_{fi} and H_c. ($H_{eff}(T)$= $H_{4f}(T)$+H_c where H_c, the field produced by the polarized conducting electrons, is assumed to be temperature independent up to 80K). The CEF parameters thus derived are A_4= 37±5 Ka_0^{-4}, A_6/A_4=

$-0.026\pm0.018\ a_0^{-2}$ and $\mu_B H_{exch} = 116\pm4$ K. Using These parameters eqQ(T) and $H_{eff}(T)$ were calculated for H_{exch} along the [001] and [110] directions and the temperature dependence of $\vec{n}$ has been obtained (see Table 1).

The three axes neutron spectrometer was recently used by Niclow et al.[20] to measure the magnon dispertion relation in single crystal $Ho_{0.88}Tb_{0.12}Fe_2$. The data were fitted by a CEF-spin wave theory and yielded $\mu_B H_{exch} = -190$K, $A_4 = 22.5\ Ka_0^{-4}$ and $A_6/A_4 = -0.045\ a_0^{-2}$. Shamir et al.[21] measured the inelastic scattering of powdered $HoFe_2$ at 4.2 K and found a peak at 100K corresponding to the first exited state of Ho^{3+} as calculated for $\mu_B H_{exch} = -150$K, $A_4 = 36Ka_0^{-4}$ and $A_6/A_4 = -0.043\ a_0^{-2}$.

Non-Cubic R-Co Alloys

Magnetocrystalline anisotropies of non-cubic R-Co alloys were treated, in a similar way, by Greedan et al.[22] and that of $SmCo_5$ by Sankar et al.[23] . The $RCo_5$and R_2Co_{17} crystallized in essentially hexagonal structure. Magnetization measurements revealed the directions of easy magnetization at the various compounds (easy axis, easy plane or cone). In the RCo_5 Compounds with magnetic rare earth elements, spin reorientations were detected. Measurements in the hard directions were also made and bulk anisotropy constants were determined. Hamiltonian (1) is applicable here too. Due to the lower symmetry more terms are retained in $\mathcal{H}_{crys}$ (Eq.4):

$$\mathcal{H}_{crys} = A_2^0\langle r^2\rangle U_2^0 + A_4^0\langle r^4\rangle U_4^0 + A_6^0\langle r^6\rangle U_6^0 + A_6^6\langle r^6\rangle U_6^6 \qquad (11)$$

The exchange field, H_{exch} in eq. 3 is either parallel to z (easy axis) or x (easy plane). Values of H_{exch} were estimated and the free energies (Eq. 5) for the two directions were calculated for various temperatures. It was found that the sign of $A^0{}_2$ determines the direction of easy magnetization. The calculated free energies were in reasonable agreement with the experiments.

REFERENCES

*Also at Materials Engineering Department,Ben-Gurion University, Beer-Sheva, Israel.

1. A.E. Clark, H.S. Belson, and N. Tamagawa, Phys. Letters, A42, 160 (1972) .
2. N.C. Koon, A.I. Schindler, and F.L. Carter, Phys. Letters, A37, 413 (1971) .
3. U. Atzmony, M.P. Dariel, E.R. Bauminger, D. Lebenbaum, I. Nowik, and S. Ofer, Phys. Rev. B7, 4220 (1973) .
4. U. Atzmony and M.P. Dariel, Phys. Rev. B13, 9 (1976) .
5. G.J. Bowden, D.St.P. Bunbury, A.P. Guimaraes, and R.E. Synder, J. Phys. C2, 1367 (1968) .

6. U. Atzmony, M.P. Dariel, and G. Dublon, Phys. Rev. B (in press) .
7. A. P. Guimaraes, Ph.D. Thesis, Manchester (1971) .
8. M. Rosen, H. Klimker, U. Atzmony,and M.P. Dariel, Phys. Rev. B8, 2336 (1973) ; J. Phys. Chem. Solids, 37, 513 (1976) .
9. U. Atzmony and M.P. Dariel, Phys. Rev. B10, 2060 (1974) .
10. U. Atzmony and G. Dublon, Proceedings of the ICM (1976) (to be published in Physica) .
11. U. Atzmony and G. Dublon, Proceedings of the Conference of the Application of Mössbauer effect, Curfu (1976) (to be published in Supp. J. Phys.).
12. D. Gignoux, F. Givord, and R. Lemaire, Phys. Rev. B12, 3878 (1975) .
13. U. Atzmony, M.P. Dariel, E.R. Bauminger, D. Lebenbaum, I. Nowik, and S. Ofer, Proceedings 10th Rare Earth Conf. Carfree, Arizona, (1973) .
14. G. Dublon, U. Atzmony, M.P. Dariel, and H. Shaked, Phys. Rev. B12, 4628 (1975) .
15. R. Yanovsky, E.R. Bauminger, D. Levron, I. Nowik, and S. Ofer, Solid State Comm. 17, 1511 (1975) .
16. A.M. van Diepen, H.W. de Wijn, and K.H.J. Buschow, Phys. Rev. B8, 1125 (1973) .
17. M. P. Dariel and U. Atzmony, Int. J. Magnetism, 4, 213 (1973) .
18. G. Dublon, M. Kroupp, M.P. Dariel, and U. Atzmony, (submitted for publication in Phys. Stat. Solidi A) .
19. G. Dublon, M.P. Dariel, and U. Atzmony, Phys. Letters A51, 262 (1975).
20. R.M. Nicklow, N.C. Koon, C.M. Williams, and J.B. Milstein, Phys. Rev. Letters, 36, 532 (1976) .
21. N. Shamir, M. Melamud, and H. Shaked (private communications) .
22. J.E. Greedan and V.U.S. Rao, J. Solid State Chem. 6, 387 (1973) .
23. S.G. Sankar, V.U.S. Rao, E. Segal, W.E. Wallace, W.G.D. Frederick, and H.J. Garrett, Phys. Rev. B11, 435 (1975) .

INFLUENCE OF CRYSTAL FIELD INTERACTION ON THE THERMAL BEHAVIOR OF $ErAl_2$*

T. Inoue, S. G. Sankar, R. S. Craig, W. E. Wallace and K. A. Gschneidner, Jr.+
Department of Chemistry, University of Pittsburgh, Pittsburgh, PA 15260
+Ames Laboratory, USERDA and Department of Materials Science and Engineering, Ames, IA 50011

ABSTRACT

Low temperature heat capacities were measured in an adiabatic calorimeter for $ErAl_2$ and $LuAl_2$ in the temperature range 4.2 - 300 K. $ErAl_2$ exhibits a λ-type anomaly in the heat capacity vs. temperature curve peaking at 10.2 K and a Schottky anomaly at nearly 23 K. Heat capacities of $LuAl_2$ were utilized to characterize the lattice and electronic contributions in $ErAl_2$. The excess entropy calculated for $ErAl_2$ is nearly R ln(2J+1) at room temperature and is suggestive of the fact that the degeneracy of the lowest ground state is completely lifted. Detailed calculations were performed to evaluate the influence of the cubic crystal field in $ErAl_2$ on the $^4I_{15/2}$ ground state multiplet of the Er^{3+} ion. Two sets of crystal field parameters which describe the heat capacity results satisfactorily were extracted from our study (B_4^0 = -1.53 x 10^{-7} ev, B_6^0 = -0.66 x 10^{-9} ev and B_4^0 = 0.48 x 10^{-7} ev, B_6^0 = -1.18 x 10^{-9} ev). It is concluded that magnetic ordering in $ErAl_2$ takes place within the Γ_8^3 quartet state. These results are discussed in the light of recent studies on other RAl_2 compounds.

INTRODUCTION

Extensive information pertaining to magnetic characteristics of rare earth-aluminum compounds has been accumulated through numerous experiments.[1] Most of the RAl_2 compounds order at low temperatures and the magnetism is considered to result from an indirect exchange between the localized 4f electrons. Below the paramagnetic to ferromagnetic transition temperature the alignment of magnetic spins gives rise to an extra contribution to the heat capacity (C_M). Thermal excitation of the 4f electrons in the crystal field levels also contribute to the heat capacity (C_{CF}) The degeneracy of f states can be lifted in general by either the crystal field or the exchange field and some of, or all of the CF levels may be involved in the (cooperative) magnetic ordering. The relative importance of the crystal field and exchange field, therefore, can greatly alter the nature of the heat capacity curve.

This was clearly evident from the heat capacity results for RAl_2 compounds reported earlier.[2,3] For example, $PrAl_2$[2] and $NdAl_2$[2] exhibit sharp λ-type anomalies at Curie temperatures which are characteristic of the cooperative phenomena. The heat capacities of $GdAl_2$[2] and $TbAl_2$,[3] however, were not λ-type but were spread over a wide range of temperatures. For $HoAl_2$,[3] a more complicated "doublet" feature of the anomaly was observed.

In this paper the heat capacity results for $ErAl_2$ and $LuAl_2$ are presented. Calculations were performed to examine the experimental results of $ErAl_2$ in the light of theory involving the influence of cubic crystalline electric field on the $^4I_{15/2}$ ground state multiplet of the trivalent erbium ion; these results are also presented.

RESULTS AND DISCUSSION

Heat capacities of $ErAl_2$ and $LuAl_2$ were measured using a calorimeter described earlier.[2] The results are shown in Fig. 1. The experimental data of $LaAl_2$ are also included in the figure for comparison.[2] In evaluating thermodynamic functions, experimental heat capacities from 1.5 to 5 K from other sources were utilized. In Fig. 1, the peak for $ErAl_2$ at nearly 10 K has the appearance of a λ-point which is characteristic of cooperative phenomena. This suggests that the magnetic ordering occurs within a few low-lying crystal field levels in a cooperative fashion. The present results for $LuAl_2$ differ considerably from the $LaAl_2$ data. The heat capacity of $LuAl_2$ is lower than those of $LaAl_2$ and of $ErAl_2$ (Fig. 1). Details of the estimation of contribution to heat capacity from non-magnetic terms in $ErAl_2$ have been reported recently.[4]

The excess entropy associated with C_{CF} and C_M is a quantity of considerable interest. For $ErAl_2$ an excess entropy of 96% of R ln(2J+1) was obtained in the present study. This is suggestive of the fact that the degeneracy of the lowest ground state of the erbium ion is completely lifted. The excess heat capacity curve shows a broad peak having its maximum at 23 K, in addition to a sharp λ-type peak at 10.2 K. The peak at 23 K is attributed to further thermal excitation among higher CF energy levels, resulting in a broad Schottky-type contribution.

The electronic ground state of the tripositive Er ion has a 16-fold degeneracy. By the influence of a cubic crystalline field in $ErAl_2$, the degeneracy is lifted into 2 doublets and 3 quartets. The following Hamiltonian was employed to treat the problem:

$$\mathcal{H}_{CF} = W\left[x\,\frac{O_4}{F(4)} + (1-|x|)\,\frac{O_6}{F(6)}\right], \qquad (1)$$

where the terms have their usual significance. The experimental excess heat capacity ΔC_p was examined by fitting the data to the energy level schemes given by eq. (1) for a variety of combinations of x and W. For the ΔC_p of $ErAl_2$ the best fits were obtained for two sets of the parameters x and OAS: +0.5, 80.3 K and -0.15, 109 K. These fits are shown in Figs. 2 and 3 along with energy schemes for each set of parameters. The mean squares of the deviations are almost the same for these two fits and the energy scheme cannot be uniquely determined by the fitting alone.

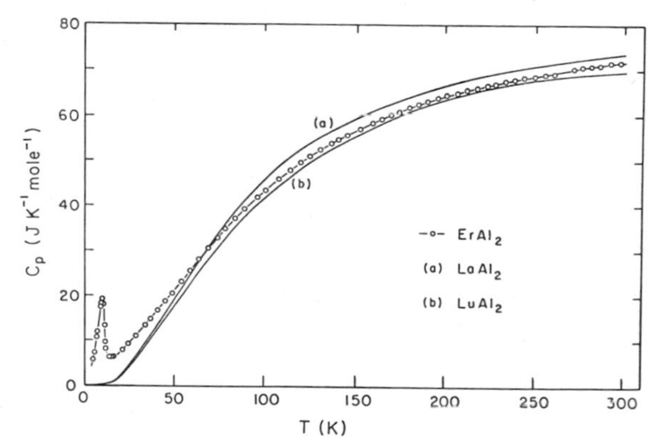

Fig. 1 Experimental heat capacity curves for $ErAl_2$, $LaAl_2$ and $LuAl_2$

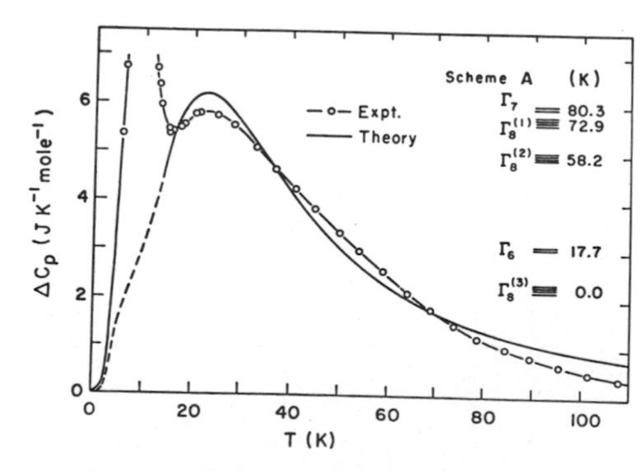

Fig. 2 Comparison of experimental and calculated crystal field heat capacities of $ErAl_2$ (Scheme A)

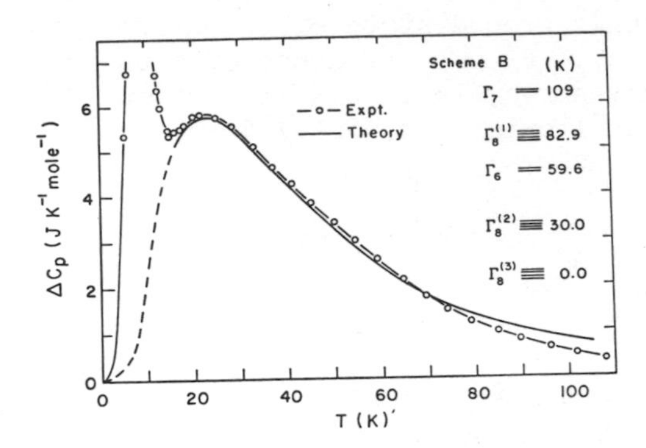

Fig. 3 Comparison of experimental and calculated crystal field heat capacities of $ErAl_2$ (Scheme B)

In order to study the multiplicity of low-lying states, the magnetic entropy was evaluated as a function of temperature. These results show that this compound acquires an entropy of nearly R ln4 at 10 K, when the ferromagnetic order is almost destroyed. This suggests that the ground crystal field state has a fourfold degeneracy. The present consideration of entropy, therefore, supports the Γ_8^3 state as the ground state for the energy schemes A and B. Among previous results for $ErAl_2$, a Kramer's doublet is reported as a ground state from Mössbauer measurements.[5] This is, however, clearly incompatible with the present result based upon the entropy.

The crystal field intensity parameters B_4^0 and B_6^0 can also be determined from the two sets of the parameters x and W [eq. (1)]. The results of the calculations are shown in Table 1.

Table 1

Crystal Field Parameters for $ErAl_2$

	(a) Evaluated from experimental data		(b) Calculated using the point charge model
	Scheme A (Fig. 2)	Scheme B (Fig. 3)	
x	+0.50	-0.15	+0.37
OAS	-80.3 (K)	-109 (K)	-11.9 (K)
W(K)	-21.2×10^{-2}	-22.5×10^{-2}	-2.86×10^{-2}
B_4^0(K)	-17.5×10^{-4}	5.60×10^{-4}	-1.75×10^{-4}
B_6^0(K)	-76.5×10^{-7}	-1.37×10^{-7}	-13.0×10^{-7}

B_4^0 and B_6^0 may be written as follows:

$$B_4^0 \sim \beta_J \, A_4^0 \, \langle r^4 \rangle$$

$$B_6^0 \sim \gamma_J \, A_6^0 \, \langle r^6 \rangle \qquad (2)$$

The signs of B_4^0 and B_6^0 are determined by the signs of A_4^0 and A_6^0. For $ErAl_2$ the following signs of (A_4^0, A_6^0) are consistent with the present C_{CF} data: (-,-) for the scheme A and (+,-) for the scheme B.

The CF effect for $ErAl_2$ has been studied also by several workers. For example, the results of magnetization measurements on a single crystal $ErAl_2$ by Purwins[6] indicate (+,-). Heer et al.[7] also reported the result (+,-) for polycrystalline $Er_cY_{1-c}Al_2$ by neutron spectroscopy, where c = 0.003 and 0.006. In a recent report Heer et al.[7] compared CF parameters for several compounds for the RAl_2 series. They showed that the signs (+,-) for (A_4^0, A_6^0) led to a consistent interpretation of the CF parameters for the RAl_2 compounds such as $NdAl_2$, $TbAl_2$, $ErAl_2$, and $Tm_{0.25}Y_{0.75}Al_2$. The above choice (+,-) is consistent with one of the results (Scheme B) obtained in the present study, although the magnitude of B_4^0 and accordingly the value of x differs from the rescaled values given by these authors.

ACKNOWLEDGEMENT

The authors wish to thank J. V. Mahoney (University of Pittsburgh) for help with many phases of this work and B. J. Beudry (Ames Laboratory, Iowa State University) for preparing the $LuAl_2$ compound.

REFERENCES

* This work was supported in part by the Army Research Office, Durham (University of Pittsburgh) and in part by the U.S. Energy Research and Development Administration under contract number W-7405-eng-82 (Ames Laboratory).

1. For example, Wallace, W. E., Rare Earth Intermetallics, p. 34, Academic Press (1973).
2. Deenadas, C., Thompson, A. W., Craig, R. S. and Wallace, W. E., J. Phys. Chem. Solids 32, 1853 (1971).
3. Hill, T. W., Wallace, W. E., Craig, R. S. and Inoue, T., J. Solid State Chem. 8, 364 (1973); Sankar, S. G., Malik, S. K. and Rao, V. U. S., ibid., July, 1976 (in press).
4. Inoue, T., Sankar, S. G., Craig, R. S., Wallace, W. E. and Gschneidner, K. A., J. Phys. Chem. Solids (communicated).
5. Wiedemann, W. and Zinn, W., Phys. Lett. 24A, 506 (1967).
6. Purwins, H. G., Z. Phys. 233, 27 (1970).
7. Heer, H., Furrer, A., Walker, E. Treyvaud, A., Purwins, H. G. and Kjems, J., J. Phys. C. 7, 1207 (1974).

MAGNETIZATION BEHAVIOUR OF $DyAl_2$ and $GdAl_2$

B. Barbara*, M.F. Rossignol*, H.-G. Purwins† and E. Walker†
*Laboratoire de Magnétisme, CNRS BP.166, F-38042 Grenoble Cedex, France
†Département de Physique de la Matière Condensée, Université de Genève, CH-1211 Genève, Switzerland

ABSTRACT

We report measurements of the magnetization of single crystals of ferromagnetic $DyAl_2$ and $GdAl_2$ between 4.2 K and 180 K in magnetic fields up to 150 kOe along the [100], [110] and [111] directions. $DyAl_2$ is strongly anisotropic with [100] as the easy direction of magnetization. $GdAl_2$ is isotropic and we observe a contribution of order of $.1\mu_B$ due to conduction electron polarization.

The magnetic properties of $REAl_2$ (RE = rare earth) intermetallic compounds have been investigated intensively. This is mainly due to the facts that all $REAl_2$ compounds crystallize in the same $MgCu_2$ Laves phase structure[1,2], that most of them order ferromagnetically [2-4] and that samples are relatively easy to prepare. Thus investigations of fundamental properties of RE ions in metallic surroundings could be carried out in a systematic way.

Magnetization measurements on polycrystals of $REAl_2$ compounds have been reported earlier [2]. Measurements on single crystals (RE = Ce, Pr, Nd, Tb, Er) have been performed also [3,5-8]. With the exception of $CeAl_2$ the magnetization and the magnetocrystalline anisotropy can be understood quantitatively in terms of the molecular field theory including a cubic crystal field [5,7,8]. In the present article we report the experimental results of $DyAl_2$ and $GdAl_2$ measured between 4.2 and 200 K with magnetic fields up to 150 kG along the [100], [110] and [111] crystallorgaphic directions. A detailed theoretical description of these results will be given elsewhere.

The measurements have been performed at the Service National des Champs Intenses at Grenoble. To measure the magnetization we used the induction method in a water cooled Bitter magnet [9]) providing a maximum static field of 150 kOe. The $DyAl_2$ and $GdAl_2$ single crystals have been prepared by the Czochralski method from 99.9 % RE and 99.999 % Al. The crystal was cut into a sphere of approximately 3 mm diameter.

The experimental results for the magnetization of $DyAl_2$ are given in fig. 1 for several temperatures as a function of magnetic field‡ applied in the three main symmetry directions. We see that $DyAl_2$ is strongly anisotropic with [100] as easy direction of magnetization. Applying the external field in the [111] direction we observe at 4.2 K a jump of the magnetization at 57 kOe. This anomaly can be understood in terms of a change of the ground state of the Dy^{3+} ion experiencing the crystal field, the molecular field and the external field. The anomaly becomes less pronounced with increasing temperature. The low temperature saturation moment in the easy direction extrapolated to zero field is $9.89\mu_B$ which is near to the free ion value of $10\mu_B$. Thus $DyAl_2$ is a system comparable to $TbAl_2$ [5]) where the crystalline electric field produces a strong magnetocrystalline anisotropy but fails to produce a seizable quenching in the presence of the exchange interaction.

In fig. 2 we give the results of the magnetization of $GdAl_2$ measured for various temperatures as a function of magnetic field along the [110] direction. We measured the same magnetization when applying the field in [100] or [111]. The saturation moment at 4.2 K is $7.10\mu_B$. This is slightly higher than the free ion value of $7\mu_B$. The small difference is probably due to conduction electron polarization. Our results on $GdAl_2$ are in good agreement with recently reported magnetization data on polycrystalline material [10]).

---

‡ By magnetic field we mean throughout this article the applied field minus the demagnetizing field.

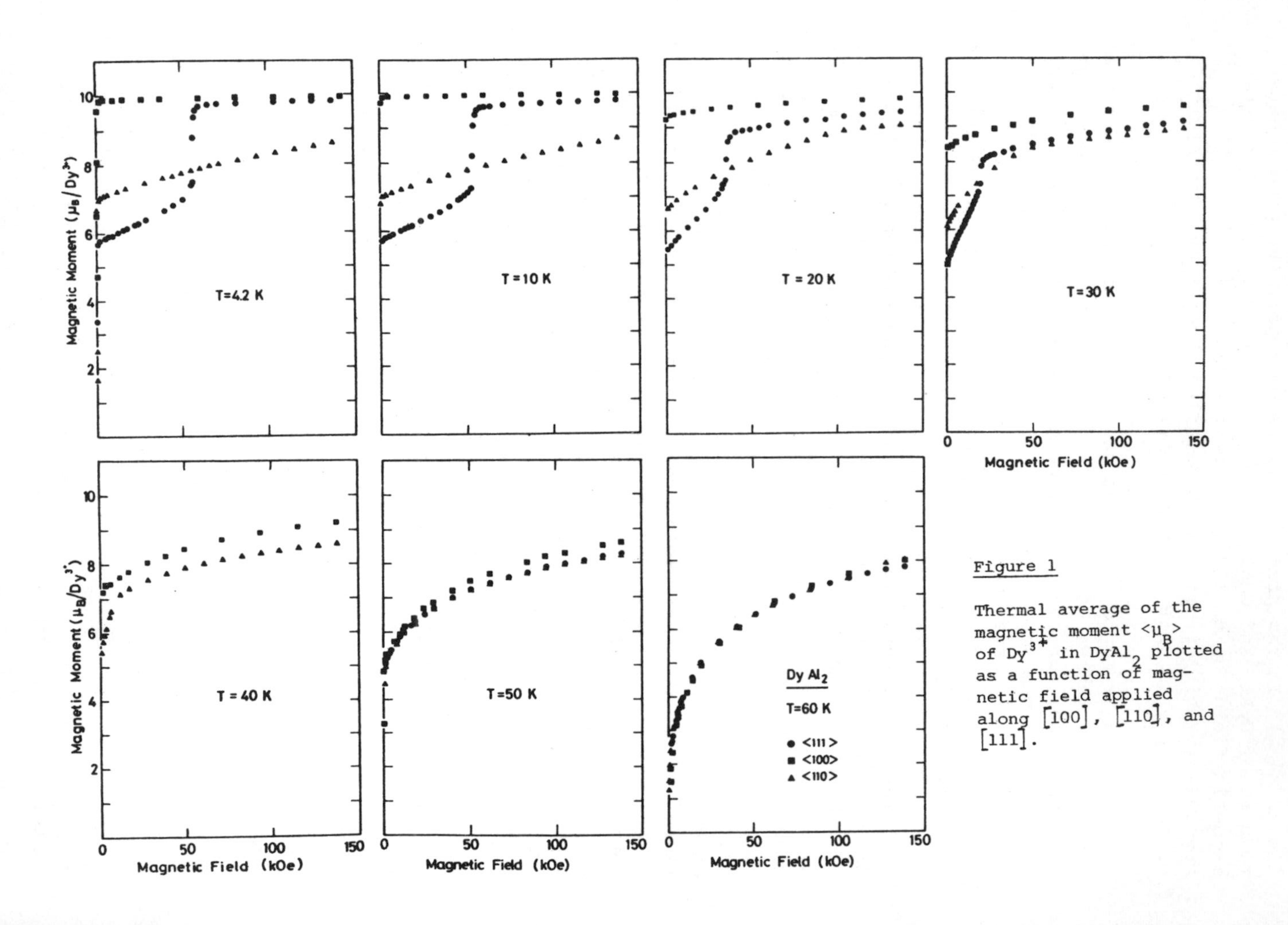

Figure 1

Thermal average of the magnetic moment $<\mu_B>$ of Dy^{3+} in $DyAl_2$ plotted as a function of magnetic field applied along [100], [110], and [111].

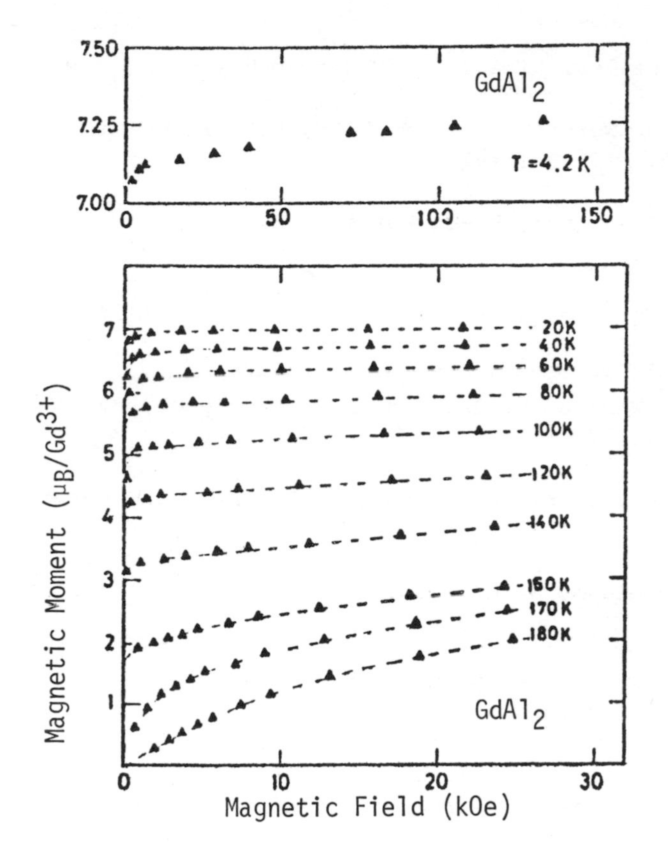

Figure 2

Thermal average of the magnetic moment $\langle\mu_B\rangle$ of Gd^{3+} in $GdAl_2$ as a function of magnetic field along |110|.

From the present work we conclude that $GdAl_2$ is an isotropic ferromagnet with isotropic exchange interaction. In contrast to $GdAl_2$, $DyAl_2$ has a strong magnetocrystalline anisotropy. Assuming that the exchange interaction is similar in $GdAl_2$ and $DyAl_2$, it is reasonable to conclude that the anisotropy observed in $DyAl_2$ is exclusively due to the interaction of the aspherical 4f charge distribution of Dy^{3+} with the crystalline electric field. In fact this model, isotropic exchange interaction plus cubic crystal field leads to a successful description of single crystal magnetization data of $REAl_2$ compounds [5,7,8]).

References

1. J.H. Wernick, S. Geller; Trans. AIME 218, 866 (1960)

2. H.J. Williams, J.H. Wernick, E.A. Nesbitt, R.C. Sherwood; J. Phys. Soc. Jap. 17 B-1 91 (1962)

3. B. Barbara, M.F. Rossignol, H.-G. Purwins, E. Walker; Solid State Commun. 17, 1525 (1975)

4. W.M. Swift, W.E. Wallace; J. Phys. Chem. Solids 29, 2053 (1968)

5. H.-G. Purwins, E. Walker, B. Barbara, M.F. Rossignol, P. Bak; J. Phys.C 7, 3573 (1974)

6. B. Barbara, J.X. Bouscherle, M.F. Rossignol; Phys.stat.sol. A 25, 165 (1974)

7. G.J. Cock, L.W. Roeland, H.-G. Purwins, E. Walker, A. Furrer; Solid State Commun. 15, 845 (1974)

8. H.-G. Purwins, E. Walker, B. Barbara, M.F. Rossignol, A. Furrer; J. Phys. C9 1025 (1976)

9. G. Aubert, H. Bartholin, D. Bloch, M. Guillot, A. Lacaze, J. Paureau, R. Pauthenet, J. Picoche, P. Rub, J.C. Vallier, A. Waintal Proc. Int. Conf. Magnetism, Moscow (1973)

10. H. Hacker, R. Gupta, L. Shepard; Phys. stat.sol. A9 601 (1972)

EFFECT OF MAGNETIC ANISOTROPY ON THE HEAT CAPACITY OF $HoAl_2$*

S. G. Sankar, S. K. Malik,[+] V. U. S. Rao and
W. E. Wallace
Department of Chemistry, University of Pittsburgh,
Pittsburgh, PA 15260
+ On leave from the Tata Institute of Fundamental
Research, Bombay, India

ABSTRACT

The heat capacity of $HoAl_2$ reveals an unexplained anomaly at 20 K. It appears that this anomaly originates from a change in the easy direction of magnetization brought about by the combined effect of crystalline electric and exchange fields. Calculations have been performed using the equivalent operator method. The exchange field acting on Ho^{3+} is determined in a self-consistent manner. Changes in the easy direction of magnetization are found for a range of reasonable crystal field parameters. The calculated Ho^{3+} magnetic moment is in agreement with experiment.

INTRODUCTION

Low temperature heat capacities of several RAl_2 compounds have been reported by Deenadas et al. (1), Hill et al. (2) and Inoue et al. (3). The destruction of ferromagnetism gives rise as expected to an extra contribution to the heat capacity. The excess heat capacity consists of a single peak in all cases, with the single exception of $HoAl_2$. The magnetic contribution to the heat capacity of $HoAl_2$ shows an unusual feature, exhibiting pronounced peaks at 20 and 28 K. The peak at 28 K corresponds to the Curie temperature; the origin of the other peak has not been clarified. It is now suggested that the low temperature peak at 20 K arises as a consequence of the change in the easy direction of magnetization due to a combined effect of crystalline electric field and exchange field.

THEORY AND CALCULATIONS

The Hamiltonian for the rare earth ion in the presence of a crystal field term $\mathcal{H}_{CF}$ and an exchange field H_{ex} is given by

$$\mathcal{H} = \mathcal{H}_{CF} + 2(g_J-1)\mu_B\, J_z\, H_{ex} \tag{1}$$

where μ_B is the Bohr magneton, g_J is the Landé g-factor and J_z is the z-component of the total angular momentum of the rare earth ion under consideration. The rare earth ion in RAl_2 compounds occupies a site of local cubic symmetry. The crystal field term for the z-axis lying along the three cubic crystallographic directions may be expressed as follows (4):

$$\mathcal{H}_{CF}^{(100)} = A_4 < r^4 > \beta(O_4^0 + 5\, O_4^{\,4}) + A_6 < r^6 > \gamma\, (O_6^0 - 21\, O_6^{\,4}) \tag{2}$$

$$\mathcal{H}_{CF}^{(110)} = -A_4 \frac{< r^4 > \beta}{4} (O_4^0 - 20\, O_4^{\,2} - 15\, O_4^{\,4}) - \frac{13}{8} A_6 < r^6 > \gamma$$
$$(O_6^0 + \frac{105}{26}\, O_6^{\,2} - \frac{105}{13}\, O_6^{\,4} + \frac{231}{26}\, O_6^{\,6}) \tag{3}$$

and

$$\mathcal{H}_{CF}^{(111)} = -\frac{2}{3} A_4 < r^4 > \beta(O_4^0 - 20\sqrt{2}\, O_4^{\,3}) +$$
$$\frac{16}{9} A_6 < r^6 > \gamma\, (O_6^0 + \frac{35\sqrt{2}}{4}\, O_6^{\,3} + \frac{77}{8}\, O_6^{\,6}) \tag{4}$$

In the above expressions the symbols have their usual significance.

In the molecular field approximation, the exchange field which the rare earth ions exert on one another can be written as

$$-2\mu_B\, H_{ex} = \mathcal{J}_{ff}\, (g_J-1) < J_z >_{av} \tag{5}$$

where $< J_z >_{av}$ denotes the Boltzmann average of the expectation value of the operator over all the levels. The exchange constant $\mathcal{J}_{ff}$ is related to the paramagnetic Curie temperature θ_P by

$$3\, k_B\, \theta_P = \mathcal{J}_{ff}\, (g_J-1)^2\, J(J+1). \tag{6}$$

From Eq. (5) we note that H_{ex} is proportional to $< J_z >_{av}$ which in turn depends implicitly on H_{ex} through the eigenfunctions and the energy eigenvalues obtained after diagonalizing the Hamiltonian given by Eq. (1). Thus, the exchange field has to be determined self-consistently for a given value of $\mathcal{J}_{ff}$, for each set of crystal field parameters $A_4 < r^4 >$ and $A_6 < r^6 >$, and for each temperature. In an actual calculation, the Hamiltonian of Eq. (1) is diagonalized for a set of parameters for an approximate value of H_{ex}. The resultant eigenvalues and eigenfunctions are utilized to evaluate $< J_z >_{av}$. This value is used to calculate H_{ex} employing Eq. (5), and this iterative process is repeated until self-consistency is achieved. These eigenvalues are used to calculate the Helmholtz free energy of the system using the expression

$$F = -k_B T \ln Z$$

where $Z = \sum_i e^{-E_i/k_B T}$.

The easy direction of magnetization at any given temperature corresponds to the one possessing the lowest free energy. The above procedure was used to calculate the temperature dependence of the free energy for Ho^{3+} ion for H_{ex} directed along the three major symmetry directions in a cube. Various combinations of $A_4 < r^4 >$, $A_6 < r^6 >$ and $\mathcal{J}_{ff}$ were used.

RESULTS AND DISCUSSION

The variation of free energy with temperature for a typical set of crystal field parameters is plotted in Figure 1. The exchange constant is calculated using Eq. (6) and the experimental value of θ_P reported by Wallace (5). The crystal field parameters are chosen on the basis of the work of Inoue et al. (3). We note from Figure 1 that the free energy along the < 110 > direction is the lowest at low temperatures, but at about 20 K the free energy along < 100 > becomes lowest; this produces a change in the easy direction of magnetization. The expected decrease in magnetocrystalline anisotropy is evident from the trends of the three curves in Figure 1.

Figure 2 shows the range of crystal field parameters for which our calculations predict a change in the easy direction of magnetization. The magnitudes of $A_4 < r^4 >$ and $A_6 < r^6 >$ have been chosen to obtain an overall splitting of 100-200 K in the ground J manifold of Ho^{3+} ion. Various experimental results on RAl_2 compounds have shown that the overall splittings in these compounds lie within this range (3,5,6). The change in the easy direction is predicted for all values of $A_4 < r^4 >$ and only negative values of $A_6 < r^6 >$, with < 110 > as the easy direction at low temperatures for all combinations. Calculations made varying $\mathcal{J}_{ff}$ for a fixed combination of $A_4 < r^4 >$ and $A_6 < r^6 >$ indicate that the transition temperature increased as a function of $\mathcal{J}_{ff}$.

Zero-field magnetization values of Ho^{3+} ion are plotted in Figure 3. We obtain a value of 9.2 μ_B at 4.2 K along the < 110 > direction. Recent magnetization measurements on single crystal

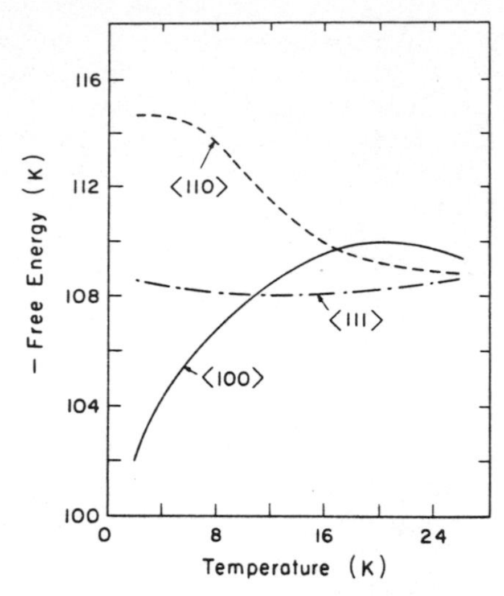

Fig. 1 Calculated free energy vs. temperature for Ho^{3+} ion in $HoAl_2$. The parameters used in the calculation are: $A_4 < r^4 > /k_B$ = 46.0 K, $A_6 < r^6 > /k_B$ = -7.7 K, $\mathcal{J}_{ff}/k_B$ = 20.0 K.

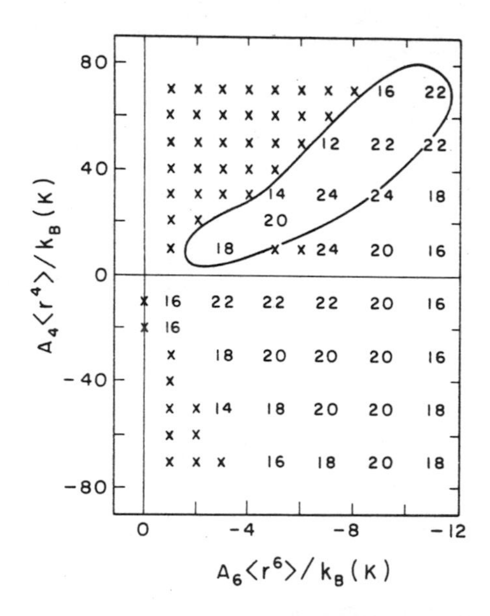

Fig. 2 Range of crystal field parameters $A_4 < r^4 >$ and $A_6 < r^6 >$ for which the calculations predict change in easy direction. Portion enclosed in the contour indicates a change from < 110 > to < 100 > direction. Portion outside the contour indicates change from < 110 > to < 111 >. The numbers indicate the transition temperature. x's define the limits.

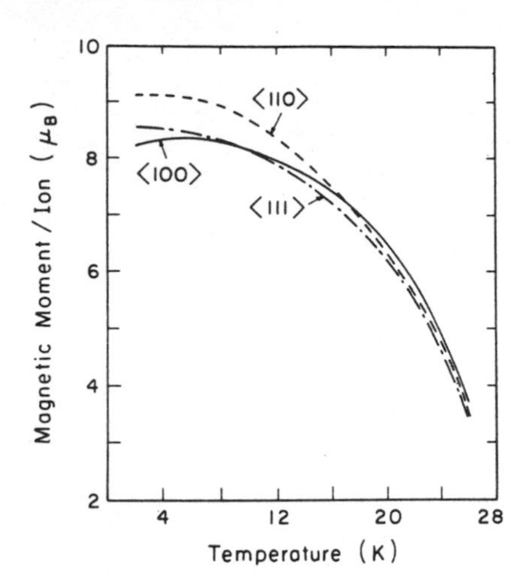

Fig. 3 Magnetic moment of the Ho^{3+} ion in $HoAl_2$ calculated as a function of temperature. The parameters are those cited in the legend of Fig. 1.

$HoAl_2$ by Barbara et al. (7) confirm the easy axis of magnetization, and the experimental value of the magnetic moment is in very good agreement with that predicted by the present calculations.

In conclusion, the best choice of crystal field parameters for $HoAl_2$ obtained in our calculations (see Fig. 1) are compatible with those obtained for other RAl_2 compounds. The origin of the peak at 20 K in the heat capacity curve reported by Hill et al. in $HoAl_2$ is thus ascribed to a change in the easy direction.

REFERENCES

* This work was supported by a grant from the Army Research Office-Durham.

1. C. Deenadas, A. W. Thompson, R. S. Craig and W. E. Wallace, J. Phys. Chem. Solids 32, 1853 (1971).
2. T. W. Hill, W. E. Wallace, R. S. Craig and T. Inoue, J. Solid State Chem. 8, 364 (1973).
3. T. Inoue, S. G. Sankar, R. S. Craig, W. E. Wallace and K. A. Gschneidner, J. Solid State Chem. (in press).
4. K. R. Lea, M. J. M. Leask and W. D. Wolf, J. Phys. Chem. Solids 23, 1381 (1962).
5. W. E. Wallace, "Rare Earth Intermetallics," Academic Press (New York), 1973, chapter 4.
6. H.-G. Purwins, E. Walker, B. Barbara, M. Rossignol and P. Bak, J. Phys. C7, 3573 (1974).
7. B. Barbara, M. F. Rossignol and J. X. Boucherle, Phys. Lett. 55A, 321 (1975).

MAGNETIC ANISOTROPY IN DILUTE MAGNESIUM MANGANESE ALLOYS

J.L. Benchimol, F.T. Hedgcock,
and J.O. Strom-Olsen
Eaton Lab, McGill University,
P.O. Box 6070, Station A,
Montreal, Quebec, Canada H3C 3G1

ABSTRACT

The paramagnetic anisotropy in single magnesium crystals containing 80, 175 and 315 p.p.m. manganese has been measured over a temperature range 300 to 1.6°K. There is observed in the paramagnetic anisotropy a term which varies as T^{-2}. This term corresponds to a concentration independent energy splitting of the manganese ion of $7.3 \pm 0.5 \times 10^{-3}$ °K. and is opposite in sign to the energy splitting observed in the Zn-Mn system.

EXPERIMENT

In previous publications we have presented[1,2] data on the magnetic anisotropy of a h.c.p. metal host containing a transition metal impurity - the particular system being zinc-manganese. We wish to report in this paper magnetic measurements on single magnesium crystals containing 80, 175, and 315 p.p.m. manganese. The magnetic anisotropy was measured directly with a torque magnetometer of sensitivity of 5×10^{-4}dyne-cm. Fig. 1 shows a typical x-y recording of the torque on a Mg-Mn sample as a function of magnetic field and the inset shows the measured torque as a function of H^2. The slope of the torque versus H^2 yields a value of $m(\chi_{\parallel}-\chi_{\perp})/2$ where the angle between the c axis and the magnetic field direction is 45° and m is the mass of the crystal. The concentrations of manganese were determined from room temperature susceptibility and residual resistivity using previously published results for the concentration dependence of these quantities[3] and the values so determined are listed in table 1. All samples were checked for the presence of ferromagnetic impurities by measuring the room temperature susceptibility as a function of field up to 15 kiloersted and by examining the angular dependence of the magnetic anisotropy. None of the samples listed in Table 1 exhibited ferromagnetism as inferred from these tests. Shown in Fig. 2 is the magnetic anisotropy $\Delta\chi$ as a function of temperature below 10°K. The values of $\Delta\chi$ for the alloys approach that for pure magnesium at higher temperatures.

ANALYSIS

As in the previously reported results on Zn-Mn, in addition to a normal isotropic Curie-Weiss paramagnetism an anisotropic term varying as T^{-2} was found. This term corresponds to an easy axis of magnetization lying parallel to the c axis of the host crystal. Whatever the physical origin of the influence of the host symmetry on the manganese ion, it can be shown that the presence of an S_z^2 term in the Hamiltonian for the localized spin causes a T^{-2} dependence of the paramagnetic susceptibility anisotropy ($\Delta\chi_i$). Let us call D the coefficient of S_z^2 where D represents the energy splitting due to the anisotropic host. In addition to DS_z^2, the Hamiltonian must also contain the Zeeman splitting, βgSH caused by the applied magnetic field. We then have

$$H = DS_z^2 - g_\perp \beta S_x H_x - g_\parallel \beta S_z H_z$$

where β is the Bohr magneton, g is assumed anisotropic because of the symmetry of the host and the two g factors, $g_\perp$ and $g_\parallel$, are defined perpendicular and parallel to the c axis. If the energy splitting is small compared to kT, the expression obtained for $\Delta\chi_i = \Delta\chi_{alloy} - \Delta\chi_{Mg}$ is

$$\frac{Nc\bar{g}^2\beta^2 S(S+1)}{3kT}\,\frac{g_\parallel^2 - g_\perp^2}{\bar{g}^2} - \frac{D(2S-1)(2S+3)}{30kT}\,\frac{2g_\parallel^2 + g_\perp^2}{\bar{g}^2}$$

where c is the atom fraction of manganese ions and

$$\bar{g}^2 = (2/3)g_\perp^2 + (1/3)g_\parallel^2.$$

Thus a graph of $\Delta\chi_i T/c$ vs T^{-1} should be linear with a slope proportional to D, independent of concentration, and with an intercept determined by $g_\parallel$ and $g_\perp$. That this result is observed is illustrated in Fig. 3. The slope of the line yields a value for D of $7.3 \pm 0.5 \times 10^{-3}$ °K while the intercept yields a value for $g_\parallel^2 - g_\perp^2/\bar{g}^2$ of

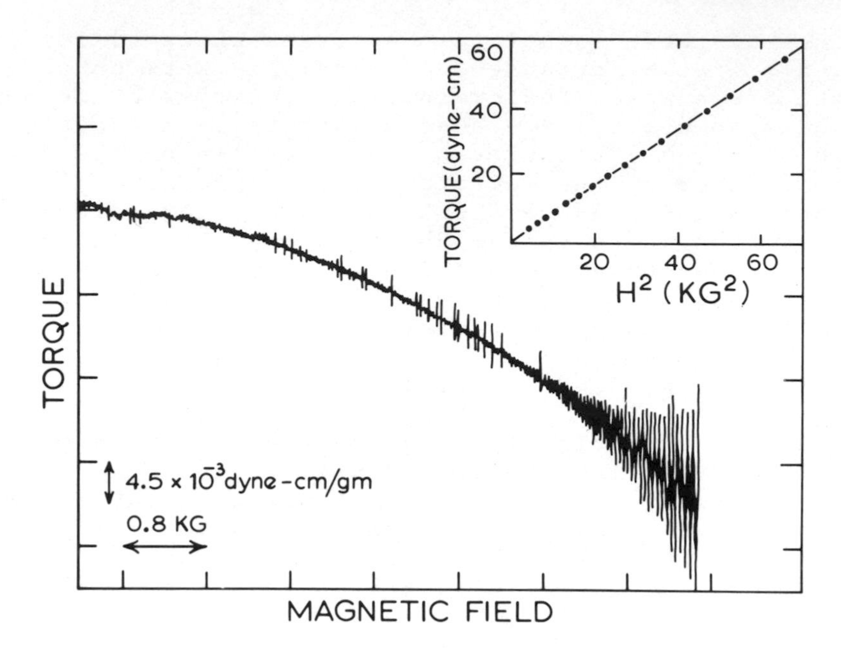

Fig.1: Typical X-Y recorder trace of the torque as a function of magnetic field for a Mg-80 p.p.m. Mn alloy at 2.4°K, showing the superposition of the de Haas - van Alphen oscillations on the static torque. The solid line represents the arithmetic mean of the torque while the inset shows the torque as function of H^2.

Sample	Concentration (ppm)				
	Nominal	from χ	Spectro-scopic	$\Delta\chi_i$ ratios	Resis-tivity
A	190	85 ± 35	80 ± 7	80	110 ± 50
B	300	73 ± 30		175	
C	420	115 ± 35		315	

Table 1: Manganese concentrations as determined by a variety of methods. The $\Delta\chi_i$ ratios were used to determine concentration assuming the concentration of sample A as determined from spectroscopic analysis was exact.

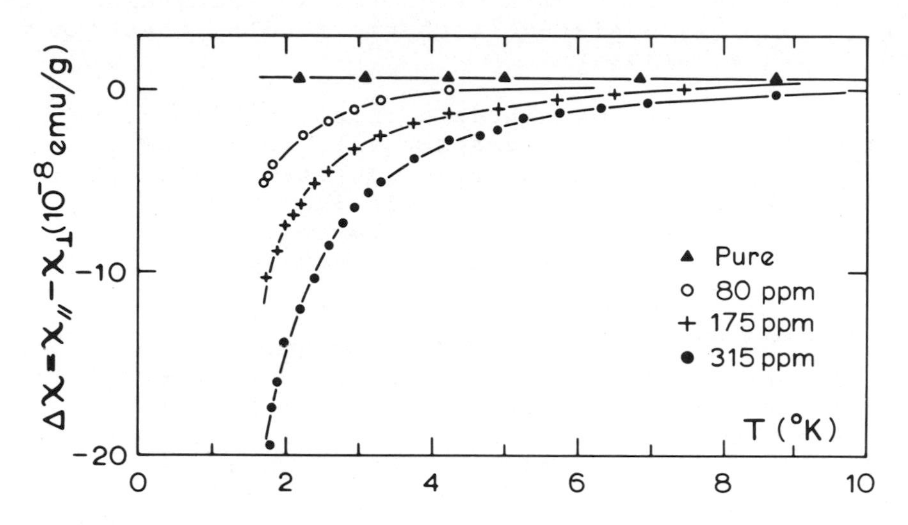

Fig.2: Static anisotropy of pure Mg and MgMn alloys as a function of temperature.

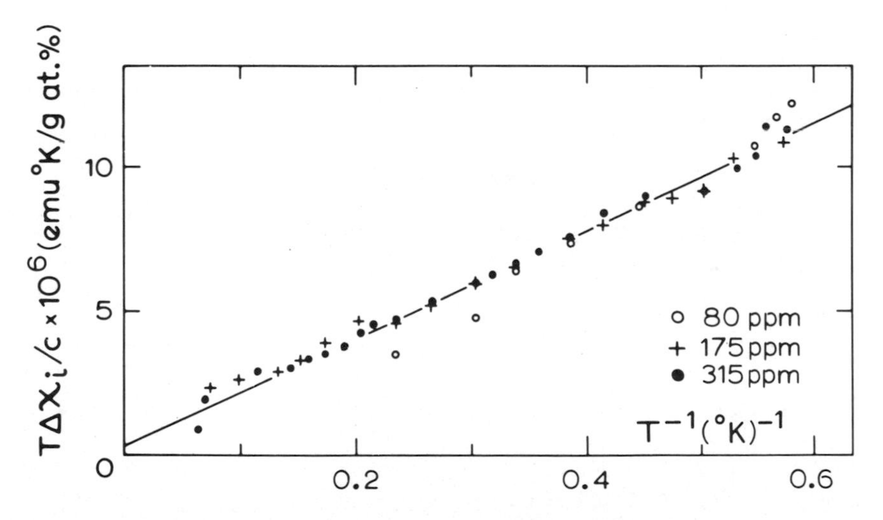

Fig.3: $(T \times \Delta\chi_i/c)$ for a number of Mg-Mn alloys as a function of inverse temperature.

0.01 ± 0.01 %. The value for D is in exact agreement with the value found by Gehman[4] from E.S.R. measurements and our g value is consistent with the g value observed in the E.S.R. experiments.

So far as the magnitude of D is concerned, we note that it is considerably smaller than the value of 8×10^{-2} °K observed in the zinc-manganese system. This might be expected since the deviation from the ideal h.c.p. structure (as inferred from the c/a ratio) is greater in the zinc host than in magnesium.

DISCUSSION

The physical origin of the energy splitting for localized moments in metals is perhaps difficult to understand in terms of conventional crystal field splitting especially for an ion such as manganese whose ground state is believed to be S-like. Smith[5] and Dixon[6] have suggested different possible origins of anisotropy, the former using spin orbit effects from the host and the latter anisotropy in the exchange interaction. The present results are consistent with either approach but the model of Smith is not yet in a sufficiently quantitative form to compare with experiment. In a forthcoming publication we will present results on the cadmium manganese system and, in the light of these new measurements, it will be possible to give a more complete discussion of the various proposed mechanisms for the explanation of these results.

REFERENCES

1. Li P.L., Hedgcock F.T., Muir W.B., and Strom-Olsen J.O. Phys. Rev. Lett., 31 29 (1973).

2. Hedgcock F.T., Lewis S., Li P.L., Strom-Olsen J.O., and Wasserman E.F., Can. J. Phys., 52, 1759 (1974).

3. Collings E.W. and Hedgcock F.T., Phys. Rev. 126, 1654 (1962)
Hedgcock F.T., Muir W.B., and Wallingford E., Can. J. Phys., 38, 376 (1960).

4. Gehman B.L., Ph.D. Thesis, 1974, University of California, San Diego.

5. Smith D.A., Journal of Magnetism and Magnetic Materials 1, 214 (1976)

6. Dixon J.M. and Dupree R., Solid State Commun., 16, 1301 (1975).

CRYSTAL FIELD EFFECTS IN Dy_2Ni_{17} FROM MÖSSBAUER SPECTROSCOPY

J. Chappert* and J.K. Yakinthos+

*DRF/Groupe Interactions Hyperfines, Centre d'Etudes Nucléaires, Grenoble, France
+Nuclear Research Center Democritos, Athens, Greece

ABSTRACT

The Mössbauer resonance of ^{161}Dy in Dy_2Ni_{17} has been recorded at 4.2 K. Two resolved hyperfine patterns corresponding to the two Dy sites are observed. This is explained by the absence of strong exchange interactions due to the small moment of the nickel atom.

INTRODUCTION

Intermetallic rare earth (R)-transition metal (M) alloys of the type R_xM_y have been much studied by X-ray, neutron diffraction and magnetization measurements. It was shown that various values of x and y lead to phases (such as RM_2, RM_3, RM_5, R_2M_7 and R_2M_{17}) whose crystallographic structures are closely related and which all derive from the well known $CaCu_5$ structure by ordered substitutions of atoms in the crystal cell [1]. Several Mössbauer studies of some of these compounds have been reported [2, 3]. It was observed that saturation values of the ^{161}Dy magnetic hyperfine fields vary at most by ∿ 10 % from one structure to the other. Also the quadrupole interactions were very similar, indicating that the main contribution to the electric field gradient (e.f.g.) at the ^{161}Dy nucleus comes from the 4f shell and that the lattice and conduction electron terms are usually small. It was in particular rather surprising to observe a unique value for the magnetic and quadrupole interactions at the ^{161}Dy nucleus for non-equivalent sites in a given compound. This is because the component ^{161}Dy spectra overlap each other closely. The situation is very different from that found in the ^{57}Fe spectroscopy where non-equivalent sites yield hyperfine interactions spread over a wide range of values. In fact the only reported Mössbauer observation of ^{161}Dy non-equivalent sites was by Crecelius _et al_. in dysprosium iron garnet which is an insulator [4].

In the course of a systematic comparison of the magnetism of the crystalline and amorphous R_xM_y alloys [3, 5], we have studied Dy_2Ni_{17} by Mössbauer spectroscopy on the ^{161}Dy resonance. This compound has a hexagonal structure (space group $P6_3/mmc$) with two Dy lattice positions (D_{3h} symmetry) called I(2b) and II(2d) in equal amount [6].

RESULTS AND DISCUSSION

The Mössbauer spectrum of polycrystalline Dy_2Ni_{17} at 4.2 K is shown in fig. 1. This is well below the Curie temperature T_C = 168 K. One notes that the spectrum is quite different from spectra reported for other Dy intermetallic compounds where two Dy sites are present. Only one hyperfine pattern was observed [1, 3]. Here the spectrum is clearly the superposition of two equally po-

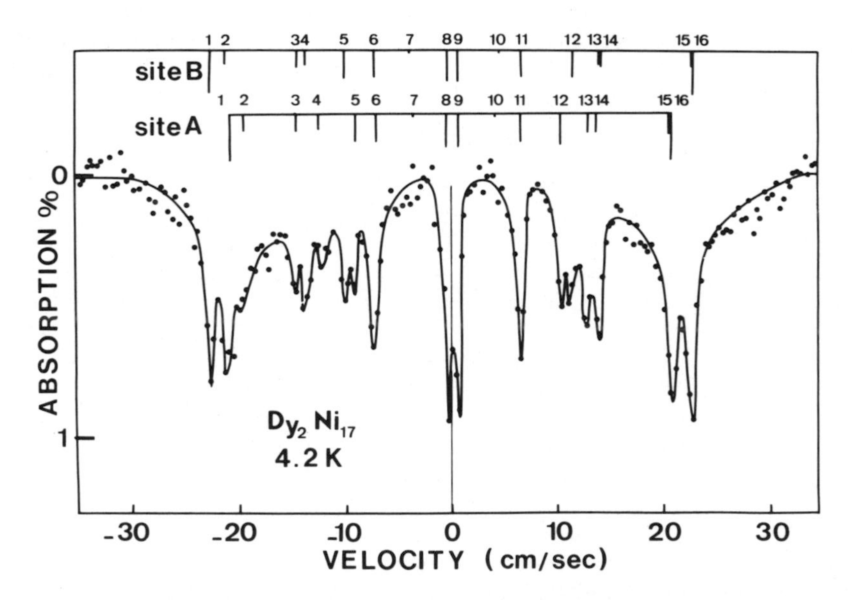

Fig. 1 Mössbauer spectrum of ^{161}Dy in Dy_2Ni_{17} at 4.2 K. The source is ^{161}Tb in GdF_3. The stick diagrams indicate the lines due to the A and B dysprosium sites.

	over-all splitting (cm/sec)	$g_o\mu_n H_{eff}$ (MHz)	μ (μ_B)	$e^2 q Q$ (cm/sec)
site A	40.8	746	8.9	8.4
site B	45.5	833	10.0	12.8

Table I

pulated 16 - line sub-spectra, whose theoretical line positions and intensities for the hyperfine interaction parameters of table I are indicated by the stick diagrams in fig. 1. The sites have been labeled A and B since it is not possible to tell from the Mössbauer data if the A and B patterns belong to I and II respectively or vice-versa. For site B the hyperfine interaction values are similar to those reported for other intermetallic compounds [2, 3]. However they are significantly smaller for site A. The values of μ are derived assuming that the magnetic moment of the Dy ion is proportional to the overall splitting and taking 10 μ_B for the moment of metallic Dy.

The magnitude of μ shows that the electronic ground state of Dy in site B is close to $|15/2>$ while in site A crystal field effects produce a mixing of states and lead to a reduction of μ. As far as the direction of the Dy moments is concerned one must consider the relative strength of the two competing interactions, the crystalline anisotropy and the exchange forces. For Dy_2Ni_{17} the crystal field interaction can be written in the simplified form :

$$\mathcal{H}_c = \alpha V_2^0 \, [3J_z^2 - J(J+1)] \qquad \text{eq. 1}$$

where α is a coefficient tabulated by Elliot and Stevens [7]. The parameter V_2^0 is defined by :

$$V_2^0 = A_2^0 \, \langle r^2 \rangle (1 - \sigma_2) \qquad \text{eq. 2}$$

where A_2^0 is a lattice sum on the neighbouring charges and σ_2 is a screening coefficient. The sign of the product αV_2^0 determines the easy direction for the magnetic moment. For Dy^{3+} $\alpha = -0.0063$ [7]. Using $\langle r^2 \rangle = 0.726$ [8] and $\sigma_2 = 0.6$ [9], we have performed a lattice sum taking into account all ions within a sphere of radius 30 Å. With the hexagonal c axis as the quantization axis we found V_2^0 (site I) = - 675 cm^{-1} and V_2^0 (site II) = 285 cm^{-1}. This calculation shows that the easy magnetization axis is perpendicular to c for site I while it is parallel to c for site II. Moreover the anisotropy of site I is bigger than that of site II. Therefore if the exchange forces tend to rotate the Dy moments, this will first occur for site II. On this basis a tentative magnetic structure would be i) μ(I) perpendicular to c and ii) μ(II) parallel or perpendicular to c. Of course only a neutron diffraction study, now in progress, will give a definite answer, since intermediate angles between parallel and perpendicular directions are possible.

A similar analysis can be made for the quadrupole interaction. It can be written as :

$$e^2 \underset{\approx}{q}Q = e^2 \underset{\approx}{q}_{Lat}(1 - \gamma_\infty)Q + e^2 \underset{\approx}{q}_{4f}(1 - R)Q \qquad \text{eq. 3}$$

where Q is the ^{161}Dy nucleus quadrupole moment, the first term is the contribution of the lattice neighbours and the second is due to the partially filled 4f shell. The parameters R and γ_∞ are the Sternheimer coefficients. One has :

$$e\, q_{Lat} = - \frac{4V_2}{\langle r^2 \rangle}$$

and $$e\, q_{4f} = - e^2 \alpha \langle r^{-3} \rangle [3J_z^2 - J(J + 1)] \; .$$

For a pure $|15/2\rangle$ state, the 4f term is 14.1 cm/sec. We calculated that the lattice terms are 3.5 cm/sec and - 1.5 cm/sec for the sites I and II respectively, taking the c axis as the quantization axis. In addition to the question of the validity of the point charge model in this metallic system, a quantitative analysis of the e^2qQ values is however difficult for three reasons : i) the real electronic state is not known. In particular it is likely that the Mössbauer pattern A comes from a mixed state. This will strongly reduce the 4f term, ii) eq. 3 involves tensors. If the spins are not collinear to the c axis it is not possible to simply add each component, iii) the charge of nickel was set equal to zero in the lattice sum. This is probably justified because in these intermetallic compounds the conduction electrons fill the 3d band of the nickel which can therefore be regarded as neutral [10]. If one sets the charge equal to + 0.2, the crystal field parameters become V_2^0 (I) = - 296 cm^{-1} and V_2^0 (II) = - 1413 cm^{-1}. The anisotropy of site II becomes very large and the Dy moments must then align in a direction perpendicular to the c axis.

Finally we may ask ourselves why the two hyperfine patterns are well resolved in Dy_2Ni_{17} while a unique spectrum is observed in the isomorphous compounds Dy_2Fe_{17} and Dy_2Co_{17} [2]. The reason is probably the strong exchange interactions which exist in the Fe and Co compounds. In Dy_2Ni_{17} the Ni magnetic moment is probably very small. Laforest et al. measured μ_{Ni} = 0.29 μ_B in Y_2Ni_{17} [11]. This suggests that, as in their Ycounterparts [12, 13], the Fe or Co atoms significantly polarize the conduction electrons at the Dy site in Dy_2Fe_{17} and Dy_2Co_{17}. This polarization must be much weaker in Dy_2Ni_{17}.

The authors gratefully acknowledge fruitful discussions with J.M.D. Coey, R. Lemaire and J. Schweizer.

REFERENCES

(1) R. Lemaire and D. Paccard, Bull. Soc. Fr. Mineral. Cristallogr., 92 9 (1969).

(2) G.J. Bowden, R.K. Day and M.Sarwar, Intern. Conf. Magnetism, Moscow, 1973 and references therein.

(3) J.K. Yankinthos and J. Chappert, Solid State Comm., 17 979 (1975).

(4) G. Crecelius, S. Hüfner and D. Quitmann, in Proceedings of the Conf. on the Applications of the Mössbauer Effect, Tihany 1969, p. 507.

(5) J.M.D. Coey, J. Chappert, J.P. Rebouillat and T.S. Wang, Phys. Rev. Letters, 36 1061 (1976).

(6) G. Bouchet, J. Laforest, R. Lemaire and J. Schweizer, C.R. Acad. Sc. Paris, 262 1227 (1966), K.H.J. Buschow, J. Less-Common Metals, 11 204 (1966).

(7) R.J. Elliot and K.W.H. Stevens, Proc. Roy. Soc., A 218 553 (1953).

(8) A.J. Freeman and R.E. Watson, Phys. Rev., 127 2058 (1962).

(9) K.C. Das and D.K. Ray, Solid State Comm., 8 2025 (1970).

(10) B. Bleaney, Proc. Phys. Soc. 82 469 (1963).

(11) J. Laforest, R. Lemaire, D. Paccard and R. Pauthenet, C.R. Acad. Sc. Paris B, 264 676 (1967).

(12) D. Givord, Thesis, Grenoble, 1973.

(13) J. Schweizer and F. Tasset, Intern. Conf. Magnetism, Moscow, 1973.

RARE EARTH WAVE FUNCTION DETERMINATION BY POLARIZED NEUTRON DIFFRACTION IN $REAl_2$

B. Barbara[x], J.X. Bourcherle[+], J.P. Desclaux[+], M.F. Rossignol[x] and J. Schweizer[+o]
x Laboratoire de Magnétisme, CNRS 166X 38042 Grenoble Cédex France
+ DRF/Centre d'Etudes Nucléaires 85X 38041 Grenoble Cédex France
o Institut Laue-Langevin 156X 38042 Grenoble Cédex France

ABSTRACT

Polarized neutron experiments have been performed on $NdAl_2$ and $CeAl_2$ compound to obtain accurate measures of form factors. From the contribution corresponding only to the 4f electrons it was possible to refine the wave function of the ground state of the rare earth ion in the crystal and molecular fields.

INTRODUCTION

Over the last 15 years the polarized neutron technique has provided very accurate measurements of magnetic form factors. In the case of 3d electrons these experiments were in very good agreement with the theory based on the non relativistic Hartree Fock wave functions, but they indicated discrepancies in the case of 4f electrons. Relativistic wave functions appeared better adapted for the rare earth case as shown for metallic gadolinium(1,2) erbium(3) and $Tb(OH)_3$(4). The tensor operator formalism due to Johnston, Lovesey and Rimmer(5,6) provides a very useful technique for calculating the magnetic form factors when the ion is spectroscopically defined as demonstrated in the case of US(7) or TmSb(8). Therefore it appears possible to go one step further and to refine from an accurately measured form factor the ground state of an atom such as for instance a rare earth ion in the crystal field.

The metallic rare earth compounds where conduction electrons are polarized by the 4f spin of the rare earth atom are interesting subjects for such studies : in this case it is in fact difficult to determine accurately the crystal field and exchange parameters from the magnetization measurements as this magnetization is the sum of two terms : the 4f moment and the conduction electrons contribution whose behaviour is essentially unknown. We have therefore applied this method in some $REAl_2$ compounds where RE is a rare earth atom.

FORM FACTOR MEASUREMENT OF $REAl_2$ COMPOUNDS

The $REAl_2$ compounds form a serie isomorphous to the cubic Laves phase $MgCu_2$[9]. They order ferromagnetically at low temperature with the exception of $CeAl_2$ whose magnetism is not fully understood at the moment. Order temperatures range up to 180 K[10]. They are good candidates for form factor measurementas there is only one crystallographic rare earth site in the cell and the crystal field acting on the rare earth atoms has a cubic symmetry.

Single crystals have been grown in tungsten crucibles using the Bridgman method. Polarized neutron measurements have been performed on the hot neutron diffractometer D5 of the Institut Laue-Langevin at several wavelengths ranging between 0.50 and 1.05 Å. The comparison between results at different wavelengths appeared to be a very suitable tool to correct extinction. The temperature of these experiments was 4.2 K and 1.5 K. A vertical field of 16.5 kOe was applied on the sample. In the case of $CeAl_2$ a cryomagnet has also been used providing a 48 kOe field.

The experiment consists in comparing, for each Bragg reflection, the two cross-sections with both orientations of the polarization $\vec{P}$ of the incident beam :

$$\frac{d\sigma}{d\Omega} = b^2 - b\,\vec{P}\,\vec{E} + \frac{1}{4}\,\vec{E}^2$$

and deducing the magnetic scattering amplitude $\vec{E}$ from the nuclear scattering length b.

MAGNETIC AMPLITUDE FORMALISM

In the Johnston, Lovesey and Rimmer formalism[5,6], the magnetic scattering amplitude $\vec{E}(\vec{K})$ is defined as a vector of spherical components

$$E(\vec{K})_Q = \frac{m}{2\pi\hbar^2} \langle\psi\,|T^1_Q\,|\psi\rangle$$

where ψ represents the electron wave function and T^1_Q the spherical components of the tensor operator interaction between neutron and magnetic electron.

When the scattering vector $\vec{K}$ is perpendicular to the magnetic moment, $E(\vec{K})_o$ is generally the only non zero component and

$$\frac{d\sigma}{d\Omega} = b^2 - bPE(\vec{K})_o + \frac{1}{4}E(\vec{K})_o^2$$

For a rare earth ion the wave function expansion implies usually only one J manifold : $\psi = \sum_M a_M |JM\rangle$ and

$$E(\vec{K})_o = 2\left(\frac{\gamma e^2}{m_e c^2}\right)(4\pi)^{1/2} \sum_{K''Q''} Y_{Q''}^{K''}(\hat{K}) \sum_{K'Q'} \{\sum_{MM'} a_M^{*} a_{M'}$$

$$[A(K''K')+B(K''K')] \langle K'Q'JM'|JM\rangle\} \langle K''Q''K'Q'|10\rangle$$

The spherical harmonics $Y_{Q''}^{K''}(\hat{K})$ show the anisotropic aspect of the form factor. The $A(K''K')$ and $B(K''K')$ terms represent respectively the orbital and spin part of the magnetic interaction. The only non zero terms are for $K'' = K' \pm 1$. They can be expressed in term of 1) quantities $A'(K'',K')$ and $C'(K'',K')$, tabulated for the rare earth in reference (11), which depend only on the 4f electron configuration in the free ion, and 2) the radial integrals $\langle j_1\rangle$ which depend on the radial part of the one electron wave function

$$\langle j_1\rangle = \int_o^\infty r^2|f(r)|^2 j_1(Kr)\, dr$$

which have been calculated for the free ion with the relativistic Dirac Fock method. This method has been described in details in (12).
Finally for each Bragg reflection (hkl)

$E_o(hkl) = \sum_{MM'} a_M^{*} a_{M'} H_{MM'}(hkl)$ where the $H_{MM'}$ are independent of the ground state of the ion. It is therefore easy to determine the set of a_M which fit better the observed magnetic amplitudes, using a least square refinement which minimizes the function $\Sigma(E_o \text{ obs} - E_o \text{ calc})^2$.

$NdAl_2$

Magnetic amplitudes have been measured for the (hk0) Bragg reflections with the field applied along the [001] easy axis of the compound. The magnetization density projection obtained by Fourier transformation of these amplitudes appeared to be the sum of two contributions : one localized on the Nd sites with a moment of $2.63 \pm 0.03\ \mu_B$ and one which represents the conduction electron polarization and which fluctuates throughout the cell[13]. On the other hand, determination of crystal field and exchange parameters have been attempted from single crystal magnetization measurements [14]. These parameters allow the calculation of the energy levels :

in the crystal field the ground state is the doublet Γ_6 which is split by the molecular field and by the 16.5 kOe applied field with an energy of about 80 K. The wave function of the ground state can thus be written as $\alpha|9/2\rangle + \beta|1/2\rangle + \gamma|-7/2\rangle$. The corresponding moment is 2.47 μ_β per Nd atom. However this determination attributes all the measured magnetization to the Nd^{3+} ion, ignoring the polarization of conduction electrons.

To meet this objection we have undertaken the least square refinement based on the form factor of the rare earth ion. For this refinement only the 23 Bragg reflections have been taken with $\sin\theta/\lambda > 0.36$ Å^{-1}. Actually, as shown by the magnetization density map, the conduction electron polarization contribution disappears for these high angle reflections and only the localized moment is taken into account. The resulting wave function, at 4.2 K, is $\psi = 0.885(1)|9/2\rangle + 0.451(4)|1/2\rangle - 0.112(15)|-7/2\rangle$ which correspond to a magnetic moment of 2.61 ± 0.01 μ_B per ND atom, in very good agreement with that obtained by the direct Fourier technique (13).

The agreement between calculated and measured magnetic amplitudes is very satisfactory and is displayed in figure 1.

$CeAl_2$

The magnetic behaviour at low temperature of $CeAl_2$ is not understood for the moment. Different models have been proposed to explain the magnetization curves(15) (figure 2) : antiferromagnetic ordering, moment reduction due to a low value of the spin correlation time or compensating conduction electron polarisation (16). Though the nature of magnetism is unknown we have measured the magnetic amplitudes of $CeAl_2$ with magnetization polarized by the field either along the [001] or the [011] direction.

In the cubic crystal field the J=5/2 manifold splits into a doublet Γ_7 and a quartet Γ_8. With the crystal field parameter W taken from specific heat measurements(17) the ground state is the doublet; the exchange and applied fields remove the degeneracy of this level. The ground state can then be written as $\psi = \alpha|3/2\rangle + \beta|-5/2\rangle$ when the field is parallel to [001] and $|\psi\rangle = \alpha'|5/2\rangle + \beta'|1/2\rangle + \gamma'|-3/2\rangle$ when it is parallel to [011]. We have refined these coefficients using the form factors measured on Bragg lines ($\sin\theta/\lambda > 0.30$ Å^{-1}) with experimental conditions corresponding to 1.5 K and 48 kOe applied field. The resulting wave functions are :

$$\psi = 0.955(15)|3/2\rangle - 0.297(51)|-5/2\rangle \text{ for } H//[001]$$
$$\psi = 0.684(8)|5/2\rangle - 0.637(14)|1/2\rangle - 0.356(29)|-3/2\rangle$$

for H//[011].

The calculated and measured amplitudes are reported on figures 3 and 4 after being normalized to the calculated moments. The agreement is very satisfactory. The form factor with the field parallel to [011] shows an unusually large dispersion of the points while

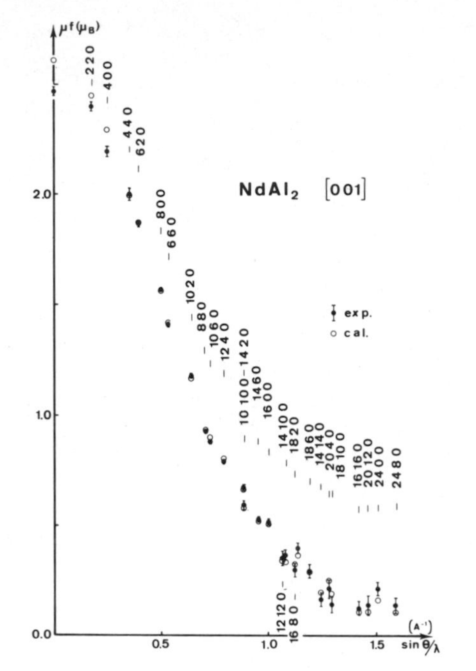

Fig. 1 : Comparison between experiment (•) and calculation (o) for magnetic amplitudes in $NdAl_2$ (H = 16.5 kOe, T = 4.2 K)

Fig. 2 : Magnetization of a $CeAl_2$ single crystal at 1.7 K (13)

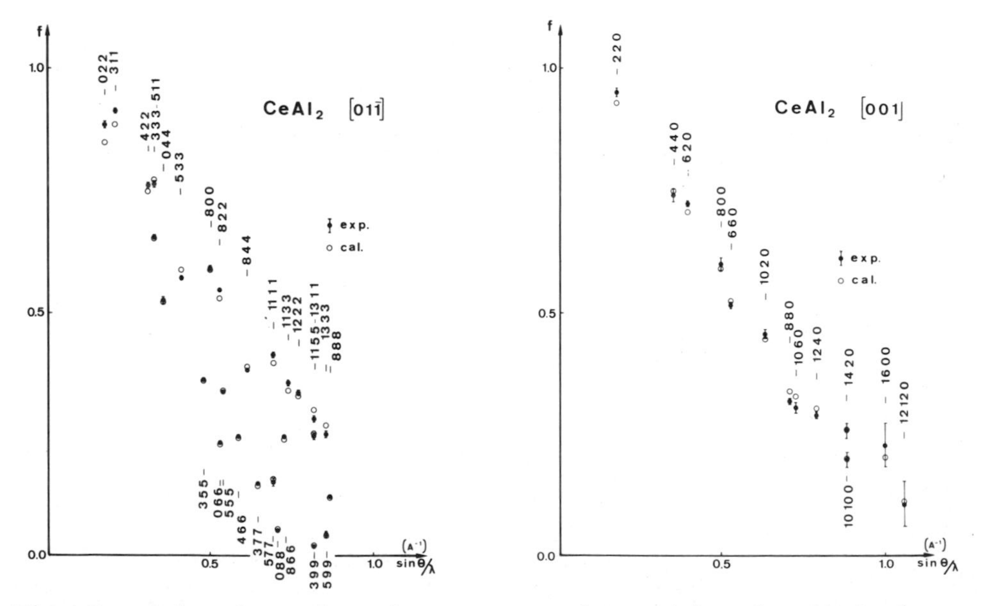

Fig. 3 and 4 : Comparison between experiment (•) and calculation (o) for form factor in $CeAl_2$ with two directions of applied field [001] and [01$\bar{1}$] (H = 48 kOe, T = 1.5 K).

this with the field parallel to (001) is smoother. It may be noted that reflections of the type (h00) and (0kk) have been measured in the two cases : the form factors are close for (h00) but very different for (0kk).

As a conclusion, in simple cases such as $REAl_2$ compounds the polarized neutron measurements of form factors are accurate enough to allow the determination of the ground state wave function. This determination is based on the shape of the rare earth 4f magnetization density and does not require a complete knowledge of the magnetism of the compounds.

REFERENCES

1. R.M. Moon, W.C. Koehler, J.W. Cable and H.R. Child, Phys. Rev. B5, 997 (1972)
2. A.J. Freeman and J.P. Desclaux, Int. J. Magnetism 3, 311 (1972)
3. C. Stassis, G.R. Kline, A.J. Freeman and J.P. Desclaux, Phys. Rev. B13, 3916 (1976)
4. G.H. Lander, T.O. Brun, J.P. Desclaux and A.J. Freeman, Phys. Rev. B8, 3237 (1973)
5. D.F. Johnston, Proc. Phys. Soc. 88, 37 (1966)
6. S.W. Lovesey and D.E. Rimmer Rep. Prog. Phys. 32, 333 (1969)
7. F.A. Wedgwood, J. Phys. C 5, 2427 (1972)
8. G.H. Lander, T.O. Brun and O. Vogt, Phys. Rev. B7, 1988 (1973)
9. J.H. Wernick and S. Geller, Trans. AIME 218, 806 (1960)
10. J.H. Williams, J.H. Wernick, E.A. Nesbitt and R.C. Sherwood, J. Phys. Soc. Japan Sup B1, 17, 91 (1962)
11. G.H. Lander and T.O. Brun, J. Chem. Phys. 53, 1387 (1970)
12. J.P. Desclaux and N. Bessis, Phys. Rev. A2, 1963 (1970)
13. J.X. Boucherle and J. Schweizer, to be published in the International Conference on Magnetism, Amsterdam, September 1976
14. B. Barbara, J.X. Boucherle and M.F. Rossignol, Phys. Stat. Sol. (a) 25, 165 (1974)
15. B. Barbara, M.F. Rossignol, M.G. Purwins and E. Walker, Sol. Stat. Comm. 17, 1525 (1975)
16. B. Cornut, Thesis, University of Grenoble (1975)
17. C. Deenedas, A.W. Thompson, R.S. Craig and W.E. Wallace, J. Phys. Chem. Solids 32, 1953 (1971)

THE COOPERATIVE JAHN-TELLER EFFECT STUDIED BY NEUTRON SCATTERING

J.K. Kjems

Research Establishment Risø

DK-4000 Roskilde, Denmark

ABSTRACT

Rare earth insulators with a degenerate or near degenerate electronic ground state have proven to be excellent model systems for the cooperative Jahn-Teller effect. Neutron scattering has been used to characterize both static and dynamic properties at the phase transitions in $TmVO_4$ (doublet, T_D = 2.1 K), $PrAlO_3$ (singlet-singlet, T_D = 151 K) and $TbVO_4$ (singlet-doublet-singlet, T_D = 33 K). Most of the experimental data can be reconciled with the RPA theory devised by Elliott et al. and such analysis can be used to give quantitative information on the electron phonon coupling in these systems. However, central peaks in the spectral response are not satisfactorily understood.

INTRODUCTION

The Cooperative Jahn-Teller Effect (CJTE) constitute a rare example of a mechanism for structural phase transitions which can be understood almost from first principles. Furthermore, the transitions often involve strain mediated forces of long range with the result that the simple RPA and mean field solutions to the microscopic Hamiltonians as a rule give very good descriptions of the observables. There has been considerable experimental and theoretical progress in this field of research over the past decade, particularly for the rare earth insulators which lend themselves conveniently to a variety of spectroscopic studies such as fluorescence, infrared, Raman and Brillouin scattering, impurity ESR, ultrasonics, X-ray and neutron diffraction and inelastic neutron scattering.

Many of these results have been summarized in the excellent review by Gehring and Gehring[1].

This paper focusses on the experimental information that has been obtained recently by inelastic neutron scattering (INS). So far, these studies have been restricted to $PrAlO_3$[2], $TmVO_4$[3] and $TbVO_4$[4] but studies of other systems are in progress.

The well known advantage of INS is the ability to probe the response throughout the Brillouin zone and it is such results which will be emphasized here.

SIMPLIFIED THEORY

In insulators the CJTE originates in the interaction between the crystal lattice and the localized orbital electronic state via the crystalline electric field. In situations where the symmetry of the crystal field gives rise to degenerate (non-Kramers) or near degenerate electronic ground states, then, at low temperatures the CJTE may lift this degeneracy by a spontaneous lattice distorsion of appropriate symmetry. The driving force is the gain in electronic energy which is linear in the distorsion whereas the balancing lattice strain energy has a quadratic dependence.

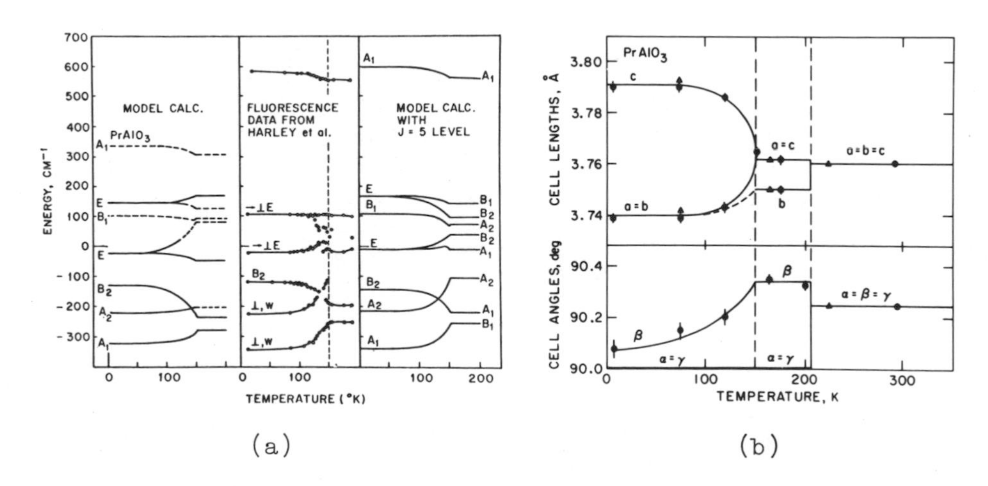

(a) (b)

Fig. 1: (a) Comparison between fluorescence data for the electronic Pr^{+3} $^3H_4$ multiplet and crystal field model calculations for $PrAlO_3$ (ref. 5).
(b) Temperature dependence of the unit cell parameters in $PrAlO_3$. Above 205 K the structure is rhombohedral, between 151 K and 205 it is orthorhombic, and the Jahn-Teller transition at 151 K results ultimately in a nearly tetragonal structure at 0 K (ref. 2).

As an example of the magnitude of the effects, Fig. 1 shows the temperature dependence of the lattice parameters and of the electronic levels of $Pr^{+3}\ {}^3H_4$ in the perovskite $PrAlO_3$ in the region of the 151 K Jahn-Teller[3] transition. Here the CJTE causes a change from orthorhombic through monoclinic to an ultimate tetragonal symmetry involving both macroscopic strain and displacements of the atoms within the unit cell. It is apparent from the fluorescence data that all the electronic levels are affected by the transition but because of the population factors it is only the two lowest levels, B_1 and A_1 of the orthorhombic phase above $T_D = 151$ K that are actively involved. In a theoretical description it suffices to consider only the interaction between these two levels and the lattice. In microscopic terms[6] this is expressed as a linear coupling between electric quadrupole operators, often in a pseudospin representation, and the normal modes of the lattice of compatible symmetry.

$$H = \sum_{n,\vec{k},p} \xi_p(\vec{k}) e^{i\vec{k}\cdot\vec{R}_n} S_n^z (a_{p,\vec{k}} + a^+_{p,\vec{k}}) + \sum_{\vec{k},p} \hbar\,\omega_p(\vec{k})(a^+_{p,\vec{k}} a_{p,\vec{k}} + \tfrac{1}{2}) + \Delta \sum_n S_n^x \qquad (1)$$

where S_n^z and S_n^x are the pseudospin operators representing two electronic states at the energies $\pm\Delta$, $a^+_{p,\vec{k}}$ is the operator for a phonon of wave vector $\vec{k}$, band p and frequency $\omega_p(\vec{k})$ and $\xi_p(\vec{k})$ denotes the corresponding Jahn-Teller coupling. The detailed solutions of this Hamiltonian are discussed in the paper by Elliot at this conference. The first step is a displaced oscillator transformation which gives

$$H = \sum_{\vec{k}} \hbar\,\omega_p(\vec{k})(\gamma^+_{p,\vec{k}} \gamma_{p,\vec{k}} + \tfrac{1}{2}) - \sum_{\vec{k}} J(\vec{k})\, \vec{S}_{\vec{k}}\, \vec{S}_{-\vec{k}} + \Delta N^{\frac{1}{2}} S_o^x$$

where

$$\gamma_{p,\vec{k}} = a_{p,\vec{k}} + \frac{\xi_p(\vec{k})}{\hbar\,\omega_p(\vec{k})} S^z_{\vec{k}} \qquad (2)$$

and

$$K_p(\vec{k}) = \frac{|\xi_p(\vec{k})|^2}{\hbar\,\omega_p(\vec{k})}, \quad J(\vec{k}) = \sum_p K_p(\vec{k}) - N^{-1} \sum_{p,\vec{k}} K_p(\vec{k})$$

This shows that the linear Jahn-Teller coupling gives rise to an effective quadrupole-quadrupole interaction, $J(\vec{k})$, which causes the spontaneous "ferroquadrupolar" transition if $J(o) > \Delta$

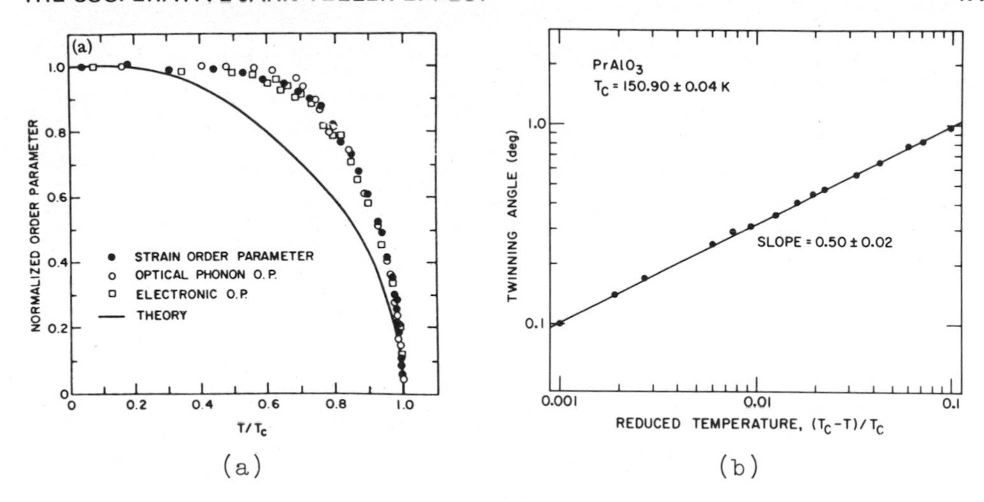

(a) (b)

Fig. 2: (a) Experimentally determined temperature variation of the reduced order-parameters in $PrAlO_3$ below T_D = 151 K. (b) Double log plot of the strain order parameter versus reduced temperature in $PrAlO_3$. The twinning angle is proportional to $\frac{c}{a} - 1$ (ref. 2).

ORDER PARAMETERS

The order parameters at the transition can be thought of as condensed normal coordinates of the Jahn-Teller coupled modes of the high temperature phase. In $PrAlO_3$ there are three such modes[2,7], the electronic transition $B_1 - A_1$ with the energy gap $2W(T)$ where $W(T_D) = \Delta$, the B_1 strain and the corresponding transverse acoustic mode, and the B_1 optic phonon. The splittings and displacements corresponding to the condensation of these modes below T_D have all been determined experimentally and the results are shown in Fig. 2. The macroscopic strain was measured by neutron diffraction using the twinning pattern[2] that evolves below T_D, the optic phonon displacement were found via the orientation of the electric field gradient tensor around a Gd^{+3} impurity as sensed by ESR[7] and the electronic energy gap was observed by fluorescence[8]. On reduced scale these three measurements give identical results and this represents an experimental proof of the assumption of a <u>linear</u> coupling between the modes. Near T_D the order parameters follow a classical powerlaw $(T_c-T)^\beta$ with $\beta=\frac{1}{2}$ as illustrated by the double logarithmic plot of the twinning angle in Fig. 2. However, the simplest mean field calculation does not give the correct temperature dependence over the whole temperature range in $PrAlO_3$. This is shown by the solid line marked "theory" in Fig. 2. In other less complicated systems such as $TmVO_4$[9] one finds a much better agreement between the experimental observations and the mean field theory for the order parameters.

MIXED MODES

The effective quadrupole coupling $J(\vec{k})$ introduces dispersion of the electronic mode, often called the vibron mode or the quadrupole exciton. In absence of coupling to the low lying acoustic phonon the RPA result for the frequency of this mode is[1]

$$\hbar^2\omega_v^2(\vec{k}) = 4\Delta(\Delta - \langle S^x\rangle J(\vec{k})) \qquad T > T_D$$

with (3)

$$\langle S^x\rangle = \tanh(\Delta/k_BT)$$

and this would be the soft mode at the transition. The interaction with the acoustic phonon and, in the limit $|\vec{k}| \to 0$, with the macroscopic strain prevents the full softning of the vibron mode and, instead, it is a mixed vibron-phonon mode that eventually becomes unstable. The frequencies, $\omega(\vec{k})$, of the hybridized modes are the solutions to

$$\hbar\omega(\vec{k})(\omega^2(\vec{k}) - \omega_v^2(\vec{k})) = \sum_p \frac{4\Delta\langle S^x\rangle\omega^3(\vec{k})K_p(\vec{k})}{\omega^2(\vec{k}) - \omega_p^2(\vec{k})} \tag{4}$$

These mixed modes have been observed by INS in $PrAlO_3$[3] (B_1 symmetry along (101)), and in $TmVO_4$[3] and $TbVO_4$[4] (B_2 symmetry along (100)). The results for $PrAlO_3$ and $TmVO_4$ are shown in Fig. 3.

As mentioned earlier there are three active modes in $PrAlO_3$ of which only the optic phonon and the acoustic phonon have an observable cross section for INS, consequently, one is only able to determine the vibron frequencies in the wave vector range where it has phonon admixture.

In the tetragonal phase[1] of $TmVO_4$ above $T_D = 2.1$ K the Tm^{+3} $^3H_6$ multiplet has a doublet ground state and, in order to observe the anticrossing, one needs to apply a magnetic field along the tetragonal C-axis[5]. The energy gap is then given by the Zeeman splitting $2\Delta = g\mu_B H$ ($g = 10.6$) and Eq. 3 and 4 are applicable. By tuning the field one is able to move the phonon-vibron crossing point out into the zone as illustrated by the measurements for different field values in Fig. 3.

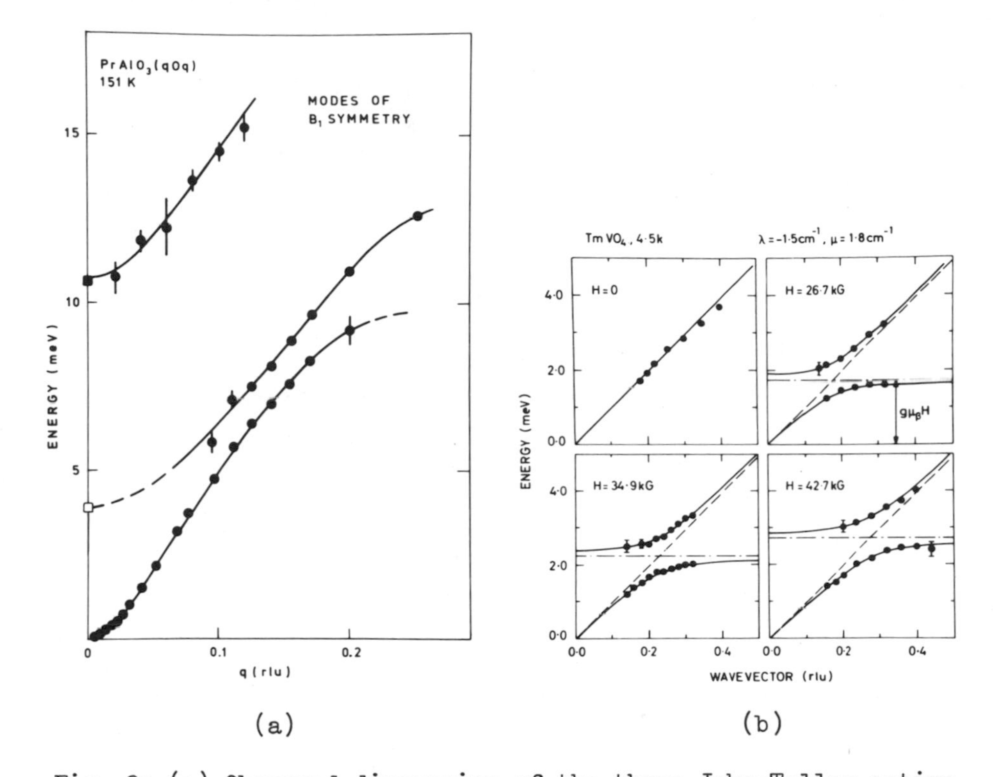

Fig. 3: (a) Observed dispersion of the three Jahn-Teller active modes of B_1 symmetry in $PrAlO_3$ along the (hoh) direction at $T = T_D$. The zone centre frequencies are determined by Raman scattering[13].
(b) The anticrossing of the acoustic phonon of B_2 symmetry propagating along (hoo) and the Zeeman split electronic ground doublet in $TmVO_4$. The magnetic field is applied along (ool). The solid lines are calculations based on Eq. 4 of the text (ref. 3).

The experimentally determined mixed mode frequencies enables one to determine the wave vector dependence of the effective interactions $K_p(\vec{k})$ by use of Eq. 4, and the results for the two systems are shown in Fig. 4.

In $TmVO_4$[3] the coupling to the acoustic phonon is found to be independent of the magnitude of $|\vec{k}|$ in the region of the Brillouin zone where the acoustic phonon has a linear dispersion, i.e. approximately half way to the zone boundary. The coupling strength agrees exceedingly well with the limiting value for $|k| \to 0$ as deduced from Raman[9] and ultrasonic measurements[10] and this result allows one to conclude that the coupling is strictly proportional

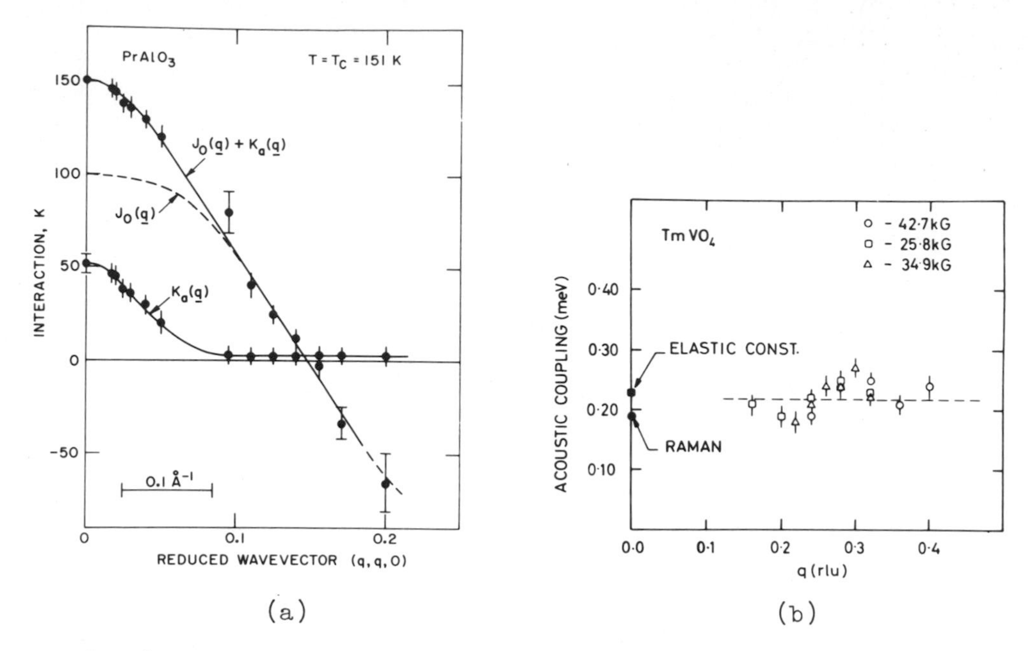

Fig. 4: Wave vector dependence of the Jahn-Teller coupling parameters for $PrAlO_3$ (a) and $TmVO_4$ (b) as determined from the experimental data in Fig. 3 by use of Eq. 4.

to the strain produced by the finite wave vector acoustic phonon in accordance with the Debye-model[1].

The interpretation of the results for $PrAlO_3$ is slightly more complicated. Here the dominating interaction is the optic phonon coupling. The acoustic coupling is found to be strongly wave vector dependent and of much larger magnitude than what was found for $TmVO_4$. This could be a result of the direct coupling between the acoustic and optic phonons in the perovskite lattice[2] which would mix the characters of these modes and thus possibly enhance the acoustic phonon-vibron interaction. The observed dispersion of the optic phonon (cubic R_{25} parentage) could then account for the rapid variation of the effective interactions with wavevector as indicated by Eq. 2.

The temperature dependences[2,3] of the mixed modes in both $PrAlO_3$ and $TmVO_4$ are found to follow from Eq. 4 when the true population factors including the otherwise neglected excited electronic states are used to calculate $\langle S^x \rangle$.

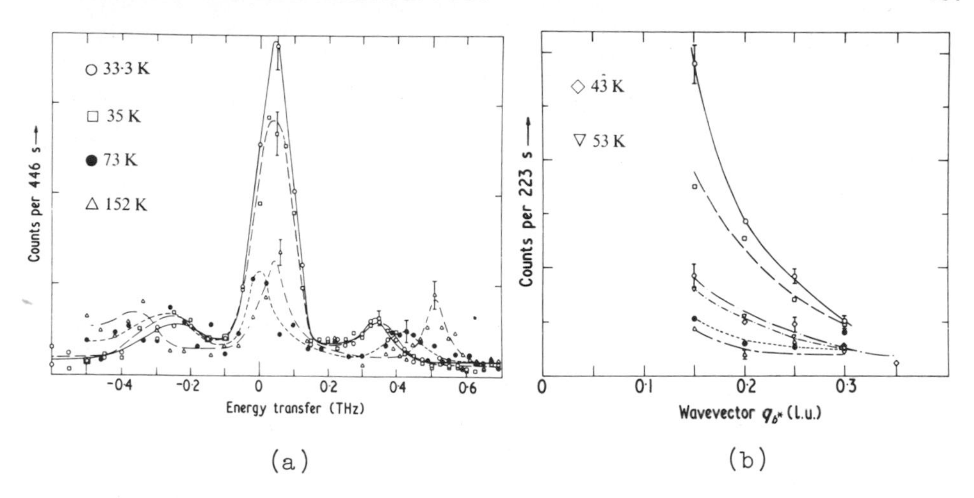

(a) (b)

Fig. 5: (a) INS response for the B_2 acoustic phonon measured at (4,0·25,0) in $TbVO_4$ at four different temperatures. The dominant feature is the peak at zero energy transfer whose width is resolution limited (ref. 4).
(b) Variation of the "central peak" intensity in $TbVO_4$ versus wavevector alonb b^* at different temperatures. The signatures correspond to those in Fig. 5(a).

In addition to the mixed mode resonances, peaks in the INS spectra at zero energy transfer have been observed near T_D in $PrAlO_3$[2] and more pronounced in $TbVO_4$[4] (T_D = 33 K). Examples of such spectra for $TbVO_4$ are shown in Fig. 5 taken at different temperatures. In this case the "central peak" is believed to originate in the peculiar electronic structure of $TbVO_4$ which consists of two singlets separated by 2.2 meV (18 cm^{-1}) and a doublet midway in energy between the singlets all of which are Jahn-Teller coupled to the lattice. In this picture the quasielastic peak corresponds to transitions within the degenerate doublet level analogous to the mechanisms that has been proposed for purely magnetic singlet ground state systems[11] with degenerate excited states, e.g. Pr_3Tl where recently a similar central peak response has been observed[12]. Given the interactions parameters one can, within the RPA, explicitly calculate[14] the full response and its temperature dependence but judging from the experience from the magnetic systems one cannot expect an equally good agreement for the critical divergencies as for the normal modes. However, for the CJTE this question is still open and awaits a detailed quantitative analysis.

The observed central peaks in $PrAlO_3$ cannot have the same electronic origin as in $TbVO_4$ since there are no degenerate electronic states in the $^3H_4$ multiplet above T_D.

SUMMARY

The information obtained so far by the INS studies of the CJTE fits exceedingly well into the picture that has immerged from other experimental studies when linked together with the RPA theory. However, the details of the critical scattering have not yet been fully understood although a recent work by Cowley[15] on the similar problem in ferroelectrics may be very relevant in this context.

The phonon quadrupole interactions found in both rare earth and actinide (UO_2) insulators may very well be significant in both rare earth and actinide metals and alloys. Hopefully, the insight which has been gained by the studies of the CJTE will be of use in the attempts to understand the often subtle manifestations of anisotropic interactions which are found in these materials.

ACKNOWLEDGEMENT

During these studies the author has benefited from his close collaboration with R.J. Birgeneau, W. Hayes and G. Shirane.

REFERENCES

1. G.A. Gehring and K.A. Gehring, Rep.Prog.Phys. 38, 1 (1975)

2. R.J. Birgeneau, J.K. Kjems, G. Shirane and L.G. van Uitert, Phys.Rev. B 10, 2512 (1974)

3. J.K. Kjems, W. Hayes and S.H. Smith, Phys.Rev.Lett. 35, 1089 (1975)

4. M.T. Hutchings, R. Scherm, S.H. Smith and S.R.P. Smith, J.Phys. C., L393 (1975)

5. K.B. Lyons, R.J. Birgeneau, E.I. Blount, and L.G. van Uitert, Phys.Rev. B 11, 891 (1975)

6. R.J. Elliot, R.T. Harley, W. Hayes and S.R.P. Smith, Proc.Roy. Soc. A 328, 217 (1972)

7. M.D. Sturge, E. Cohen, L.G. van Uitert, and R.P. van Stapele, Phys.Rev. B 11, 4768 (1975)

8. R.T. Harley, W. Hayes, A.M. Perry, and S.R.P. Smith, J.Phys.C. 6, 2382 (1973)

9. R.T. Harley, W. Hayes, and S.R.P. Smith, J.Phys.C. 5, 1501 (1972)

10. R.L. Melcher, E. Pytte, and B.A. Scott, Phys.Rev.Lett. 31, 307 (1973)

11. S.R.P. Smith, J.Phys.C. 5, L157 (1972)

12. J. Als-Nielsen, J.K. Kjems, W.J.L. Buyers and R.J. Birgeneau (to be published)

13. R.T. Harley, W. Hayes, A.M. Perry and S.R.P. Smith, J.Phys.C. 8, L123 (1975)

14. W.J.L. Buyers, in: Magnetism and Magnetic Materials, AIP Conference Proc. 27 (1975)

15. R. Cowley, Phys. Rev. Lett. 36, 744(1976).

LOCAL JAHN-TELLER EFFECT AT A STRUCTURAL PHASE TRANSITION

K.-H. Höck and H. Thomas

Institut für Physik, Universität Basel

Klingelbergstrasse 82, CH-4056 Basel

ABSTRACT

We investigate the critical enhancement of the local Jahn-Teller (JT) effect of a JT-impurity ion produced by the coupling to the soft mode of a structural phase transition of the host crystal.

EPR-spectra of JT-impurities in crystals have commonly been interpreted in terms of a quasimolecular model /1/ which takes into account the vibronic coupling to a single mode only. Recent investigations /2,3/ have shown that the properties of a JT-impurity become modified due to the vibronic coupling to the phonon continuum of the host crystal (multimode JT-effect). Within the framework of harmonic lattice theory the effects are expected to be ordinarily rather small, because the induced JT-distortion extends only over a few lattice distances. However, if the JT-distortion couples to the soft mode of a structural phase transition of the host crystal, the range of distortion increases as the critical temperature is approached, and one obtains a strong, temperature-dependent enhancement of the local JT-effect (Fig. 1).

In order to study this effect, we consider the specific problem of an impurity ion with a doublet electronic ground state of E_g-symmetry at a lattice site of cubic symmetry (O_h) in a non-JT-crystal, which undergoes a displacive ferrodistortive phase transition with an

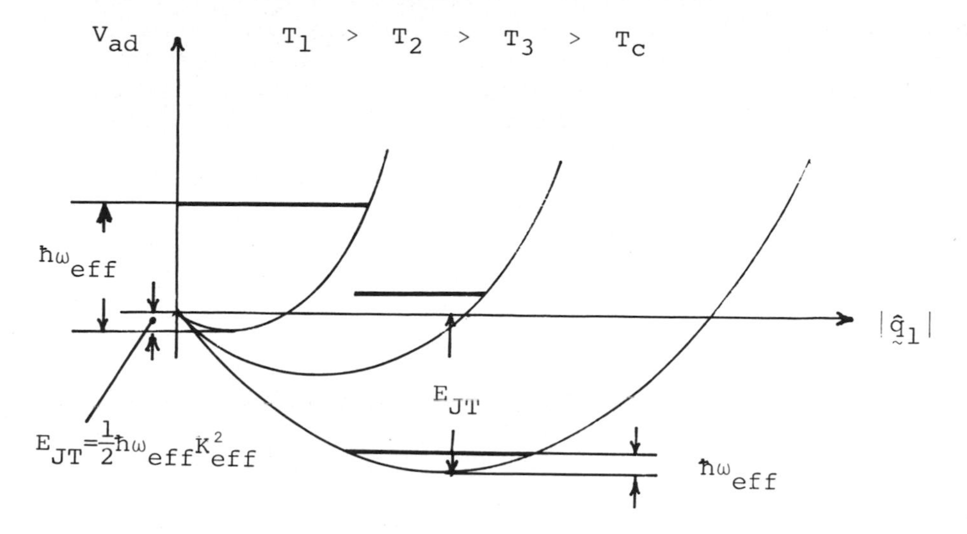

Fig. 1: Critical enhancement of the local JT-effect as the temperature approaches the transition temperature

E_g-type soft optical-phonon branch in the high-temperature phase.

We treat the multimode JT-problem

$$(1) \quad H = \sum_q \hbar\omega_q(T) \{\tfrac{1}{2} \underset{\sim}{p}_q^2 + \tfrac{1}{2} \underset{\sim}{q}_q^2 + K_q(T) \underset{\sim}{q}_q \cdot \underset{\sim}{\sigma}\} \;.$$

$\underset{\sim}{q}_q = (q_{q,\theta}, q_{q,\varepsilon})$ = dimensionless coordinates of the E_g-type symmetry-adapted phonon modes

$$\underset{\sim}{\sigma} = (\sigma_\theta, \sigma_\varepsilon) \quad \text{with} \quad \sigma_\theta = \begin{pmatrix} -1 & 0 \\ 0 & 1 \end{pmatrix} \quad \sigma_\varepsilon = \begin{pmatrix} 0 & 1 \\ 1 & 0 \end{pmatrix}$$

where we have described the host crystal by an effective harmonic Hamiltonian with temperature-dependent frequencies $\omega_q(T)$ of the soft optical phonon branch. In the mean-field-approximation for small wave vectors q, the spectrum is given by

$$\omega_q^2(T) = \omega_o^2(T) + v \cdot q^2$$

with the soft optical phonon frequency

$$\omega_o^2(T) \propto T - T_c$$

$K_q^2(T) \propto \frac{1}{\omega_q^{\,3}(T)}$ is a dimensionless JT-coupling constant.

We adopt the single-effective-frequency approximation described by Halperin and Englman /2/: The multimode Hamiltonian (Eq. 1) is replaced by an effective single-mode Hamiltonian containing the coupling to a kind of effective interaction mode which is strongly affected by the JT-coupling, and which is determined such that the approximation to the multimode coupling is optimized:

$$(2) \quad H = \hbar\omega_{eff}(T) \; \{\tfrac{1}{2}\, \hat{\underset{\sim}{p}}_1^2 + \tfrac{1}{2}\, \hat{\underset{\sim}{q}}_1^2 + K_{eff}^2(T)\; \hat{\underset{\sim}{q}}_1 \cdot \underset{\sim}{\sigma}\} + H'$$

The effective interaction mode $\hat{\underset{\sim}{q}}_1$ is given by a linear combination of the original symmetry-adapted phonon modes $\underset{\sim}{q}_q$, and extends over the correlation distance of the optical phonon displacements. For a soft optical phonon branch, this correlation distance diverges at the critical point, thus giving rise to the enhancement of the local JT-effect. Due to this enhancement a weak JT-effect ($K_{eff} < 1$) changes into a moderately strong JT-effect ($K_{eff} > 1$) close to the transition temperature. This is shown in Fig. 2, which depicts the temperature variation of the effective JT-coupling-constant $K_{eff}(T)$. The variation from a weak to a moderately strong JT-effect, when the temperature approaches the critical temperature, is reflected in a characteristic temperature-dependence of the reduction factors p and q, which measure the influence of the JT-effect on the EPR-spectrum. The results of the calculation of the reduction factors p and q are shown in Fig. 3.

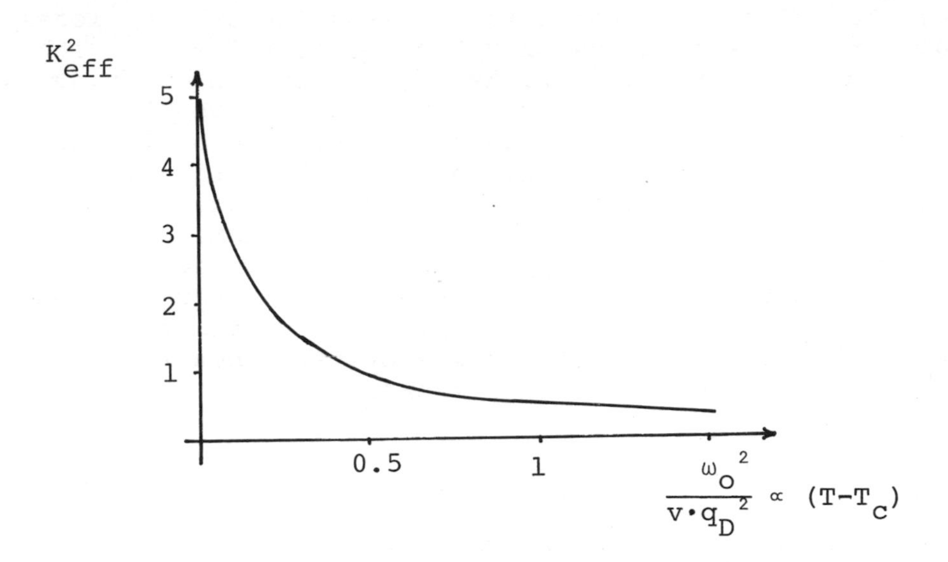

Fig. 2: Temperature variation of the effective JT-coupling constant $K_{eff}(T)$

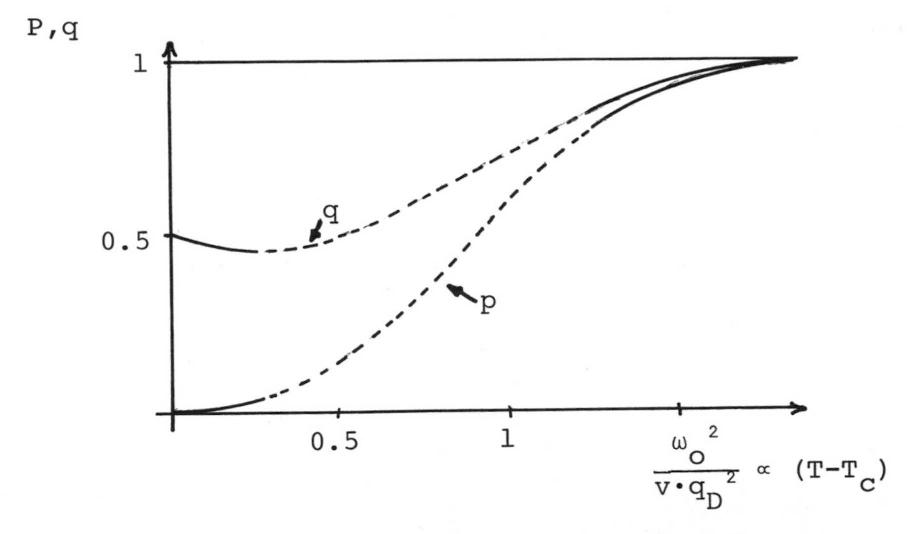

Fig. 3: Temperature variation of the reduction factors p and q as a consequence of the critical enhancement of the local JT-effect

We have further calculated the spatial extension of the effective interaction modes $\tilde{\hat{q}}_1$ around the local JT centre. We find that for $T \neq T_c$, it falls off with an Ornstein-Zernicke law. At $T = T_c$, on the other hand, the spatial dependence is governed by a power law depending on the value of the effective JT coupling constant $K_{eff}(T=T_c)$. It should be noted that this is the distortion field associated with the optical phonon coordinate only. Because of the lattice-dynamical coupling to the acoustical phonon coordinates there will be in addition an elastic strain field falling off with a power law already for $T \neq T_c$. This contribution is, however, not expected to show a critical behaviour in our model of a soft optical phonon.

REFERENCES

/1/ F.S. Ham, in Electron Paramagnetic Resonance ed. by S. Gschwind (Plenum, New York, 1972)

/2/ M.C.M. O'Brien, 1972 J.Phys.C: Solid State Phys. 5, 2045

/3/ R. Englman, B. Halperin, 1973 J.Phys.C: Solid State Phys. 6, L219

SPIN WAVES AND CENTRAL MODES IN CRYSTAL FIELD SYSTEMS

T.M. Holden, W.J.L. Buyers and E.C. Svensson
Atomic Energy of Canada Ltd., Chalk River, Ontario,
Canada, and
H.-G. Purwins*
Dépt. de Physique, Université, Geneva, Switzerland

ABSTRACT

For magnetically ordered intermetallic compounds such as $PrAl_2$, $Pr_3T\ell$ and TbSb neutron inelastic scattering experiments show that the spin-wave spectrum consists of several branches of excitations. These correspond at T=0 to allowed transitions between the ground state and excited states of the lowest multiplet of the rare-earth ion in the exchange field existing in the material. A good description of the spin-wave spectrum can be obtained within the random phase approximation with crystal field terms and interionic interactions. At elevated temperatures excited-state spin waves are observed which correspond to transitions between excited states of the multiplet. Close to the ordering temperature and ordering wavevector the spectrum is dominated by elastic scattering with less important contributions from inelastic transitions. Generalization of the spin-wave theory to finite temperatures leads to a qualitatively correct description of these effects and insight into the nature of the 'central' mode.

INTRODUCTION

Crystal-field effects and exchange interactions dominate the magnetic behaviour of materials containing rare-earth ions. The effective crystal field lifts the free-ion degeneracy giving rise to a multiplet of levels characteristically spaced over several THz in frequency, and in special cases producing non-magnetic singlet or doublet ground states. An exchange interaction between the rare-earth ions mixes the wave functions of the members of the multiplet and raises the degeneracy of the crystal-field states. A moment may thus be induced into an otherwise non-magnetic state. The most effective spectroscopic probe of these states is neutron inelastic scattering and considerable advances have been made in the decade in which this technique has been applied to materials containing rare-earth ions. The low temperature spin-wave spectra are more complicated than for spin-only magnets since there are usually several possible dipole transitions from the ground state to excited states of the lowest multiplet. As an example of a system showing a complex spectrum with interplay between crystal field and exchange effects, recent results on $PrAl_2$[1] are examined in the framework of random phase approximation theory for the following Hamiltonian.

$$H = H_{cf} + \sum_{jj'} \sum_{kk'} J\begin{pmatrix} jj' \\ kk' \end{pmatrix} \underset{\sim}{S}\begin{pmatrix} jk \end{pmatrix} \cdot \underset{\sim}{S}\begin{pmatrix} j'k' \end{pmatrix} \qquad (1)$$

where $\underset{\sim}{S}$ is the total angular momentum on site j and sublattice k and $J\begin{pmatrix} jj' \\ kk' \end{pmatrix}$ is an isotropic exchange parameter. The crystal field part, for a quantization direction along a cube edge of a cubic material is

$$H_{cf} = \sum_{jk} B_4^0 \left\{ O_4^0\,(jk) + 5\, O_4^4\,(jk) \right\} + B_6^0 \left\{ O_6^0(jk) - 21\, O_6^4(jk) \right\} \qquad (2)$$

Where B_n^0 are crystal field parameters and O_n^m are operator equivalents tabulated, for instance, by Hutchings[2]. Other terms may be necessary such as the effective coupling between ions via macroscopic strains in the crystal[3,4], if there is a known macroscopic strain. In some cases e.g. DySb[5], measurements of the velocity of sound indicate that additional quadrupole-quadrupole interactions are important and show which terms dominate. In general it is recognized that many anisotropic exchange interactions between ions are present in some degree[6].

II LOW TEMPERATURE SPIN-WAVE SPECTRA

Several experiments have been reported recently on rare-earth compounds which allow the exchange and crystal field effects to be separated (see for example refs. 7, 8, 1, 4). $PrAl_2$ is interesting in that consideration of crystal field and isotropic exchange provide a good qualitative description of the spectrum, yet there are enough discrepancies to suggest that other interactions are also present. $PrAl_2$ is a metallic ferromagnet ($T_c = 32.1 \pm 0.1$ k) which crystallizes in the cubic Laves phase with two Pr ions per unit cell. The overall effect of the Pr and Al neighbours is to give a cubic crystal field which splits the J=4 multiplet (see Fig. 1) giving a Γ_3 doublet ground state both components of which are non-magnetic.

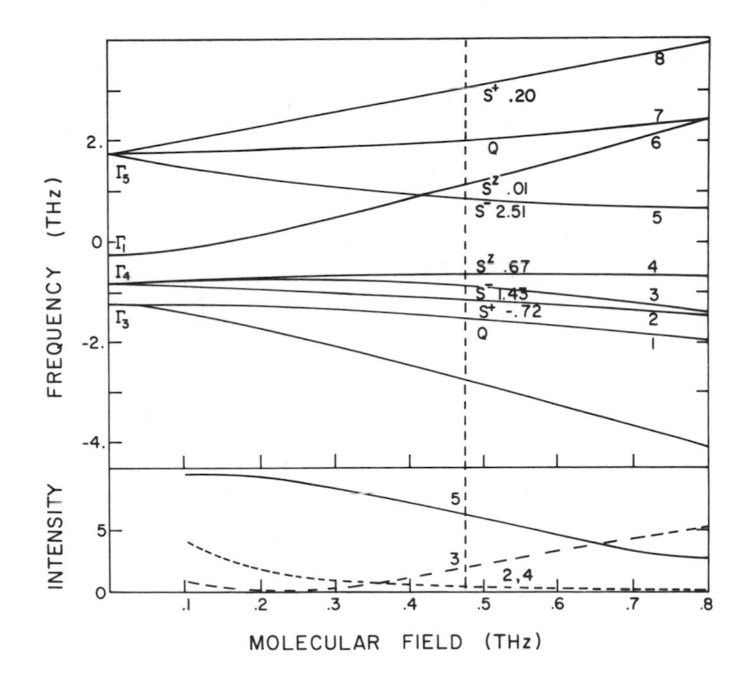

Fig. 1: Energy levels of the Pr^{3+} ion in a crystal field appropriate to $PrAl_2$ as a function of molecular field. The crystal-field-only states are denoted by their symmetry labels Γ_n. The broken vertical line gives the molecular field for $PrAl_2$ at T=0. The lower portion gives the intensity of the strongest transitions as a function of field.

The exchange, which is three times larger than that necessary to induce magnetic order at T=0, gives rise to magnetic ordering of the induced moment type. The easy axis of magnetization is the cube edge[9] and the measured magnetic moment per ion, 2.88 ± 0.05 μ_B, is below the free-ion moment (3.2 μ_B) for Pr.

The measurements of the spin-wave spectrum were made with a triple-axis crystal spectrometer at the NRU Reactor, Chalk River operated in the constant-momentum transfer (constant-$\underset{\sim}{Q}$) mode of operation with fixed scattered neutron energy. The dispersion relation for $PrAl_2$ for modes propagating in the [00ζ] and [ζζ0] directions is shown in Fig. 2. Near ζ = 0.5 in the "acoustic"

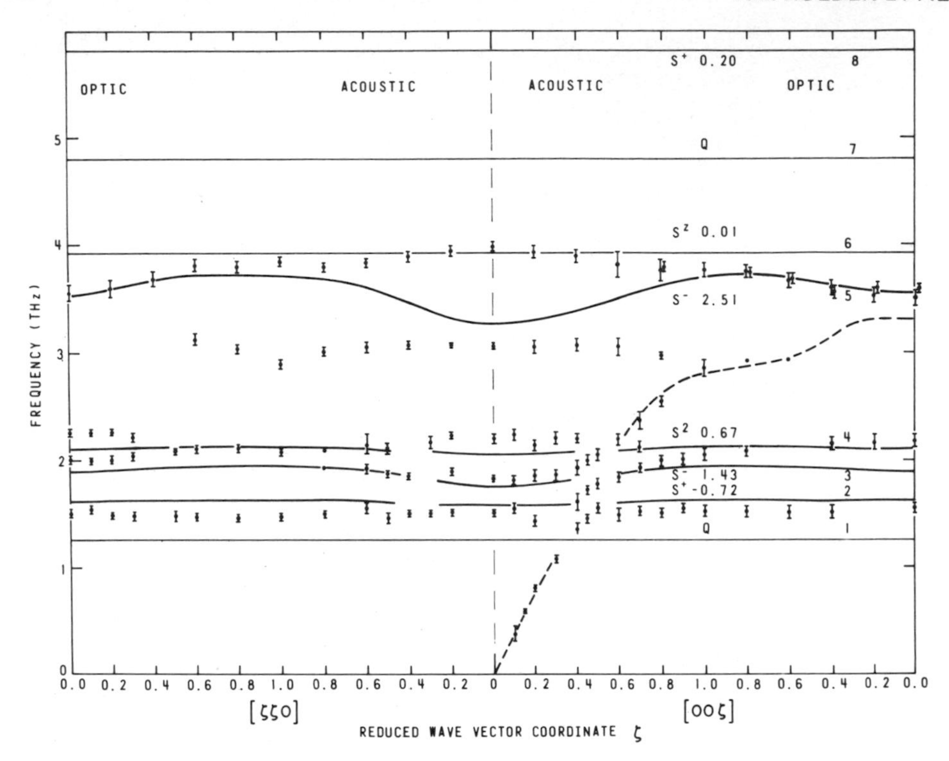

Fig. 2: Spin-wave dispersion relation for $PrAl_2$ in the [00ζ] and [ζζ0] directions. Solid lines show the best fit of the multilevel spin-wave model. ([1] and Purwins, Holden, Buyers and Svensson to be published.)

zone the [00ζ] TA phonon branch interacts with the two lowest observed magnetic branches. The dashed curve shows the dispersion relation for the TA phonon branch at 300K. A similar interaction occurs near $\zeta = 0.4$ in the [ζζ0] direction. Magnon-phonon interactions frequently occur in magnetically ordered crystal field systems when the vibrational motion of the neighbours modulates the crystal field acting on the rare-earth ion.

An interaction, whose origin is unknown, splits the strongest magnetic-dipole mode of the spectrum. At $\zeta = 0$ (acoustic zone) two modes are observed at 3.07 ± 0.03 and 3.99 ± 0.04 which are thought to correspond to the same single-ion transition. Although at $\zeta = 0$ the lower mode is twice as intense as the upper mode, the intensity of the lower mode falls rapidly as ζ increases and the intensity transfers to the upper mode. At the zone boundary $\zeta = 1.0$ in the [00ζ] direction, the lower mode has almost the same frequency as the TA phonon mode, and its extension into the optic zone follows the phonon frequencies but its intensity disappears beyond ζ =0.6 in the optic zone. The interaction may be a complicated magnon-phonon effect, or it may be due to higher order coupling mechanisms.

These experiments require for their interpretation the inclusion of all transitions from the ground state to excited states of the ground manifold and the theoretical treatment follows that of Buyers et al.[10]

The single-ion dynamical response to a probe of frequency ω is

$$g^{\alpha\alpha^*}(\omega,T) = \sum_{n,m} \frac{(f_n - f_m)\; S^{\alpha}_{nm}\; S^{\alpha^*}_{mn}}{h\omega - E_n + E_m} \tag{3}$$

Here f_n and E_n are the population and frequency of the nth level and S^{α}_{nm} (α is +, - or z) is the matrix element between the n and m levels in the self-consistent molecular field. The single-ion response reduces at T=0 to a set of transitions from the ground state to excited states with strengths given by the square of the matrix element involved. The perturbation of the J=4 multiplet for Pr ions in the molecular field which obtains in $PrAl_2$, is shown in Fig. 1. In this field we would expect a strong transverse S^- transition to the third excited state, flanked by two weaker transverse S^+ and longitudinal S^z transitions, together with a very strong S^- transition to the fifth state. This scheme is in reasonable accord with what is observed at a zone boundary where off-diagonal mixing terms are weakest. The relative intensities of the three lowest dipole modes are not even approximately correct unless the molecular field exceeds = 0.5 THz (see the lower part of Fig. 1). This observation effectively puts a lower limit on the magnitude of the exchange interaction.

When the inter-ion parts of the exchange are included the full transverse and longitudinal response functions are given by

$$G^{+-}(\omega,q,T) = \frac{g^{+-}(\omega,T)}{1 - J(q)\, g^{+-}(\omega,T)} \tag{4}$$

$$G^{zz}(\omega,q,T) = \frac{g^{zz}(\omega,T)}{1 - 2\,J(q)\, g^{zz}(\omega,T)} \tag{5}$$

where J(q) is the Fourier transform of the exchange. The neutron inelastic scattering cross-section is proportional to the imaginary part of the response and the frequencies are given by the poles of the response functions (4) and (5).

The best least squares fit of this model to the experimental results for $PrAl_2$, excluding data in the neighbourhood of the interactions referred to above, is shown by the solid curves in Fig. 2 which gives a good overall description of the results. It is surprising that exchange interactions which are three times that required to induce magnetic order give rise to such flat dispersion relations for the three lowest dipole modes. However, as a general rule, there is appreciable dispersion of the corresponding mode only when the matrix element S^{α}_{nm} is large.

Discrepancies exist, however, between theory and experiment; the positions of the lowest three dipole modes are not exactly reproduced. In addition the theoretical relative strengths of the longitudinal S^z and transverse S^- modes at $\zeta = 0$ (optic zone) near 2 THz disagree with experiment. Apart from these detailed discrepancies, the strong perturbation of the intense mode between 3 and 4 THz poses the most serious difficulty in $PrAl_2$. If the perturbation is a magnon-phonon interaction the splitting is large compared with the 0.2 THz splittings observed for the lower branches. We are lead to the conclusion that the principal features of the results can be explained without difficulty, although the details which are not understood may indicate that other interactions must be included.

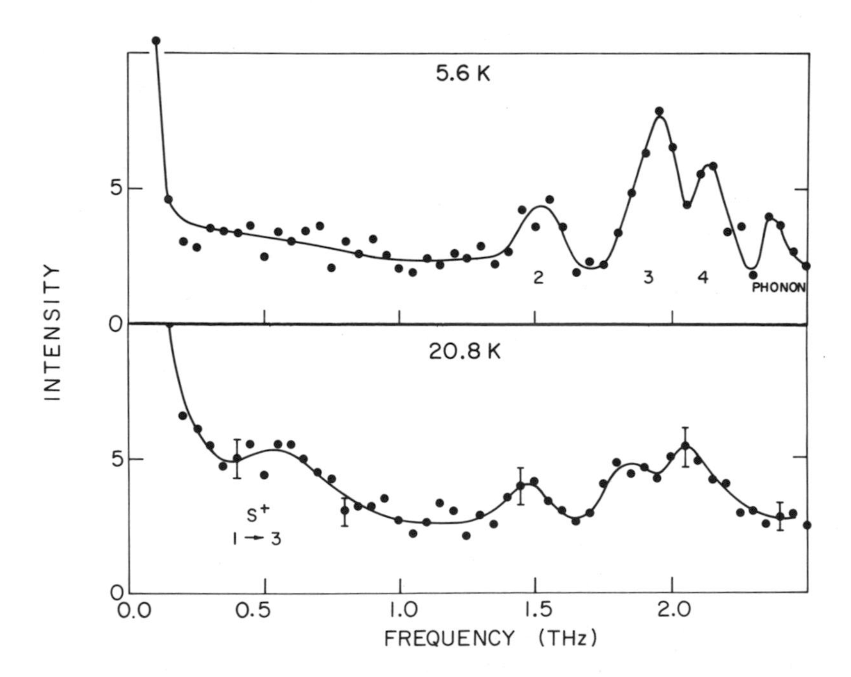

Fig. 3: Temperature dependence of the neutron scattering in $PrAl_2$ at a wavevector transfer $2\pi/a$ [2,2,0.7] showing the excited-state spin wave between the first and third excited states at 0.55 THz.

III TEMPERATURE EFFECTS

As the temperature is raised within the ordered phase of a crystal field system, the excited states of the manifold become populated and excited-state spin waves (Fig. 3) can occur, corresponding to transitions between excited states of the manifold. These were first observed directly in $TbAl_2$[7] and were also observed in $PrAl_2$ although the nature of the excited-state spin wave is different in the two cases. In $TbAl_2$ the frequency of the single-ion excited-state transition falls in the band of ground-state spin waves, and although the matrix element for the transition is low the measured intensity is appreciable due to the admixture of the ground state component. The situation is analogous to resonant-mode scattering. In $PrAl_2$ the excited-state spin wave is observed at 0.55 ± 0.10 THz and corresponds to a transition between the first and third excited states. This mode falls outside the band of ground-state spin waves and is observed through its intrinsic dipole strength. This case is analogous to local mode scattering.

Examination of equations (3)-(5) in the region close to the critical temperature for magnetic ordering[11] gives insight into the predominant modes of excitation near a second-order phase transition in crystal field systems. A second order magnetic phase transition corresponds to the divergence of the longitudinal susceptibility at the ordering wave-vector (q=0 for a ferromagnet). There are two components[12] to the longitudinal single-ion susceptibility g^{zz} (0,T) corresponding to Van Vleck contributions and long wavelength zero frequency modes. Within the random phase approximation the latter component is entirely elastic since it originates in terms in (3) in which m=n above T_c, and from static fluctuations in moment below T_c. It can be shown[11] that the contribution to the dynamical response and hence the neutron scattering from the inelastic Van Vleck terms does not diverge at T_c. Correspondingly the crystal field modes soften in frequency but do not fall to zero in general. In fact a central peak from the zero frequency component grows in intensity and finally diverges at T_c. This account of the phase transition was first given by Smith[13]. The RPA introduces no damping contributions from exchange fluctuations so that the picture of undamped elastic and inelastic modes emerging from the simple theory will only be accurate when the exchange is weak relative to the crystal field.

Experimental evidence for central modes near T_c in crystal field systems is only just becoming available. Preliminary results of Als-Nielsen, Kjems and Buyers (Proceedings of the International Conference in Magnetism, 1976, to be published) on $Pr_3T\ell$ exhibit the softening of the inelastic transitions to a finite frequency at T_c accompanied by an elastic peak which has a maximum at T_c. High resolution experiments (0.04 THz FWHM) on

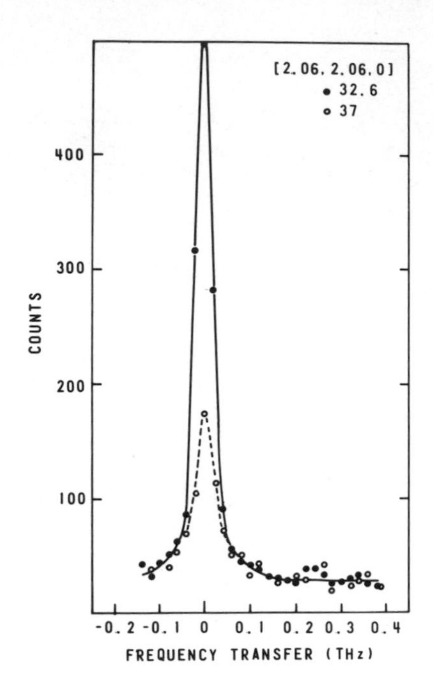

Fig. 4: Temperature dependence of the neutron scattering in $PrAl_2$ just above T_c close to the ordering wavevector for a ferromagnet showing predominantly elastic response.

$PrAl_2$ (Fig. 4) (Purwins, Holden, Buyers and Svensson to be published) show that the major component of the magnetic scattering just above T_c is strictly elastic as opposed to a rather broad distribution corresponding to exchange broadened crystal field levels. This also suggests that the divergent contribution comes from the elastic central component.

REFERENCES

1. H.-G. Purwins, W.J.L. Buyers, T.M. Holden and E.C. Svensson. AIP. Conference Proceedings No. 24, 259 (1976).

2. M.W. Hutchings, in Solid State Physics, edited by F. Sertz and D. Turnbull (Academic, New York, 1964), 16, 227.

3. P. Bak and P.A. Lingard, J. Phys. C 6, 3774 (1973).

4. A. Furrer, W.J.L. Buyers, R.M. Nicklow and O. Vogt, Phys. Rev. B. 14 (1976) to be published.

5. L.F. Uffer, P.M. Levy and H.H. Chen, AIP Conference Proceedings No. 10, 553 (1973).

6. P. Fulde, Z. Physik B 20, 89 (1975).

7. W. Buhrer, M. Godet, H.-G. Purwins and E. Walker. S.S. Comm. 13, 881 (1973). H.-G. Purwins, J.G. Houmann, P. Bak and E. Walker, Phys. Rev. Lett. 31, 1585 (1973).

8. T.M. Holden, E.C. Svensson, W.J.L. Buyers and O. Vogt, Phys. Rev. B10, 3864 (1974).

9. H.-G. Purwins, E. Walker, B. Barbara, M.F. Rossignol and P. Bak. J. Phys. C 7, 3573 (1974).

10. W.J.L. Buyers, T.M. Holden and A. Perreault, Phys. Rev. B. 11, 266 (1975).

11. W.J.L. Buyers, AIP. Conference Proceedings No. 24, 27 (1975).

12. M.W. Lines J. Phys. C. 7, L287 (1974).

13. S.R.P. Smith J. Phys. C 5, L157 (1972).

* Present Address: Laboratorium für Festkörperphysik
ETH-Hönggerberg
8093 Zürich, Switzerland

SPIN DYNAMICS OF A KONDO SYSTEM IN THE PRESENCE OF CRYSTAL FIELDS+

M. Loewenhaupt[x] and F. Steglich[*]

[x]Institut für Festkörperforschung der KFA Jülich
D-5170 Jülich
[*]II. Physikalisches Institut der Universität zu Köln
D-5000 Köln 41

ABSTRACT

Neutron scattering experiments were performed on single crystals of $(Ce_xLa_{1-x})Al_2$ with x varying between 0.07 and 0.70 in the temperature range from 3 to 300 K. The energy spectra of the magnetically scattered neutrons were analysed in terms of two quasielastic lines due to the relaxation of the isolated Ce^{3+} ions in the Γ^7 and Γ^8 levels and of an inelastic line due to the crystal field transition between the levels. The temperature dependence of the intensities and linewidths is discussed.

INTRODUCTION

Recent neutron scattering[1] and NMR[2] experiments show that the spin relaxation rates of isolated 3d-Kondo impurities exhibit a temperature dependence which deviates appreciably from a simple Korringa law, but which agrees well with the results of the Kondo theory[3]. On the other hand, in Kondo systems with 4f-impurities complications arise due to crystal field (CF) effects. From theoretical arguments[4] a Kondo temperature is expected which increases strongly with increasing degeneracy of the CF levels.

In this paper we report on neutron scattering experiments on the $(Ce_xLa_{1-x})Al_2$ system, which has been extensively studied by other experiments in the limits of small Ce-concentration ($x < 0.01$) and low temperature ($T < 10$ K)[5]. The large number of anomalies, which have been detected for this system, has been explained in

terms of:
(1) a Kondo effect of a low lying Γ^7 doublet of the Ce^{3+} ions with only weak admixture of an excited Γ^8 quartet and
(2) an extremely weak RKKY interaction between the magnetic ions. Due to the second fact one expects single impurity behaviour even for high Ce-concentrations. From the energy spectra we hoped to seperate two quasielastic and an inelastic line which will then allow to deduce the Kondo temperatures of the two CF-levels and the splitting energy, as well.

EXPERIMENT

$(Ce_xLa_{1-x})Al_2$ single crystals with x=o, o.o7, o.15, o.4o and o.7o were prepared using the Czochralski method (for details see Ref.[6]) with its axis parallel to the [1oo] direction.
The experiments were performed on the diffuse scattering apparatus D7 at the HFR Grenoble. The incident neutron energy was 3.53 meV. The energy of the scattered neutrons was analysed using a time-of-flight method. For calibration of the neutron intensity a vanadium reference scatterer was used. The experiments on the single crystals were performed in two different ways:
(1) the samples with the small Ce concentrations of only 7% and 15% were measured against the pure $LaAl_2$ single crystal under identical conditions; the magnetic scattering was then decuced from the difference of the TOF-spectra of the doped and the pure sample.
(2) the samples with the high Ce concentrations of 4o% and 7o% were measured against the normal background of the experimental set-up. We then subtracted the $LaAl_2$ spectra of (1) to obtain the magnetic scattering. To give an idea of the relative intensity and structure of the nonmagnetic contributions we show in Fig. 1 for T=3oo K the TOF-spectra of $Ce_{o,4}La_{o,6}Al_2$, and $LaAl_2$ and the difference of the two spectra. Nonmagnetic scattering is due to
(a) elastic incoherent scattering around channel 1o6 and
(b) inelastic coherent phonon scattering in the range of channel 35 to 55 at the energy gain side. The intensity of (b) however decreases strongly with decreasing temperature.

RESULTS AND ANALYSIS

Fig. 2 shows typical magnetic TOF-spectra of the crystal with 4o% Ce concentration at 2o K, 6o K and 2oo K which we obtained by subtracting the pure $LaAl_2$ spectra. At low temperatures ($T \sim 2oK$) there is only a narrow quasielastic line due to the slow relaxation of the Γ^7 ground state. The excited Γ^8 level is not yet populated. So it does not contribute to quasielastic scattering and also not to inelastic scattering at the energy gain side. With increasing temperature ($T \sim 6o$ K) the excited level gets populated which allows

the observation of an inelastic line due to the CF transition $\Gamma^8 \rightarrow \Gamma^7$ and an additional quasielastic line due to the relaxation of the excited level. For high temperatures (T$\sim$ 2oo K) the lines are extremely broadened which makes a separation of quasielastic and inelastic contributions nearly impossible.
To analyse the data we use the expression for the double differential cross section of a crystal field split ion given by [7], allowing for two modifications: (1) the interaction of the CF levels with the conduction electrons is taken into account by finite linewidths of the two quasielastic lines and the inelastic line, (2)

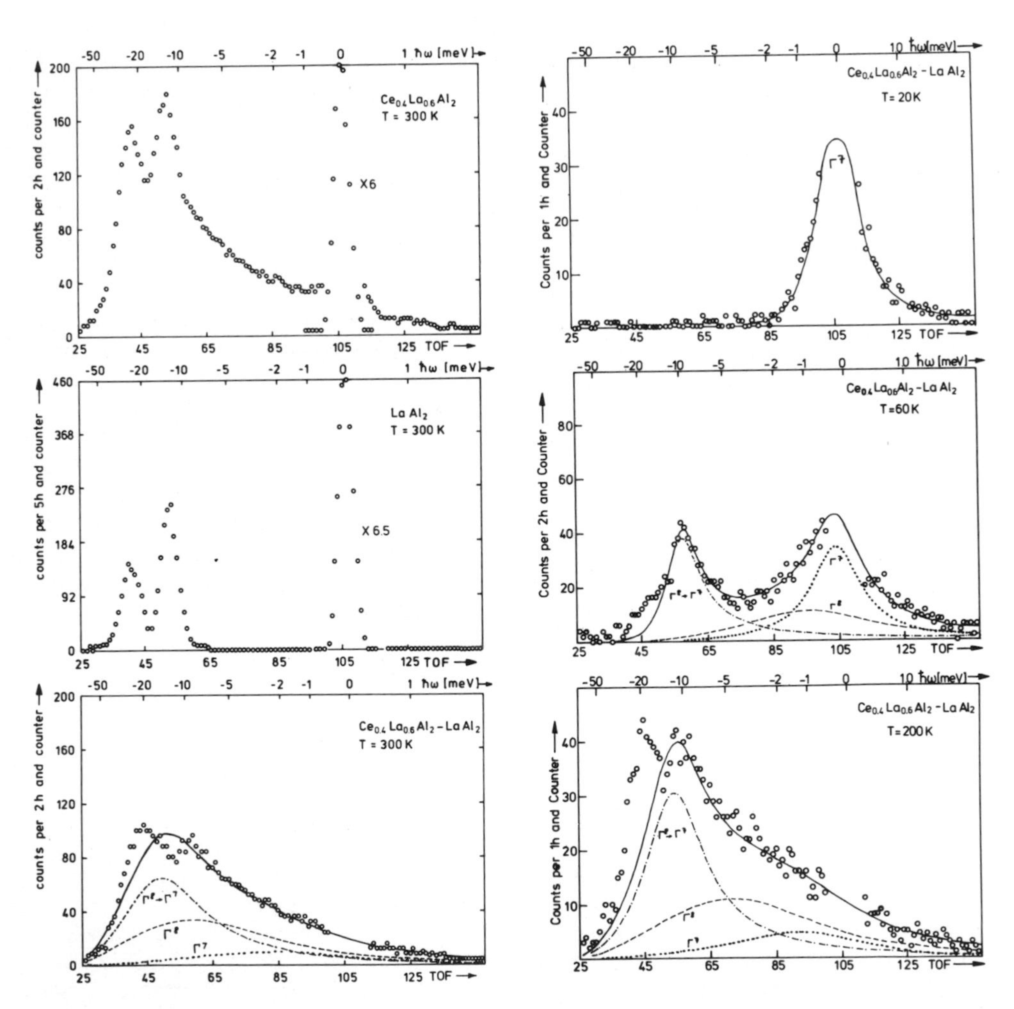

Fig. 1 TOF-spectra at T=3oo K

Fig. 2 Difference of TOF-spectra at T=2oK, 6oK and 2ooK.

the intensities are reduced according to the reduction of the measured static susceptibility[8]. Details of the data evaluation will be given elsewhere[9]. A reasonable fit to the data is obtained by using $\Delta = (9\pm1)$ meV for the crystal field splitting with a Γ^7 ground state and a Γ^8 excited state which was also observed in $CeAl_2$[10]. The different contributions (dashed lines) to the total magnetic scattering (full line) are indicated in Fig. 2. To deduce the different linewidths from the measured spectra we proceed as follows. If we try to fit the data with only one quasielastic line (same linewidth for Γ^7 and Γ^8) we get small values for the linewidth for temperatures up to 6o K, then a steep increase around 1oo K and correspondingly high values for the room temperature measurements. As the main contribution to the quasielastic scattering is due to the Γ^7 ground state at low temperatures and due to the Γ^8 excited state at high temperatures we attach the small linewidth to the Γ^7 state and the large linewidth to the Γ^8 state. With this consideration the analysis of all of our data with different linewidths for the quasielastic and inelastic lines yields the temperature dependence of the half linewidths as shown in Fig. 3. But we have to note that the results have not been obtained by a least-squares method and that the error bars are confined to our special type of data evaluation.

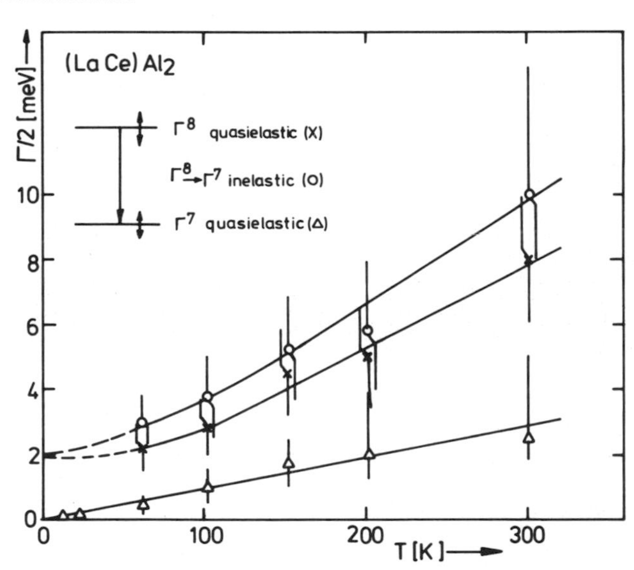

Fig. 3

Temperature dependence of the quasielastic and inelastic linewidths for the CF levels in $(\underline{La}Ce)Al_2$.

DISCUSSION

For the Γ^7 doublet ground state of the isolated Ce ions in $(Ce\underline{La})Al_2$ we observe a Korringa relation for the temperature dependence of its relaxation rate measured by its half linewidth of the quasielastic line (Fig.3). This was expected since the Kondo temperature of the $\Gamma 7$ level is rather small, T_K=o.36 K[5]. The slope of the experimental curve however, exceeds the "normal" Korringa slope (as calculated with N(o).J=o.o8 from other experiments[5]) approximately by a factor of 3. This increase in slope was also observed for Kondo systems with 3d-impurities[1,2]. For the $\Gamma 8$ quartet excited state we observe considerably higher relaxation rates than for the $\Gamma 7$ level and an intercept of approx. 2 meV for the half-linewidth if extrapolating to $T \rightarrow o$. This indicates a Kondo temperature of about 25 K for the excited level, two orders of magnitude larger than for the ground level but still smaller than the CF splitting. The temperature dependence of the linewidth of the inelastic line is in agreement with calculations of Fulde and Peschel[11].

We note that these experiments provide for the first time a direct proof that the spin relaxation rates increase rapidly with the degeneracy of the magnetic impurity state - in qualitative agreement with theoretical expectations[4].

We acknowledge stimulating discussions with Prof. W. Schmatz and the valuable help of Dr. W. Just during the neutron scattering experiments.

REFERENCES

+ Work performed within the research program of the Sonderforschungsbereich 125, Aachen-Jülich-Köln

1. M. Loewenhaupt, W. Just: Phys. Lett. 53A, 3o5 (1975)

2. H. Alloul: preprint

3. W. Götze, P. Schlottmann: J. Low Temp. Phys. 16, 87 (1974)

4. B. Cornut, B. Coqblin: Phys. Rev. B5, 4541 (1972)

5. F. Steglich: Z. Phys. B23, 331 (1976)

6. J.H. Moeser: Dissertation, Köln, unpublished (1976)

7. R.J. Birgeneau: J. Phys. Chem. Sol. 33, 59 (1972)

8. M.B. Maple: Appl. Phys. 9, 179 (1976)

9. M. Loewenhaupt, W. Schmatz, to be published

1o. E. Walker, H.-G. Purwins, M. Landolt, F. Hulliger: J. Less-Common Metals 33, 2o3 (1973)

11. P. Fulde, I. Peschel: Adv. Phys. 21, 1 (1972)

BEHAVIOUR OF CRYSTAL FIELD PARAMETERS IN ErM AND HoM EQUIATOMIC COMPOUNDS, IN RELATIONSHIP WITH BAND STRUCTURE

D. Schmitt, P. Morin and J. Pierre

Laboratoire de Magnétisme, C.N.R.S.,

166X, 38042-Grenoble-Cedex, France

ABSTRACT

New results on CEF parameters in holmium compounds are obtained by neutron spectroscopy. They give informations on the origin of CEF in intermetallics, and are used to interpret the magnetic properties.

INTRODUCTION

Among the intermetallic rare earth compounds, the equiatomic RM phases of CsCl cubic structure provide a simple model to study the variations of crystal field (CEF) parameters in isomorphous series and thus to understand the origin of crystal field in alloys:

1 - the 4f distribution of the rare earth ion influences the coulombic term, mostly through its radial expansion and the exchange term with the band in a much more complex manner,

2 - the CEF parameters vary with the valency and the electronegativity of the alloyed metal and the spatial distribution of conduction electrons in the solid.

We present in the following a review of the different parameters obtained on Ho and Er compounds by neutron spectroscopy and their consequences on physical properties such as anisotropy and magnetostriction.

EXPERIMENTAL RESULTS

Neutron spectroscopy experiments were performed on IN7 time of flight-spectrometer at the Institut Laue-Langevin at Grenoble, experimental details may be found in earlier papers[1] dealing with

Er compounds with Cu, Ag, Zn, Mg and Pd. We describe here some new results obtained on the isomorphous compounds with holmium. The rare earth was generally substituted by yttrium in order to reduce the magnetic interactions and to study the paramagnetic state at temperatures low enough to populate only a few levels. The results were analyzed within the classical formalism, starting with Lea, Leask and Wolf's Hamiltonian[2] :

$$\mathcal{H}_c = W\,x\,\frac{O_4}{F_4} + W(1-|x|)\,\frac{O_6}{F_6} = A_4\,\beta\langle r^4\rangle\,O_4 + A_6\,\gamma\langle r^6\rangle\,O_6$$

and the neutron cross section formula derived by Trammell[3] ; assuming a gaussian shape for the transitions. As examples, we present only the spectra concerning the Ho compounds with Cu and Zn due to the space lack.

$Ho_{0.25}Y_{.75}Cu$ compound

Spectra taken at 15 and 70 K are given in figure 1. The left side spectra were obtained with an incident energy E_o = 26.6 meV. At lower temperature we observe two transitions at 3.9 and 7.7 meV, whereas a third transition appears at 12.2 meV at higher temperatures.

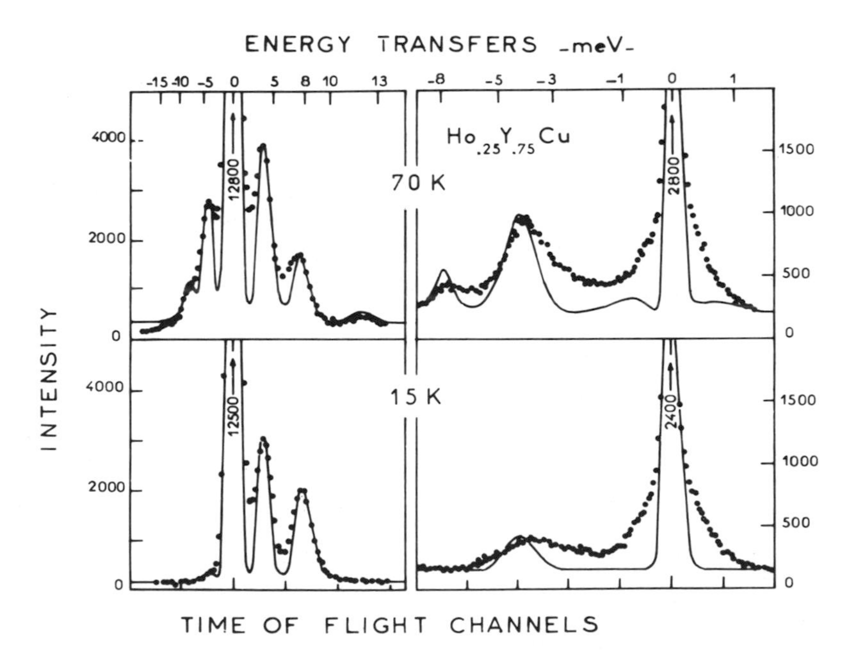

Fig. 1 : Neutron spectra for $Ho_{.25}Y_{.75}Cu$. The incident energy is 26.6 meV at left, and 6.65 meV at right. Continuous lines are fits with W = 0.40 K, x = 0.34.

The two first transitions are seen as well on the energy loss spectrum with E_o = 6.65 meV (right side of figure 1). From energy and intensity ratios, the CEF scheme is determined without ambiguity W = 0.40 ± 0.02 K (0.0345 meV) x = 0.34 ± 0.02, the ground level being the $\Gamma_5^{(1)}$ triplet.

$Ho_{.15}Y_{.85}Zn$

The spectrum taken at 15 K with E_o = 22.69 meV shows two transitions at 6.1 and 8.5 meV (figure 2), whereas some other energy transfers appear at 68 K. The best agreement is obtained with W = 0.36 ± 0.02 K, x = 0.08 ± 0.02, the fit at 68 K could be improved by taking larger transitions widths than assumed here (0.5 meV).

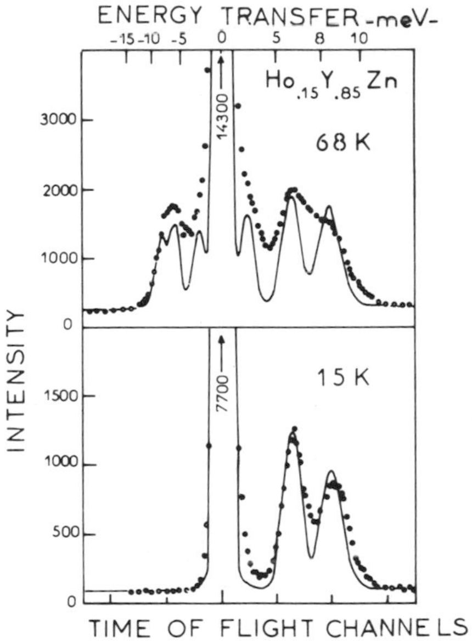

Fig. 2 : Neutron spectra for $Ho_{.15}Y_{.85}Zn$ (E_o = 22.69 meV). The fit is for W = 0.36 K, x=0.08.

All the results are summarized in Table I together with previous results on RRh compounds[4]. For $Ho_{.2}Y_{.8}Ag$, we obtained qualitatively the same results than Furrer[5]. For ErPd, the results lead to some ambiguity but only one possibility is coherent with the neighbouring compounds. For $Ho_{.2}Y_{.8}Mg$, only one composite transition has a large intensity ; one will see later how the magnetocrystalline anisotropy helps to choose the good solution. The variation of parameters in Ho and Er series is very similar, however, it exists a significant shift toward more positive values for the $A_4\langle r^4\rangle$ parameter in Ho compounds.

CONSEQUENCES ON MAGNETIC STRUCTURES AND ANISOTROPY

The precise knowledge of CEF parameters allows to predict the direction and anisotropy of the moment in the ordered state ; taking into account the Heisenberg exchange besides the one-ion crystal field Hamiltonian. However, we shall see that, in certain cases, other terms such as magnetoelastic coupling or anisotropic exchange should be introduced.

Table I : Lattice and crystal field parameters in Ho and Er compounds

HoM	a (Å)	$A_4<r^4>$ (K)	$A_6<r^6>$ (K)	ErM	a (Å)	$A_4<r^4>$ (K)	$A_6<r^6>$ (K)
Rh	3.377	-125	-18	Rh	3.361	-123	-18.8
				Pd	3.445	-117	-11.8
Cu	3.444	- 68	-15	Cu	3.430	- 84	-15.1
Ag	3.593	- 67	-12	Ag	3.574	- 82	-10.3
Zn	3.547	- 14	-18	Zn	3.531	- 36	-17.8
Mg	3.761	+ 42	-13	Mg	3.748	+ 3.6	-10.8

HoZn compound

Previous studies on this ferromagnet (T_C = 75 K) by magnetic measurements on single crystals and specific heat showed a moment rotation between twofold and threefold axes at T_R = 23 K[6]. That was interpreted including only Heisenberg exchange and led to a first estimation of CEF parameters. The actual values are in fact very different, and require a reinterpretation of the properties. Taking W = 0.36 K and x = 0.08, we derive a much too high rotation temperature ; the agreement is much better with addition of second order terms : i) a magnetoelastic term arising from a rhomboedral distortion actually observed : $\mathcal{H}_d = Wy(O_2^2 + 2\sqrt{2}\ O_2^1)$, with the quantization axis taken along a fourfold axis. ii) a biquadratic exchange which may be written with cubic symmetry as :

$$\mathcal{H}_b = K\left[<O_2^0>_T\ O_2^0 + 3<O_2^2>_T\ O_2^2\right]$$

Parameters as y = -0.04 and K = 6.5×10^{-4} K satisfactorily describe the behaviour with temperature. Such terms should also explain the behaviour of DyZn, where, in spite of cubic CEF parameters, the moment orders along a fourfold axis with a huge magnetostriction.

HoMg and ErMg compounds

These compounds order at 21 and 7 K respectively. Their magnetization has been studied on single crystals[7]. The magnetic order is not ferromagnetic, but due to low interactions, the moment is aligned in the high applied fields (150 kOe), and is strongly anisotropic : CEF parameters deduced from these data are in good agreement with neutron spectroscopy results.

ANALYSIS OF C.E.F. PARAMETERS

The crystal field is built from several contributions :

a - Effective charges borne by ligands were evaluated by augmented-plane wave method (APW). This contribution was found rather small ($|A_4\langle r^4\rangle| < 10$ K, $A_6\langle r^6\rangle \simeq -1$ K).

b - Coulombic terms arise from the anisotropic distribution of band electrons within central or ligand spheres. The large d character around the rare earth ion contributes to the 4th order term. Within APW model, it was found roughly constant and positive for Ag, Zn, Mg compounds ($A_4\langle r^4\rangle_c = +150$ K). The variation of $A_4\langle r^4\rangle$ seems closely related to the progressive localization of d electrons from Rh to Zn and their absence in Mg ; it follows rather well the electronegativity differences between elements, as previously shown on pnictides[8].

c - Covalency and overlap effect may also give a contribution, particularly to the 6th order term.

d - From the CEF parameters of RMg compounds, we may expect that anisotropic exchange terms are nearly opposite to coulombic ones, as for rare earths diluted in noble metals[9].

Considering the large magnitude of different terms, the difference of radial expansion $\langle r^4\rangle$ between holmium and erbium may account for a variation of 10 to 20 K for $A_4\langle r^4\rangle$.

We shall extend our determinations to other compounds in order to see the influence of anisotropic exchange terms. Elastic constant measurements[10] have shown the occurence of biquadratic exchange in Tm compounds.

REFERENCES

1. P. Morin, et al. Phys. Rev. B 9, 4932 (1974)
   P. Morin, J. Pierre, D. Schmitt, J. Physique, 37, 611 (1976)
2. K.R. Lea, J.M. Leask, W.P. Wolf, J. Phys. Chem. Solids, 23, 1381 (1962)
3. G.T. Trammell, Phys. Rev., 92, 1387 (1953)
4. R. Chamard-Bois et al., Solid. Stat. Commun., 13, 1549 (1973)
5. U. Tellenbach, A. Furrer, A.H. Millhouse, J. Phys. C, 8, 3833 (1975)
6. P. Morin, J. Pierre, J. Chaussy, Phys. Stat. Sol. a, 24, 245 (1974)
7. R. Aléonard, P. Morin, J. Pierre, D. Schmitt, J. Phys. F, 6, 1361(1976)
8. E. Bucher, J. Maita, Solid Stat. Comm., 13, 215 (1973)
9. H.C. Chow, Phys. Rev. B, 7, 3404 (1973)
10. P. Morin, A. Waintal, B. Lüthi, Phys. Rev. B, 13, xxx (1976)

CRYSTAL FIELD IN LIQUID RARE EARTH METALS

A.H. Millhouse* and A. Furrer

Institut für Reaktortechnik ETHZ, EIR
CH-5303 Würenlingen, Switzerland
*Permanent Address: Hahn-Meitner Institut für Kernforschung, D-1000 Berlin 39

ABSTRACT

Neutron inelastic scattering measurements have been performed on liquid neodymium and ytterbium. The results demonstrate that a well-defined crystal field of uniaxial symmetry exists in liquid Nd, while no crystal-field transitions have been observed in liquid Yb. The results for Nd and those previously reported for liquid Ce and Pr (1) are correlated in terms of reduced crystal-field parameters.

INTRODUCTION

A large number of experiments have now been performed which demonstrate the power of the neutron inelastic scattering technique in determining the splittings of 4f-electron ground-state multiplets in rare earth crystalline solids. The results of these experiments have demonstrated that in metallic systems, where conduction electron shielding occurs, only the first few neighbouring shells of ions are important in determining the strength of the crystalline electric field at a particular ionic site.

In liquid rare earth metals whose temperatures are not too far above the melting points it is expected that the local surrounding of an ion is not too much different from that in the solid state. The big difference between the liquid and the solid is that in the liquid the ions are not localized at a particular site but can move or diffuse through the sample. If the ions in the liquid remain localized for a time which is long compared to the transit time of a neutron in the scattering experiment, then one would expect that the instantaneous picture of the liquid seen by the neutron is very similar to that in an amorphous system.

Calculations have been carried out for an amorphous rare earth alloy to examine the local crystal field within clusters of ions of the type expected to exist in such a material (2). The calculations revealed a dominant and well defined uniaxial component with the local axes varying randomly from site to site. The crystal-field Hamiltonian is of the form

$$\mathcal{H}_{cf} = - D J_z^2 \quad , \qquad (1)$$

where D is the uniaxial crystal-field parameter and z varies randomly from ion to ion.

In previous publications (1) we presented neutron inelastic scattering results of measurements on liquid La, Ce, and Pr which demonstrate that the crystal field is well defined and that the crystal-field potential has uniaxial symmetry. These results have now been extended by performing measurements on liquid Nd and Yb.

EXPERIMENTAL RESULTS AND ANALYSIS

The data were obtained using the triple-axis spectrometer at the reactor Diorit in Würenlingen operating in the constant-Q mode, and both neutron energy gain and neutron energy loss processes have been observed. The incident neutron energy was fixed at 14.96 meV and the beam passed through a pyrolitic graphite filter. The samples were sealed in evacuated Ta cylindrical containers of 1 cm diameter and 4 cm length. Johnson-Matthey vacuum melted rare earths were used.

Energy spectra of neutrons scattered from liquid Yb at 900°C are shown in Fig. 1. The energy spectra vary smoothly as a function of energy transfer and are very similar to the energy spectra observed for "non-magnetic" La (1). There is no evidence for crystal-field transitions at all.

Energy spectra of neutrons scattered from liquid Nd at 1100°C are shown in Fig. 2. In addition to the elastic line five inelastic peaks are observed. The peaks are more pronounced on the energy gain side than on the energy loss side because of cross-section and resolution effects. By considering the Q-dependence of the scattering ($\hbar Q$: momentum transfer) we conclude that peak number 5 is not of magnetic origin, while the peaks 1,2,3, and 4 at 0.8 meV, 1.9 meV,

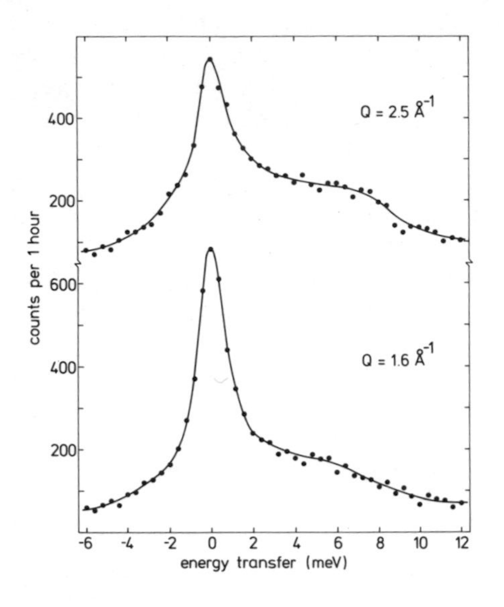

Fig. 1:
Energy spectra of neutrons scattered from liquid Ytterbium at 900°C. The solid line is drawn as a guide to the eye.

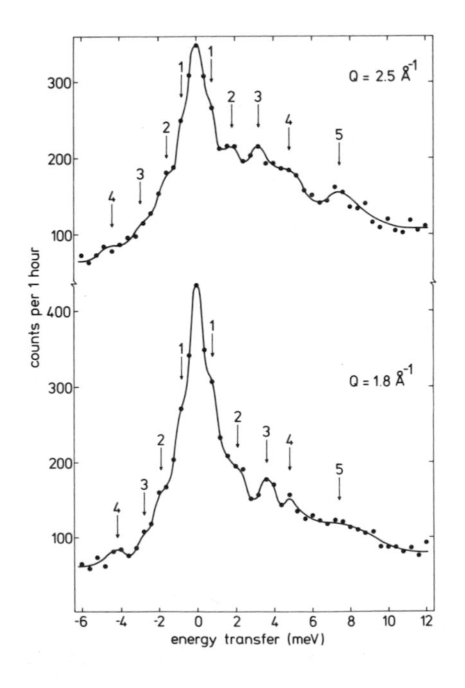

Fig. 2:
Energy spectra of neutrons scattered from liquid neodymium at 1100°C. The solid line is drawn as a guide to the eye.

3.2 meV, and 4.5 meV are interpreted as crystal-field transitions. Due to the fact that the half-widths are somewhat larger than the separation of the peaks, the lines are only partly resolved.

Under the action of the uniaxial potential given in Eq. (1) the $^4I_{9/2}$ ground-state multiplet of Nd^{3+} will split into five doublets having pure Kramers degenerate wavefunctions, $|\pm m\rangle$, $m = 1/2,\ldots,9/2$. The allowed transitions between these levels give rise to four inelastic peaks at energies 2D, 4D, 6D, and 8D, which is in good agreement with the observations. For the uniaxial crystal-field parameter the value $|D| = 0.51 \pm 0.05$ meV is obtained. Analysing the intensities of the crystal-field transitions is difficult because of the unknown non-magnetic background, but Fig. 2 suggests that the intensities are gradually decreasing from peak 1 to peak 4 which is also in agreement with the predictions of the uniaxial model.

DISCUSSION

Our measurements on Nd and our previous results for Ce and Pr demonstrate that the crystal field is well defined in liquid rare earth metals, supporting the liquid structure model in which an ion sits at a site in a cluster for a time long compared to the transit time of the interacting neutron, and then jumps to another site (jump diffusion model). The clusters are of such nature that the electrostatic field produced has a dominant and well defined uniaxial component.

The lack of crystal-field transitions for Yb is interpreted as due to the Yb ions being in the 2+ state. This is in agreement with the analysis of a recent neutron diffraction study of the structure factor for liquid Yb (3).

A comparison of the uniaxial crystal-field parameters for Ce, Pr, and Nd can be made by calculating the reduced crystal-field parameters

$$D' = \frac{DR^3}{\langle r^2\rangle \chi_2} \quad , \qquad (2)$$

where χ_2 is a reduced matrix element (4), and R should be chosen as the position of the first maximum in the radial distribution function (3,5) which is approximately R=3.5 Å. For $\langle r^2\rangle$ we used the non-relativistic radial integrals tabulated by Freeman and Watson (6). The results are listed in Table I. The reduced crystal-field parameters D' turn out to be significantly different for each rare earth

Table I: Uniaxial crystal-field parameters D and reduced crystal-field parameters D' of liquid rare earth metals for which experimental information on the crystal-field splitting is available.

Rare earth	D (meV)	D' (10^{-4} meV cm)	Reference
Ce	1.74 ± 0.08	0.39 ± 0.02	(1)
Pr	1.1 ± 0.1	0.74 ± 0.07	(1)
Nd	0.51 ± 0.05	1.2 ± 0.1	this work

element which means that the effective charge distribution in liquid rare earth metals is dependent on the particular rare earth element.

REFERENCES

(1) A.H. Millhouse and A. Furrer, Phys. Rev. Lett. 35, 1231 (1975): A.H. Millhouse and A. Furrer, Magnetism and Magnetic Materials - 1975, AIP Conference Proceedings, Vol. 29 (New York, 1976), p.257

(2) D. Sarkar, R. Segnan, E.K. Cornell, E. Callen, R. Harris, M. Plischke, and M.J. Zuckermann, Phys. Rev. Lett. 32, 542 (1974); R.W. Cochrane, R. Harris, and M. Plischke, J. Non-cryst. Solids 15, 239 (1974)

(3) H. Rudin, P. Fischer, and G. Meier, Institut für Reaktortechnik ETHZ, Report No. AF-SSP-95 (1976), p. 41

(4) K.W.H. Stevens, Proc. Phys. Soc. A65, 209 (1952)

(5) R. Bellissent and G. Tourand, J. Physique 36, 97 (1975)

(6) R.J. Freeman and R.E. Watson, Phys. Rev. 127, 2058 (1962)

CRYSTAL FIELDS IN ACTINIDE INTERMETALLICS - PROSPECTS AND PROBLEMS*

G. H. Lander

Argonne National Laboratory

Argonne, Illinois 60439, U. S. A.

ABSTRACT

In spite of the complications found in actinide (5*f*) systems, the results of optical spectroscopy on ionic 5*f* materials can be interpreted in terms of well-localized crystal-field states. From these studies we anticipate the electronic structure of lanthanide and actinide systems to be similar. Surprisingly, therefore, neutron inelastic experiments performed to date on metallic actinide systems have found no evidence for well-localized states. These experiments are discussed. On the other hand, neutron elastic experiments to measure the magnetization density show, particularly in the case of USb, that the ground-state free-ion wave functions are strongly modified by the crystal field. We suggest that the absence of discrete crystal-field excitations may be a result of strong interaction between the 5*f* and 6*d* electrons.

I. INTRODUCTION

For ionic materials containing either 4*f* or 5*f* electrons, optical spectroscopy has for many years provided detailed information about the quantum energy levels of the systems. Although the analysis of actinide (5*f*) spectra is complicated by the presence of roughly comparable Coulomb, spin-orbit, and crystal-field interactions, computational methods for unravelling the spectra are

*Work supported by the U. S. Energy Research and Development Administration.

available.[1] Because of the extended nature of the 5*f* electrons, the crystal-field interactions are about a factor of 5 greater in the actinides than in the corresponding lanthanide systems.

Of course, for metallic materials photon spectroscopy is not applicable, but the technique of neutron inelastic scattering has proved most successful in characterizing the crystal-field parameters in a host of lanthanide systems--as the Proceedings of this Conference so aptly demonstrate. It seems reasonable to expect, therefore, that neutron inelastic scattering will be as useful in investigating actinides, provided that neutron spectroscopy can cover a sufficient energy range. The importance of determining these parameters directly, apart from their intrinsic interest, is that with them we can build a model to interpret other physical measurements, such as magnetization, resistivity, and specific heat. Up to the present time, however, <u>no crystal-field transitions have been observed in any actinide compound</u>. This presents a major puzzle that is not understood. One of the purposes of this paper is to tabulate the experiments so far (Sec. II) and suggest possible reasons for their failure (Sec. IV). In Sec. III we describe neutron <u>elastic</u> experiments that suggest the existence of well-defined crystal-field states, at least in the limit of a long time average.

II. NEUTRON INELASTIC SCATTERING

The experiments performed to date on actinide systems are given in Table I. With the possible exception of UPd_4, the magnetic scattering in these systems is either absent or spread over a wide range of energy transfer. A number of remarks can be made about Table I.

1. Except for $Np_{0.02}Y_{0.98}$, all experiments have been performed with uranium compounds. Since the inelastic cross sections are small, large amounts of material are needed (usually at least 20 g of polycrystalline sample) and, even apart from the handling problems associated with transuranium isotopes, this seems unlikely to change in the near future. However, by analogy with cerium (the first lanthanide element with 4*f* electrons), the uranium systems may prove more difficult to understand than the heavier actinides, in which trivalent localized 5*f* behavior is anticipated.
2. Exchange effects are known to cause the crystal-field states to broaden in energy. Experiments on materials that order magnetically have been performed above T_N (or T_C), except for the triple-axis experiments on single crystals of UN and USb. However, even in dilute alloys,[5,6] or in materials such as USn_3, which does not order at all, no sharp excitations have been seen. We assume from these comparisons that, although exchange broadening may be important, it does not play the central role in suppressing the crystal-field transitions.

3. The failure to find evidence for spin-wave excitations below T_N in either UN or USb constitutes a major mystery. The crystal size of USb was rather small (0.04 cm^3) and intensity may be a limiting factor here. However, in the case of UN the crystal volume was 1.5 cm^3 and the complete phonon spectrum, including the optical mode at ∿50 meV were readily seen with a triple-axis spectrometer.
4. Finally, we note that the "failures" appear to be equally spread over various neutron laboratories and, since the authors are unsure of the significance of their null result, most of the work remains unpublished!

Table I: Brief description of experiments performed on actinide systems to search for crystal-field excitations with neutron inelastic scattering. The samples were polycrystalline unless otherwise noted.

Compound	Magnetic Properties	Remarks	Ref.	Laboratory
UC	Nonmagnetic	No visible magnetic	2	AERE (UK)
UN	AFI T_N = 53 K	No visible magnetic	2	AERE (UK)
UP	AFI T_N = 125 K	Broad magnetic	2	AERE (UK)
US	F T_C = 178 K	Broad magnetic	2	AERE (UK)
USe	F T_C = 160 K	Some (?) magnetic	2	AERE (UK)
USn_3	No ordering	No visible magnetic	3	ANL (USA)
UN	AFI T_N = 53 K	Single crystal (∿1.5 cm^3), no magnetic excitations seen with 3-axis.	4	AECL (CAN.)
$U_{0.15}Th_{0.85}Se$	No ordering	No visible magnetic	5	ILL (FR.)
UTe	F T_C = 104 K	Broad magnetic (?) at ∿45 meV	5	ILL (FR.)
$Np_{0.02}Y_{0.98}$	No ordering	No visible magnetic	6	BNL (USA)
UPd_4	AF T_N = 30 K	Possible transition at 5 meV	7	EIR (SWITZ.)
USb	AFI T_N = 241 K	Single crystal (0.04 cm^3). No magnetic excitations seen with 3-axis	8	ANL (USA) and AECL (CAN.)

III. NEUTRON ELASTIC MAGNETIC SCATTERING

A. General

An important physical consequence of the electrostatic potential arising from the neighboring ions is that certain directions become energetically favorable for the electron orbits at the central ion. In the absence of a crystal-field potential the electron orbits are governed by the Coulomb correlation energy and the way in which $\vec{L}$ and $\vec{S}$ couple together to give a resultant $\vec{J}$. The shape of the charge and magnetization densities is therefore directly related to the ground-state wave functions in the solid. Elastic magnetic scattering presents a method for determining the magnetization density, and hence the ground state. In general, these experiments are more tedious than straightforward crystal-field spectroscopy. First, they require single crystals. Second, because the dominant component of the magnetization density is spherical, accurate measurements are needed. The measurement of aspherical densities has led to the identification of the crystal-field ground state in a number of transition metal systems.[9]

The reduction of experimental data is relatively easy with *d* electron compounds, in which the orbital angular momentum is quenched. The Fourier transform of the experimental data then represents the magnetization density directly. However, if a large orbital moment is present, as is the case for most *f* electron systems, then the magnetization $\vec{M}(\vec{r})$ is a vector quantity. The orbital current density essentially contributes a term which is perpendicular to $\vec{J}$ in the magnetization density. In general, therefore, the experimental data (by necessity a scalar quantity) cannot be Fourier inverted to give the vector $\vec{M}(\vec{r})$. Under certain conditions the only component measured may be $M_z(\vec{r})$, which may then be Fourier inverted. Unfortunately, such an experimental arrangement may also prevent the observation of the maximum asphericity in the system.[10]

The neutron elastic magnetic cross section is sensitive to both the angular (discussed above) and radial part of the 5*f* wave function. The radial part is of interest because it is related to the extent of localization of the *f* electrons. However, we shall not discuss this aspect here, but simply state that relativistic calculations of the single electron radial distribution function are in good agreement with experimental data,[11] and that evidence for spatial delocalization of the 5*f* electrons has been found in certain neptunium compounds.[12]

Three neutron experiments have been performed on 5*f* electron compounds to determine the magnetization densities. Wedgwood[13] measured the magnetic cross section from the ferromagnet US, and we have recently measured both the paramagnetic[14] and antiferromagnetic[15] cross section from UO_2. The magnetization density in US exhibits rather little asphericity, so that a precise identification of the crystal-field ground state is difficult. We certainly can exclude a number of configurations,[11] but a band structure approach may be more appropriate for US. For UO_2 a rather novel internal distortion was found.[16] This effect, together with the complex domain configuration, precludes an accurate determination of the aspherical magnetization density. In contrast, the neutron cross section from USb exhibits considerable anisotropy,[17] suggesting that a unique identification of the ground state may be possible in this case.

B. Theory

To calculate the elastic neutron cross section we use the tensor-operator method.[18] The magnetic scattering is defined as a vector $\vec{E}$ with spherical components E_Q given by

$$(2\pi\hbar/m)\ E_Q = \langle\psi_e | T_Q^K\ (e,\vec{\kappa}) | \psi_e\rangle \tag{1}$$

where the electron wave functions are presented by ψ_e, and $T_Q^K(e,\vec{\kappa})$ defines the neutron-electron interaction through a rank one tensor. The electron wave function is constructed in the $|SLJM\rangle$ basis and matrix elements between these and the spin and orbital operators are determined with the aid of standard angular momentum coupling techniques. The results may be expressed as coefficients of terms $Y_{Q''}^{K''}(\Theta,\Phi)\ \langle j_i(\kappa)\rangle$. The spherical harmonics $Y_{Q''}^{K''}$ are functions of the angle Θ between the moment direction $\vec{\mu}$ (or $\vec{J}$) and the scattering vector $\vec{\kappa}$ ($|\vec{\kappa}| = \kappa = 4\pi\sin\theta/\lambda$), and the angle Φ between the scattering vector $\vec{\kappa}$ and a fixed direction in the plane perpendicular to $\vec{\mu}$. The functions $\langle j_i(\kappa)\rangle$ are smoothly varying functions of κ and represent the radial distribution of the $5f$ electrons.[11] Since K'' and Q'' can have even values up to 8, and i have values up to 6, a large number of coefficients are involved.[14] However, both the functions $\langle j_4\rangle$ and $\langle j_6\rangle$ and their coefficients are small, and the most sensitive region experimentally involves $\langle j_2\rangle$. (Recall that $\langle j_0\rangle$ gives the spherical component of the density distribution.) Using Eq. (1) we can construct an effective form factor $f'(\vec{\kappa})$ such that

$$f'(\vec{\kappa}) = \langle j_0\rangle + c_2\langle j_2\rangle + c_4\langle j_4\rangle + c_6\langle j_6\rangle \tag{2}$$

where the c_i coefficients are functions of Θ and Φ, but not of κ. Specifically the problem reduces to one of considering the coefficient $c_2(\Theta,\Phi)$. Furthermore, if $\vec{\mu}\,||\,[001]$, then c_2 is independent[19] of the angle Φ. We have, therefore, a relatively simple way of sensing the quadrupole shape profile, provided that the experimental data can be measured with sufficient accuracy. In practice, the simplest procedure is to reduce both the experimental and theoretical cross sections to an effective form factor and examine these as a function of κ. To focus on the ashpericity alone, we consider pairs of reflections that occur at the same κ. This method eliminates $<j_0>$ and, assuming c_4 is small,

$$\Delta f = f'(\vec{\kappa}_1) - f'(\vec{\kappa}_2) \approx \{c_2(\Theta_1) - c_2(\Theta_2)\} <j_2(\kappa)> \tag{3}$$

where $|\vec{\kappa}_1| = |\vec{\kappa}_2| = \kappa$, and $\Theta_1 < \Theta_2$. The two different shapes, oblate and prolate, are illustrated in Fig. 1, together with their respective values of Δf. Note that we have drawn the shapes axially symmetric, whereas for $\vec{J}||[001]$ only four-fold symmetry will be present about the [001] axis.

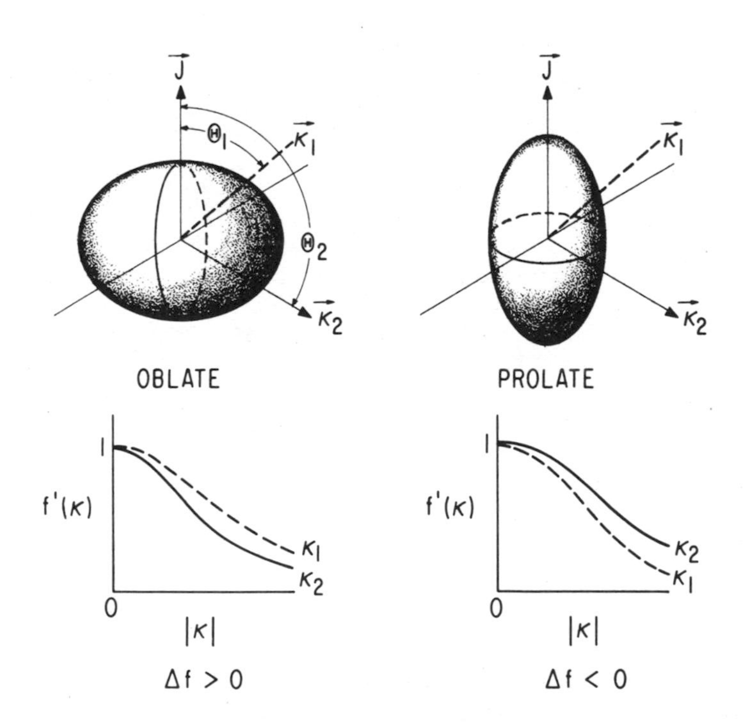

Fig. 1: Schematic representation of magnetization densities and magnetic form factors for oblate and prolate densities. Δf is defined in Eq. (3).

C. Experimental Analysis for Uranium Antimonide

The UX compounds, all with the NaCl crystal structure, with X = N, P, As, Sb, S, Se, and Te have been extensively studied over the last decade.[20] The compound USb (a_o = 6.197 Å) orders antiferromagnetically at 241.2 K with the type I magnetic structure, in which ferromagnetic (001) planes are stacked in the alternating sequence + - + -. The moments are aligned parallel to [001]. This simple magnetic structure, together with the large ordered moment of 2.8 μ_B, allowed accurate integrated intensities to be obtained. If uranium is trivalent in USb, the ionic configuration is f^3 : $^4I_{9/2}$, assuming that pure Russell-Saunders coupling applies. The crystal-field states are then two quartets $\Gamma_8^{(1)}$ and $\Gamma_8^{(2)}$ and a Kramers doublet Γ_6. With an exchange field applied, the ground-state wave

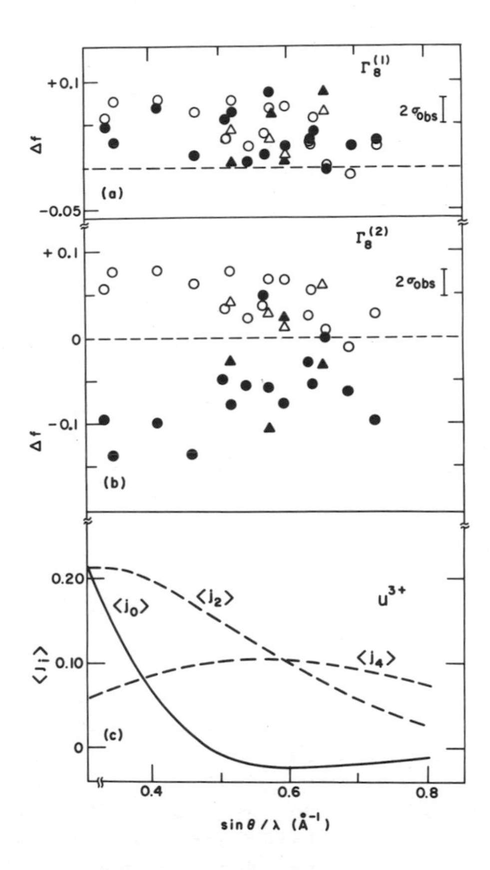

Fig. 2: Anisotropy in the form factor of USb. Δf is defined in Eq. (3). The open points are experimental values. The calculated values for $\Gamma_8^{(1)}$ and $\Gamma_8^{(2)}$ are shown in (a) and (b), respectively. The radial integrals $\langle j_i \rangle$ are plotted for this region of sin θ/λ in (c).

functions for the two quartets are

$$
\begin{aligned}
\Gamma_8^{(1)} \quad \psi_e &= 0.97|7/2\rangle - 0.25|-1/2\rangle - 0.01|-9/2\rangle \\
\mu_{sat} &= 2.36\ \mu_B \\
\Gamma_8^{(2)} \quad \psi_e &= 0.79|9/2\rangle - 0.59|1/2\rangle - 0.14|-7/2\rangle \\
\mu_{sat} &= 2.12\ \mu_B\ .
\end{aligned}
\tag{4}
$$

In NdSb, the analogous $4f^3$ compound, the $\Gamma_8^{(2)}$ wave function is the ground state. The quadrupole shape components are quite different for the $\Gamma_8^{(1)}$ and $\Gamma_8^{(2)}$ states, as is illustrated by a comparison between experimental and theoretical values of Δf in Fig. 2. Clearly, the $\Gamma_8^{(1)}$ state, which is oblate in shape, is a much better fit to the experimental results from USb. As Eq. (4) shows the dominant $|M\rangle$ component is $|7/2\rangle$, rather than the $|9/2\rangle$ of the free-ion and $\Gamma_8^{(2)}$ state. This is a surprising result, anticipated neither by point-charge estimates nor by comparison with the lanthanide series. The change in sign (as compared to the $4f$ compounds) for the crystal-field parameters V_4 and V_6 probably reflects the interaction between the $5f$ and $6d$ (and possibly also the $7s$) electrons.

IV. DISCUSSION

At the present time our understanding of the actinides is still at a primitive level, especially as compared to that of the lanthanide series. One problem, to which we referred earlier, is that the unavailability of transuranium isotopes has meant that the vast majority of work has been performed on compounds of uranium. For the transplutonium elements both theory and the limited experimental data available suggest that trivalent localized $5f$ electron behavior occurs.[21] Uranium is the analogue of cerium or neodymium, and the behavior is correspondingly more complex. For the light actinides the $5f$ wave functions are sufficiently extended in real space to be sensitive to the nearest neighbor atoms. In addition, they interact, and probably hybridize with, the $6d$ electrons.[22] This situation does not occur to any appreciable extent for the $4f$ electrons, since they are sufficiently well localized to be insensitive to their environment. It is this important difference between the $5f$ and $4f$ electrons that probably prevents the observation of localized excitations. A simple conceptual way to view this interaction is that it prevents the time correlation between two localized f electron states, either on the same ion (crystal-field excitation) or between neighboring ions in the ordered state (spin waves). Such a conceptual model at least has the virtue that it allows one to reconcile the <u>presence</u> of strong crystal-field effects in the elastic scattering (and other measurements of static or long-time average properties) and the <u>absence</u> of such effects in experiments that investigate the dynamical response of the system.

At least two other possible explanations exist. First, these compounds may exhibit interconfiguration fluctuations. In the absence of direct evidence, for example from X-ray photoemission experiments, this idea remains speculative, although such effects no doubt will be found in some actinide systems.[23,24] Indirect evidence for more than one valence state per actinide species has recently been inferred from neutron experiments[25] on $AmFe_2$.

Second, we ask whether the crystal-field splittings are simply too large in energy to be observed with neutron spectroscopy? The term $\langle r^4 \rangle / R^5$ in the crystal-field potential is about a factor of five larger in the actinides than in the lanthanides, so that excitations should still be visible with neutron spectroscopy. Of course, if the crystal-field potential is very large the experiments become difficult. For example, we are presently planning an experiment at Argonne National Laboratory to search for the $\Gamma_5 - \Gamma_3$ transition in UO_2, which according to calculations,[26] has an energy of $\sim$180 meV. In the neutron experiments the momentum transfer must be as small as possible to minimize the intensity loss associated with the decrease in the magnetic form factor. To conserve both energy and momentum we must use high-energy incident neutrons. The present proposal is to use an incident beam of energy $\sim$400 meV, a scattering angle of $\sim 10°$, and a momentum transfer of ~ 3.9 Å^{-1}.

In conclusion, much work needs to be done on metallic actinide systems before the true nature (or absence) of crystal-field and spin-wave excitations is understood. As with so many properties of this fascinating series, the behavior tends to be intermediate between that of the essentially itinerant behavior of the *3d* series and the localized electron behavior of the *4f* series; thus presenting a formidable challenge to the theorists. The purpose of this paper has been to give a progress report (very little) and to whet the appetites of both experimentalists and theorists involved in considerations of excitation spectra.

ACKNOWLEDGMENTS

I would like to thank T. M. Holden and W. J.L. Buyers for their advice and efforts with the UN and USb experiments. I am indebted to B. R. Cooper for a number of discussions, and to J. F. Reddy and O. Vogt for sample preparation. I am particularly grateful to the ETH, Zurich, and to the National Science Foundation for support to attend the Conference.

REFERENCES

1. B. G. Wybourne, *"Spectroscopic Properties of Rare Earths"*, Interscience, N. Y., 1965, p. 198 et. seq.

2. F. A. Wedgwood, J. Phys. C 7, 3203 (1974).
3. G. H. Lander, T. O. Brun, B. W. Veal, and D. J. Lam, *"Proceedings of the Montreal Conference on Crystalline Electric Fields in Metals and Alloys"*, (Univ. of Montreal, 1975, ed. R.A.B. Devine) p. 480.
4. W. J. L. Buyers, T. M. Holden, E. C. Svensson, and G. H. Lander, Abstract for Inter'l. Magnetism Conf., Amsterdam, Sept. 1976.
5. C. H. DeNovion, J. L. Boutard, F. A. Wedgwood, and A. Murani, Inst. Laue-Langevin, Annual Report, Dec. 1975, p. 195.
6. S. Shapiro, G. H. Lander, and P. Bak, March 1976 (unpublished).
7. A. Furrer and A. Murasik, Progress Report, EIR, Wurenlingen, Dec. 1975, p. 32.
8. T. M. Holden, G. H. Lander, and S. K. Sinha, (May 1976 unpublished).
9. W. Marshall and S. W. Lovesey, *"Theory of Thermal Neutron Scattering"* (Oxford U.P., London, 1971), Ch. 7.
10. O. Steinsvoll, G. Shirane, R. Nathans, M. Blume, H. A. Alperin, and S. J. Pickart, Phys. Rev. 161, 499 (1967).
11. A. J. Freeman, J. P. Desclaux, G. H. Lander, and J. Faber, Jr., Phys Rev. B 13, 1168 (1976).
12. B. D. Dunlap and G. H. Lander, Phys. Rev. Letters 33, 1046 (1974); A. T. Aldred, B. D. Dunlap, and G. H. Lander, Phys. Rev. B 14, (1976).
13. F. A. Wedgwood, J. Phys. C 5, 2427 (1972).
14. G. H. Lander, J. Faber, Jr., A. J. Freeman, and J. P. Desclaux, Phys. Rev. B 13, 1177 (1976).
15. J. Faber Jr., and G. H. Lander, Phys. Rev. B 14, (1976).
16. J. Faber Jr., G. H. Lander, and B. R. Cooper, Phys. Rev. Letters 35, 1770 (1975).
17. G. H. Lander, M. H. Mueller, D. M. Sparlin, and O. Vogt, Phys Rev. (to be published).
18. See Ref. 9, Chapter 6.
19. T. O. Brun and G. H. Lander, Phys. Rev. Letters 29, 1172 (1972); G. H. Lander, T. O. Brun, and O. Vogt, Phys. Rev. B 7, 1988 (1973).
20. G. H. Lander, AIP Conf. Proc. 29, 311 (1976).
21. *"The Actinides: Electronic Structure and Related Properties"*, ed. by A. J. Freeman and J. B. Darby (Acad. Press. N.Y. 1974), Vol. I, Ch. 1 and 2, Vol. II, Ch. 6.
22. A. J. Arko, M. B. Brodsky, G. W. Crabtree, D. Karim, D. D. Koelling, L. R. Windmiller, and J. B. Ketterson, Phys. Rev. B 12, 4102 (1975).
23. J. M. Robinson, AIP Conf. Proc. 29, 319 (1976), ibid. 34, (to be published 1976).
24. H. J. van Daal, K. H. J. Buschow, P. B. Van Aken, M. H. van Maaren, Phys. Rev. Letters 34, 1457 (1975).
25. G. H. Lander, A. T. Aldred, B. D. Dunlap, and G. K. Shenoy, Abstract for Inter'l. Magnetism Conf., Amsterdam, Sept. 1976.
26. H. U. Rahman and W. A. Runciman, J. Phys. Chem. Solids 27, 1833, (1966).

ENHANCED COULOMB AND EXCHANGE INTERACTION CRYSTAL FIELDS IN THE MODIFIED ZENER MODEL OF MAGNONS IN METALLIC FERROMAGNETS

Edward Siegel*

Energy Laboratory
Public Service Electric and Gas Company
Maplewood, New Jersey 07040

ABSTRACT

The Bartel[1] treatment of magnons in metallic ferromagnets within the modified Zener model of itinerant-localized (s-electron or hole) exchange is generalized to include ten-fold d-electron (hole) spin-orbital degeneracy. As such one can ab. initio include enhanced intra-cell Coulomb repulsive crystal fields and exchange attractive crystal fields. This treatment includes itinerant-itinerant (d-electron or hole) Coulomb and exchange interactions as well as itinerant-localized (d-electron or hole) interaction; Hund's rule coupling is included automatically. Also ab. initio the ferromagnet specificity is included via the d-bandfilling and d-bandwidth inherent in the degnerate theory; the possibility of finite temperatures is also included. Acoustic and optic ferromagnon branches are recalculated and corrected and the position of the Stoner particle-hole continuum boundary is recalculated at zero and finite temperatures.

****

Bartel[1] has recently treated the modified Zener model[2](MZM) for ferromagnetic transition metals following the original suggestion of Herring[3]. Previously Arai and Parrinello[4] and Bartel[5] had developed the MZM. Briefly the MZM is described as follows - a lattice of metal atoms, some containing x d-electrons and some containing x+1 d-electrons. Bartel stressed that the existence of an acoustic ferromagnon branch <u>does not</u> depend upon the coupling of the itinerant d-electrons to the localized d-electron spins. Acoustic ferromagnons exist in the MZM in the J=0 or $\bar{S}$=0 limit ie. in the pure Hartree

* Present Address: Molecular Energy Research Institute, Westwood, N.J.

model limit with no Fock type exchange term present! This statement is strictly true. We shall see that, in the Siegel-Kemeny TDHM[12] it is also strictly true. The extra electron is treated as itinerant and couples to the localized spin situated on a lattice site via Hund's rule. At each site the x d-electrons are coupled according to Hund's rule to yield the localized spin of maximum multiplicity.

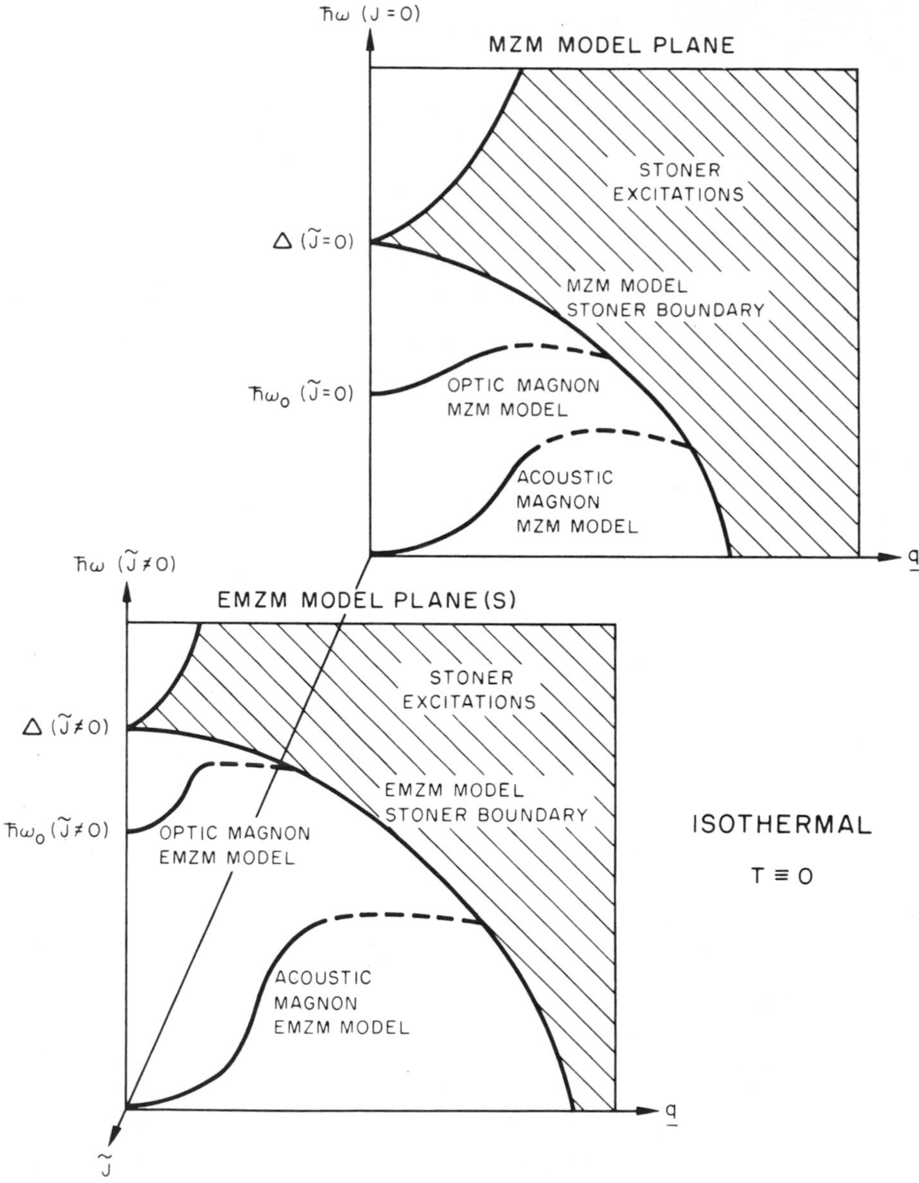

Figure 1. Evolution of acoustic and optic ferromagnons from MZM to EMZM.

We study the ferromagnon dispersion relations at frequencies and and wavevectors large enough to the onset of the electron-hole pair Stoner continuum at temperatures near the Curie temperature. We perform a model calculation of the imaginary part of the transverse dynamic susceptibility $\mathrm{Im}\chi^{+-}(q,\omega)$, related to the transverse inelastic neutron scattering cross section. Utilized in this analysis are the relation to the transverse spin Green's function $G_s(\underline{q},\omega)$ and the decoupling procedure in the R.P.A. as developed by Bartel[5],

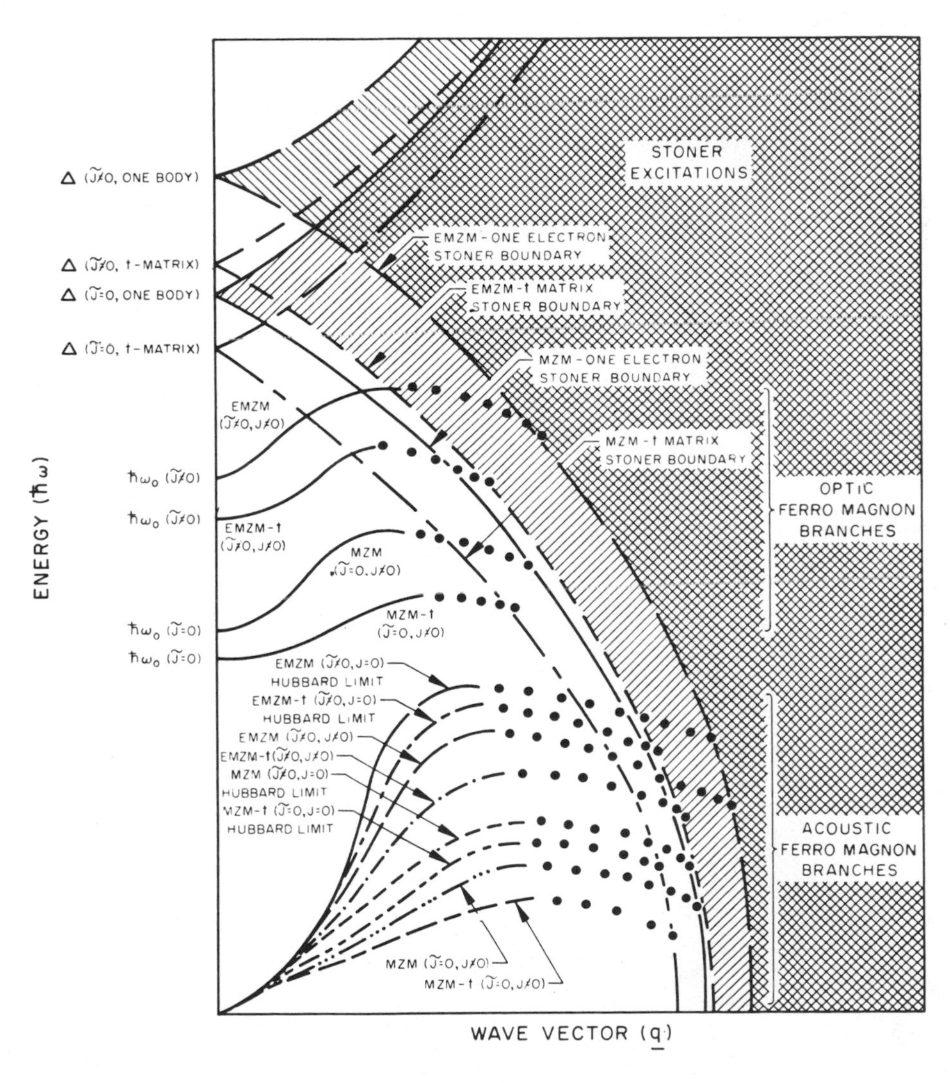

Figure 2. Acoustic and optic ferromagnon branch dispersion relations compared.

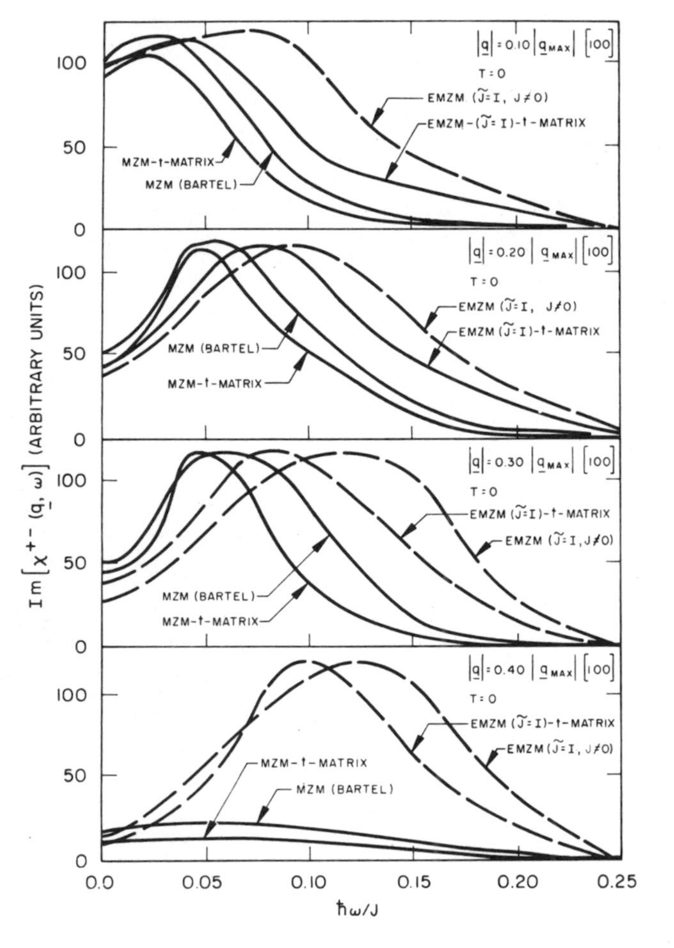

Figure 3. Acoustic ferromagnon branch.

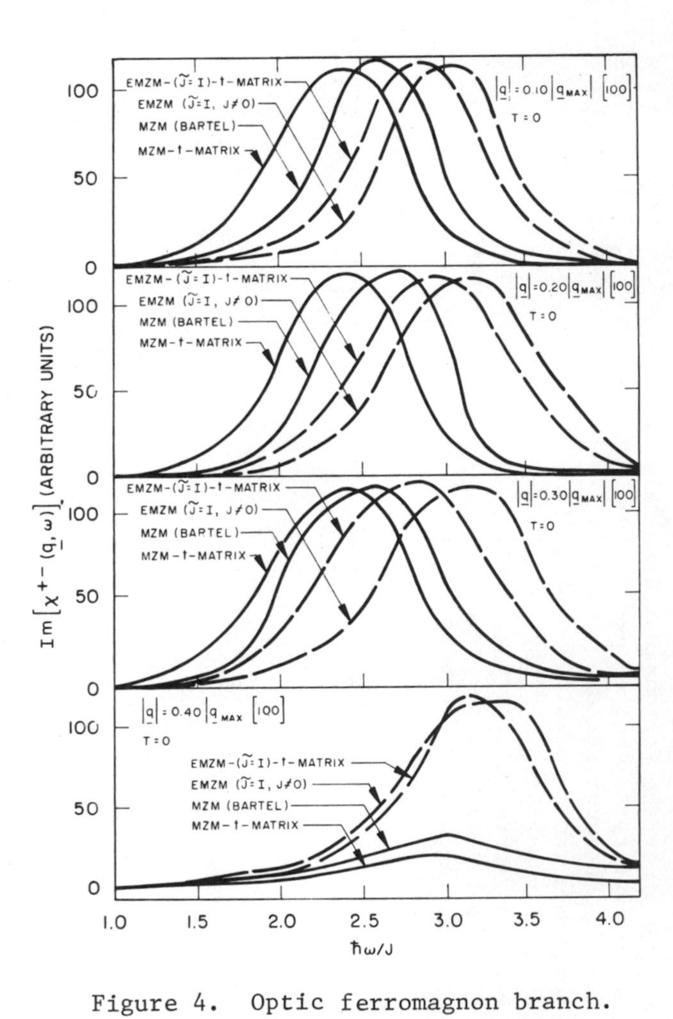

Figure 4. Optic ferromagnon branch.

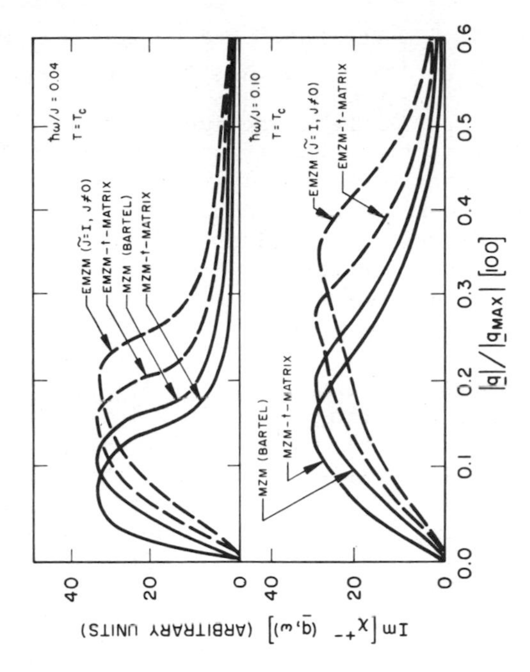

Figure 6. Optic ferromagnon branch.

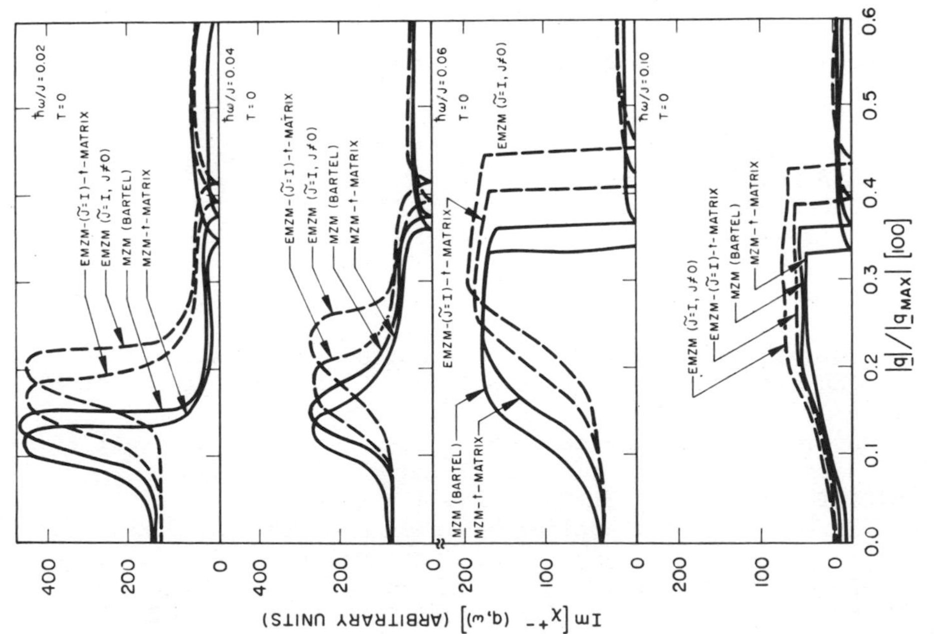

Figure 5. Acoustic ferromagnon branch.

Izuyama, Kim and Kubo[6] and Doniach and Wohlfarth[7]

$$\chi^{+-}(\underline{q},\omega) = -(2\pi)^{-1}\chi^{o}(\underline{q},\omega)/[1-I_{eff}(\omega)\chi^{o}(\underline{q},\omega)] \quad (1)$$

where I_{eff} is the effective electron-electron interaction and $\chi^{o}(\underline{q},\omega)$ is the enhanced spin susceptibility. The resonances of the singularity of $\chi^{+-}(\underline{q},\omega)$ are of two types: ferromagnon when q and ω values do not produce a branch cut in $\chi^{o}(q,\omega)$ or single particle-hole (Stoner) excitations when a branch cut exists. The MZM and its d-electron counterpart, the TDHM[12], possess two ferromagnon branches (optic and acoustic)

$$h\omega^{AM}(\underline{q}) = D^{AM}q^2 \text{ and } h\omega^{OM}(\underline{q}) = h\omega_o + D^{OM}(Ih\omega_o/J\Delta - 1)q^2 \quad (2)$$

where J is the MZM localized-itinerant exchange interaction and Δ is the Stoner exchange d-band splitting. Utilizing various expressions for D[7-11] we calculate the ferromagnon dispersion relations for the optic and acoustic branches in the d-electron MZM (EMZM), a t-matrix correction to it (EMZM-t) and a t-matrix correction to the MZM (MZM-t) by studying how D varies as we vary the model. An extensive summary of the calculation will be presented elsewhere[13]. In the figures, J is the MZM localized-itinerant exchange while $\tilde{J}$ is the itinerant-itinerant Hund's rule d-d exchange. We see that the position of the Stoner continuum and the exchange splitting Δ vary with model change ie. with J. Also the imaginary part of the transverse dynamic susceptibility changes as the model varies. The EMZM has enhanced local Coulomb and exchange crystal fields over the MZM, these being due to d-electron degeneracy. The t-matrix inclusion only partially cancels this trend. The MZM contains only an artificially included local-itinerant exchange field (J) which is automatically included in the EMZM within the $\tilde{J}$ crystal field.

REFERENCES

1. L.C. Bartel, Phys. Rev. B8, 11, 5316 (1973).
2. C. Zener, Phys. Rev. 82, 403 (1951).
3. C. Herring in Magnetism, Ed. by G. Rado and H. Suhl, Academic Press (1966).
4. T. Arai and M. Parrinello, Phys. Rev. Letters 27, 1226 (1971).
5. L.C. Bartel, Phys. Rev. B7, 3153 (1973).
6. S. Doniach and E.P. Wohlfarth, Proc. Roy. Soc. A296 (1967).
7. T. Izuyama, D. Kim and R. Kubo, Jnl. Phys. Soc. Japan 18, 1025 (1963).
8. S. Doniach, Lecture Notes, Mount Tremblant Summer School (1973).
9. C. Herring, Phys. Rev. 85, 1003 and 87, 60 (1952).
10. E.D. Thompson, Adv. in Phys. 14, 213 (1965).
11. A. Kutsuki and E.P. Wohlfarth, Proc. Roy. Soc, A345, 231 (1969).
12. E. Siegel and G. Kemeny, Phys. Stat. Sol.(b)50,593 (1972); ibid, (b)55,817 (1973); Ph.D. Thesis, Michigan State Univ. (1970); T. Moriya, in Proceedings of Fermi School, Course XXXVII, Academic Press (1967).
13. E. Siegel, Jnl. of Mag. and Mag. Mtls. (1977).

IONIC ENERGY LEVELS OF 3d IMPURITIES IN NORMAL METALS*

John A. Gardner

Department of Physics, Oregon State University
Corvallis, Oregon 97331
U.S.A.

INTRODUCTION

Finding an appropriate zeroth order theoretical framework for the 3d electrons of dilute transitional impurities in metals has proven to be difficult. Usually the 3d electrons hybridize strongly with conduction electrons, so the standard crystal field approximation is not obviously appropriate. Quite commonly, 3d states are treated as "virtual bound states" or scattering resonances in the conduction band, but this one-electron framework has proven inadequate to describe a number of important experimental results. There is now a good deal of evidence that even in the presence of quite strong 3d-conduction electron (k·d) interaction, ionic quantum numbers in many instances are still good, and orbital and crystal field energy level splittings are of the same magnitude as those commonly found in non-metallic solids.[1,2]

The experimental evidence supporting this conclusion includes transport properties, ESR in bottlenecked impurity states, neutron scattering, etc., but the most direct evidence arises from magnetic susceptibility and nuclear hyperfine measurements, primarily nuclear magnetic resonance. Unfortunately there is little undisputed optical information about impurity ionic energy level structure. For "magnetic" impurities where this structure should be clearly resolvable, oscillator strengths are found to be vanishingly small. The reason for this peculiarity is probably associated with the large k·d interaction but is presently not understood.

Because of the strong interaction with conduction electrons, the 3d ionic energy levels are coupled and lifetime broadened. The ground state level width can be estimated from magnetic properties and varies from a fraction of a °K (Mn in noble metal hosts) to one or two eV (Co in polyvalent metal hosts) or more. When the line width becomes as large as several eV, the ionic "ground state" may become a superposition of several coupled low-lying ionic levels. Even in this extreme limit however, there is experimental evidence[1] that intraionic interactions are still much too strong to justify a one-electron resonant scattering approximation.

MAGNETIC PROPERTIES OF 3d IMPURITIES

Since impurity properties are more or less symmetric about Mn in the 3d series, it is commonly accepted[3] that the 3d ground state has the doubly ionized atom configuration (although there appear to be some exceptions; for example the Cr ground state apparently is sometimes $3d^5$ rather than $3d^4$).[4] As shown below, crystal field splitting of the Hund's rule ground state can in principle be deduced from magnetic and hyperfine data which are sensitive to orbital and spin quantum numbers of the ground and low-lying excited levels.

At low temperature, or if the first excited ionic level lies well above the ground state, the magnetic susceptibility will be given by

$$\chi = \chi_{VV} + \chi_{dyn} \tag{1}$$

where χ_{VV} is the temperature independent Van Vleck susceptibility. The dynamic susceptibility χ_{dyn} arises from population differences when the (originally degenerate) ground state levels are split by a magnetic field. In the small field limit,

$$\chi_{dyn} = C/T \tag{2}$$

when the k·d interaction energy V_{kd} is negligible, but the form can be qualitatively different when V_{kd} is large. A phenomenological expression

$$\chi_{dyn} = (1-\delta)C/(T+T_K) \tag{3}$$

is found to be appropriate in most cases, but the precise form is not important to the present discussion. T_K is a measure of the ionic level width. Both the level broadening and the relative g shift δ are caused by the k·d interaction, but the relationship among T_K, δ, and V_{kd} is not well understood.

Hyperfine coupling of the 3d shell with the impurity nucleus causes a Knight shift given by[5]

$$\kappa = \kappa_0 + \alpha\chi_{spin} + \beta\chi_{orb} \quad (4)$$

where κ_0 is the (small and roughly constant) contact term observed in all metals, α is the core-polarization coupling constant, and β is the orbital hyperfine coupling constant. In general α is negative and of order -50 kG/μ_B, while β is positive and an order of magnitude larger. Since κ is very sensitive to orbital magnetic moment on the impurity, comparison of κ and χ yields information on the relative orbital and spin contributions to the susceptibility.

We shall denote the fractional orbital contribution to χ_{VV} and χ_{dyn} by f_{VV} and f_{dyn}, respectively. If the lowest excited ionic energy level is well above the ground state, f_{VV} and f_{dyn} are temperature independent, and Eq. 4 may be written,

$$\kappa = \kappa_0 + [\alpha(1-f_{VV}) + \beta f_{VV}]\chi_{VV} + [\alpha(1-f_{dyn}) + \beta f_{dyn}]\chi_{dyn} \quad (5)$$

$$= \kappa_1 + k\chi \quad (6)$$

where $\kappa_1 = \kappa_0 + [\alpha(f_{dyn}-f_{VV}) + \beta(f_{VV}-f_{dyn})]\chi_{VV}$ (7)

and $k = \alpha(1-f_{dyn}) + \beta f_{dyn}$ (8)

are constants.

The relationship given in Eq. 6 is found to hold in several impurity systems including CuMn[6], AuV[7], and possibly AuCo[8,9]. In CuMn, both χ and $\kappa-\kappa_0$ approximately follow a Curie law, and from the experimental k and Curie constant, it is clear that the Mn ionic ground state is $3d^5(^6S)$. T_K is a small fraction of a °K in this system. For AuV, κ_1 is found to be 1.8%, and k is negative. This result is consistent with a ground A_{2g} orbital singlet arising from crystal field splitting of the $3d^3(^4F)$ term. The crystal field parameter 10Dq is of order 0.3 eV, and T_K is about 300°K. The magnitude of the spin-orbit coupling cannot be determined. For AuCo, κ_1 is small, and k is approximately 40 mole/cm^3.[9] Preliminary analysis by Dupree, et al.,[8] suggests a spin-orbit splitting of about the same size as the free ion and a crystal field splitting 10Dq of order 0.3 eV.

The relationship given in Eq. 6 also holds for "nonmagnetic" Co in the polyvalent liquid hosts, Sn, Al, and Al_xCu_{1-x}.[1,10] For these alloys, χ_{VV} is negligible, and χ_{dyn} is dependent on temperature and host composition, presumably through the quantities δ and T_K given in Eq. 3. It is found that κ vs χ is linear, indicating that the Co ground state is the same in all these host metals. The linewidth varies from roughly 0.3 to 2 eV in the various hosts and apparently washes out spin-orbit and crystal field splittings. The ground state does not appear to arise from a single ionic

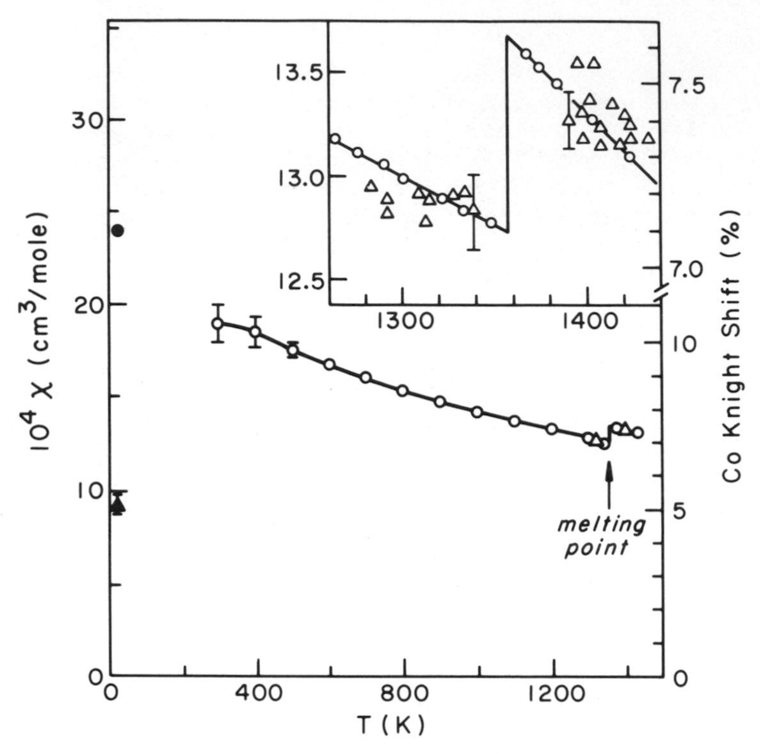

Fig. 1: Susceptibility χ (circles) and Knight shift κ (triangles) of Co impurities in Cu vs temperature. χ and κ scaled to coincide at high temperature. Knight shift data taken from (open triangles) Ref. 11 and (closed triangle) S. Wada and K. Asayama, Ref. 12. The low temperature susceptibility point (closed circle) is from R. Tournier and A. Blandin, Phys. Rev. Lett. 24, 397 (1970).

term, but rather is a superposition of the $3d^7(^4F)$ term and another having small f_{dyn}, possibly $3d^7(^4P)$ or $3d^6(^5D)$.

Eq. 6 does not hold for CuCo.[11] This can be seen from Fig. 1 which shows κ and χ vs temperature for Co in copper. In this impurity system, experimental Knight shift data have not been obtained at intermediate temperature, because the Co solubility is very small. The susceptibility is also difficult to measure, and the data shown are taken for a sample containing only 0.10 at. % Co. Below 600°K, small ferromagnetic Co precipitates formed, and the data with large error bars were found by extrapolating χ to infinite magnetic field in order to correct for the magnetic clusters. Precision of other points is smaller than the symbols, but there is an over-all scale uncertainty of approximately 10%. The inset in Fig. 1 shows susceptibility and Knight shift data near the melting point where Co solubility is appreciable. The representative error bars in the inset apply to all the Knight shifts shown.

Within experimental uncertainty, the high temperature data are consistent with $T_K \approx 1500°K$ (linewidth about 0.1 eV) and a Co $3d^7(^4F)$ ground state, which implies negligible χ_{VV} and $k \approx 50$ mole/cm^3. For this ground state, $\kappa - \kappa_0$ (here $\kappa_0 \approx 0.5\%$) and χ should be everywhere proportional, but it is obvious that at low temperature κ falls well below the high temperature extrapolation. It is likely that at low temperature, crystal fields split the ^4F ground state and reduce f_{dyn}. The nonlinearity of $\kappa - \kappa_0$ vs χ may arise either because the crystal field strength is very small ($10Dq \lesssim 1000°K$) or greatly reduced by vibrational motion at high temperature. It is unfortunate that experimental work in the intermediate temperature transition region is so difficult. This is the only 3d impurity system where this kind of anomaly is known to occur.

There are only two other 3d impurity-metal systems for which enough experimental information exists to deduce ionic energy levels. Those are MoCo and WCo where Narath[2] has shown that in both, Co has an A_{2g} ground state arising from crystal field ($10Dq \approx 0.6$ eV) and spin-orbit perturbations of the $3d^7(^4F)$ or possibly $3d^8(^3F)$ term. This experimental work was facilitated by relatively low values of T_K and a near-cancellation between the orbital and spin hyperfine fields.

REFERENCES

* Research supported by the National Science Foundation and the Oregon State University Graduate Research Fund.

1. J.A. Gardner, Phys. Rev. B14, in press
2. A. Narath, Phys. Rev. B13, 3724 (1976)
3. L.L. Hirst, Z. Physik 241, 9 (1971)
4. S. Sotier, thesis, Technical University of Munich, 1973 (unpublished)
5. A.J. Freeman and R.E. Watson, Magnetism (G. Rado and H. Suhl, eds.) Vol. IIA, Academic Press, N.Y. 1965
6. K. Kume and K. Mizoguchi, Sol. St. Commun. 16, 675 (1975)
7. A. Narath, Sol. St. Commun. 10, 521 (1972)
8. R. Dupree, R. Walstedt, and W.W. Warren, Jr., Bull. Am. Phys. Soc. 21, 328 (1976)
9. J.A. Gardner, "Susceptibility of Co impurities in Au", to be published
10. J.A. Gardner and C. Ardary, Sol. St. Commun. 19, 143 (1976)
11. J.A. Gardner, J. Bremer, and J. Bensel, "Magnetic properties of dilute Co impurities in Cu", to be published
12. S. Wada and K. Asayama, J. Phys. Soc. Japan 30, 1337 (1971)

THERMAL CONDUCTIVITY OF ANTIFERROMAGNETIC $FeCl_2$*

M.D. Tiwari

Department of Physics, Birla Postgraduate College,
Garhwal University, Srinagar, Garhwal, U.P., India

ABSTRACT

Modified Callaway's Model for thermal conductivity has been employed to study the thermal conductivity results of antiferromagnetic $FeCl_2$ in the temperature range 1.2-80 K. The experimental results are interpreted in terms of boundary, point defect and umklapp scatterings and exchange crystal-field energy levels of Fe^{2+} ions in the crystalline lattice of $FeCl_2$. The resonant character of the thermal conductivity has been explained on the basis of magnon-phonon interactions. A reasonably good fit with the experiment has been obtained.

INTRODUCTION

Laurence (1) reported experimental thermal conductivity results on $FeCl_2$ over the temperature range 3-90 K. Like other magnetic substances (2) $FeCl_2$ has also a sharp dip in the variation of its thermal conductivity with temperature. The anomalies arising in other substances (2) have been explained either in terms of scattering of phonons by critical fluctuations in the spin density (3,4) near the magnetic transition or as a result of phonon-induced electronic excitations. The magnetic properties of $FeCl_2$ studied by many other techniques show that there are a strong ferromagnetic exchange interaction among the intralayer ferrous ions and a weak antiferromagnetic interlayer interaction which gives rise to an antiferromagnetic order up to T_N = 23.5 K. We have studied here the results of $FeCl_2$ in terms of boundary, point defect, umklapp scattering, phonon-induced electronic excitations among the trigonal crystal and exchange-field split energy levels of Fe^{2+} ions in the crystalline lattice of $FeCl_2$ and magnon-phonon interactions. The theoretical computations show good agreement with the experiment.

THEORY AND RESULTS

The magnetic dipole excitation spectra (5) of Fe^{2+} ions in $FeCl_2$ at T=10 K showed that there are two groups of phonons whose energies are centered at 31 and 44 cm^{-1} (ground state triplet transitions) with a resonant type of scattering described by a relaxation time τ_r. The frequency dependence of τ_r used here is of the form

$$\tau_r^{-1} = S_i \frac{x^2}{(x_i^2 - x^2)^2} \quad . \tag{1}$$

Here S_i is the spin-phonon coupling parameter and the subscript i denotes the particular phonon induced electronic transition. x is given by $\hbar\omega/k_BT$.

The two transitions are not simultaneously involved at all temperatures, but as T_N is approached from the low temperature side, the exchange field decreases causing an eventual merger (5) of the J_z=+1 level with the J_z=-1 level at T_N, and thereby giving absolutely no possibility of oberving the 44 cm^{-1} resonance above T_N. The merger is not a sudden process but is continuous with decreasing exchange field. The experiments show that at ∿23 K the level J_z=+1 crosses the level J_z=0 and finally merges with the J_z=-1 level at T_N. This means that instead of talking of a fixed 44 cm^{-1} transition one must also take into account its variation with temperature. The 44 cm^{-1} transition which is found at T=10 K must be recognized as a 31 cm^{-1} transition at ∿23 K where the levels cross over. However, to simplify our calculations we took a temperature independent transition at 44 cm^{-1} below T_N while above T_N the 31 cm^{-1} transition associated with the paramagnetic state was taken and the relaxation-rate term was defined by

$$\tau_r^{-1} = \tau_{r,44}^{-1}\ \theta(23.5-T) + \tau_{r,31}^{-1}\ \theta(T-23.5)$$

where θ(x) is a step function which takes into account that below and above T_N the 44 and 31 cm^{-1} transitions, respectively, are active.

Callaway's (6) integral for the thermal conductivity at temperature T is given by

$$K(T) = \frac{k_B T^3}{6\pi^2 \hbar^3} \int_0^{\theta_D/T} \tau_c \, v \, \frac{x^4 e^x}{(e^x-1)^2} \, dx \quad , \tag{2}$$

where v is the phonon group velocity and the combined relaxation time τ_c is given as

$$\tau_c^{-1} = \tau_B^{-1} + \tau_{pt}^{-1} + \tau_{ph}^{-1} + \tau_r^{-1} \quad . \tag{3}$$

The last term at the right hand side is given by Eq. (1). The other three terms arise due to boundary effect, point defect and phonon-phonon scattering:

$$\tau_B^{-1} = v/L$$

$$\tau_{pt}^{-1} = A\,\omega^4$$

$$\tau_{ph}^{-1} = B\,\omega^2\,T^3 \quad .$$

L is the characteristic length of the crystal, A and B are adjustable parameters. We have taken the results for magnon-phonon interactions as given by Laurence and Petitgrand (7). They have also explained the thermal conductivity results theoretically but their calculations do not satisfy the experimental results around the first maximum.

We find that we can successfully account for the experimental results in the whole temperature range 1.2-80 K. Our computed results are shown in Fig. 1. The various parameters used in the analysis are given in Table I. The S_i parameters for both transitions are adjusted so that one gets the same value of the thermal conductivity at T=23.5 K. The value of L given by theory does not give good agreement with the experiment and therefore we treat L as an unknown parameter to which we assign a value that gives the best fit to the experimental results.

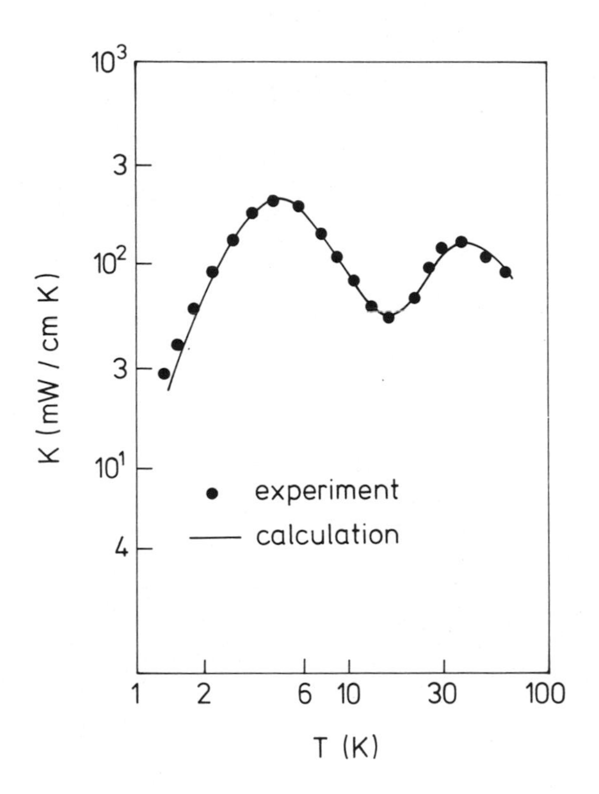

Figure 1: Thermal conductivity of $FeCl_2$.

Table I: Parameters used in the analysis.

$A = 3.2 \times 10^{-44}\ sec^3$	$S_{44} = 3.28 \times 10^4\ sec^{-1}$
$B = 1.63 \times 10^{-12}\ deg^{-4}$	$S_{31} = 1.71 \times 10^4\ sec^{-1}$
L = 0.018 (0.54 theoretical)	θ_D = 162 K

ACKNOWLEDGEMENTS

The author is grateful to Drs. Bal K. Agrawal, F.W. Sheard, and Prof. G.S. Verma for their interest in the present work. Prof. L.J. Challis is thanked for encouragement and the help in various ways. The computations were performed on IBM 7044/1401 computors installed at I.I.T., Kanpur.

REFERENCES

* Work supported by the University Grants Commission, New Delhi

1. G. Laurence, Phys. Letters 34A, 308 (1971)
2. G.S. Dixon and D. Walton, Phys. Rev. 185, 735 (1969); J.E. Rives, D. Walton, and G.S. Dixon, J. Appl. Phys. 41, 1435 (1970); M.J. Metcalfe and H.M. Rosenberg, J. Phys. C 5, 450 (1972)
3. K. Kawaski, Prog. Theor. Phys. (Kyoto) 29, 801 (1963)
4. H. Stern, J. Phys. Chem. Solids 26, 153 (1965)
5. R. Alben, J. Phys. Soc. Japan 26, 261 (1969)
6. J. Callaway, Phys. Rev. 113, 1046 (1959)
7. G. Laurence and D. Petitgrand, Phys. Rev. B 8, 2130 (1973)

CRYSTAL FIELD EFFECTS ON HYDROGEN IN TRANSITION METALS

J. Hauck and H.J. Schenk[+]

Institut für Festkörperforschung
[+]Institut für Chemie
KFA Jülich, D-5170 Jülich, W.-Germany

ABSTRACT

The partial loss of the cohesive metal-metal bonding in interstitial solid solutions of hydrogen in transition metals is compensated by Coulomb interactions. The hydrogen atoms occupy octahedral interstices at high electronegativity of the metal (PdH, NiH, β-VH). Otherwise tetrahedral sites are preferred by crystal field energy (δ-VH, β-TaH, NbH_2, etc.). The metal lattice can get distorted by the repulsion of neighbouring atoms, which can be shown by Madelung calculations on β, γ, ζ-NbH and NbH_2.

The transitions metals V, Nb, Ta, Cr, Ni, Pd and the high temperature modification of Ti, Zr and Hf are forming hydrides, which can be considered as interstitial solid solution of hydrogen in the metal lattice[1]. With increasing hydrogen content the lattice of the metal atoms gets expanded thus decreasing the cohesive metal-metal interaction (fig. 1). The hydrogen atoms can gain on electron density within its nearest surrounding or can also loose electron charge to the metal. The Coulomb energy of the lattice is increased at a high exchange of electrons and compensates the loss of the metal-metal cohesive energy. The kind of electron distribution can be estimated by the electronegativity or the electron work function, which are decreasing in the series Pd>Ni>(H)>Cr>V>Ta>Nb>Ti>Hf>Zr[2]. Also the relative volume expansion shows this relation and might be due to an increasing transfer of electrons from the metal to the hydrogen atoms, which shortens the metal-metal bonding distances at constant H-content (fig. 1). The resulting effective charge of the hydrogen atoms should be important for the preference of octahedral or tetrahe-

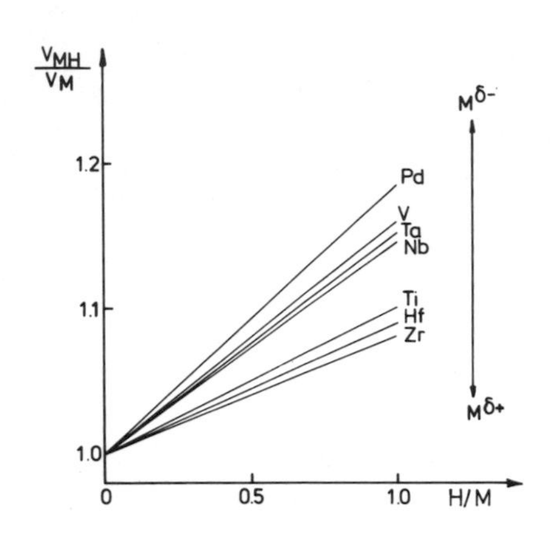

Fig. 1: Approximate relative volume expansion of transition metal hydrides

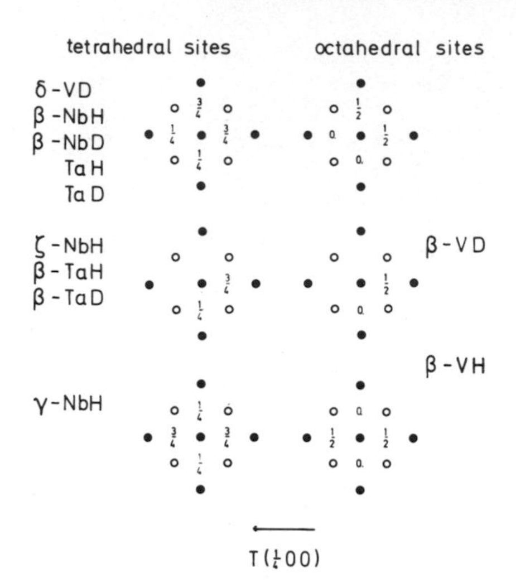

Fig. 2: Ordering of hydrogen on tetrahedral or octahedral sites in closely related crystal structures of V-, Nb- and Ta-hydrides (open circles represent metal atoms at x=0, full circles at x=0.5, the numbers indicate the projection of the hydrogen atoms at the (b,c) plane

dral interstices. Some structures with ordered hydrogen distribution in bcc metals are shown in fig. 2. The octahedral sites can be transformed to tetrahedral sites of related structures by the translation $\frac{1}{4}$ 0 0 perpendicular to the plane of paper. In fcc metal lattice the octahedral interstices of the NaCl-structure (PdH, NiH) have a similar relation to the tetrahedral interstices of the ZnS-structure with the translation $\frac{1}{4}\ \frac{1}{4}\ \frac{1}{4}$. The site preference in related structures should mainly be determined by crystal field energy.

The electron density at tetrahedral and octahedral interstices of transition metals will be different because of the nonspherical distribution of the d-orbitals. The t_{2g} orbitals are overlapping with the orbitals of neighboring metal atoms and give rise to conduction bands. The e_g orbitals are pointing to octahedral interstices in bcc and fcc lattice. Though the actual electron concentration is unknown , one can show by different models, that the electron density will be higher at octahedral than at tetrahedral interstices[3]. Therefore, hydrogen with a low effective electron concentration will be attracted to octahedral sites in Pd, Ni and probably also Cr, whereas hydrogen atoms with a high electron density are repelled to tetrahedral sites in Nb, Ta, Ti, Zr and Hf. Vanadium has the highest electronegativity among the elements of the fourth and fifth transition metal group. At low H concentrations and low temperatures (β-VH, β-VD) hydrogen occupies octahedral sites. At higher H-content the electrons get more localized and the tetrahedral sites will be preferred (δ-VD, α-VH, VH_2).

At a given site preference the ordering of the hydrogen atoms in the metal lattice will largely be determined by the Coulomb interaction between neighbouring atoms. In a qualitative approach for the Coulomb lattice energy the sum of partial Madelung constants PMC contributing to the Madelung part of the lattice energy (MAPLE)[4] was calculated. Similar results should be obtained with the Ewald-Fuchs-potential of Coulomb interaction by which the electrostatic energy of pure metals was calculated or by some other kind of screened potential[5].

For hydrogen in bcc metal lattice at H/M = 1 the highest Madelung energy was obtained for γ-NbH-structure and about 1 % less for β-NbH, whereas all other distributions yielded lower values. γ-NbH is pseudocubic, whereas β- and ζ-NbH are orthorhombic which is in accordance with experimental data. In β-NbH the neighboring hydrogen atoms with $T(0\,\frac{1}{2}\,\frac{1}{2})$ have a stronger repulsive interaction than those with $T(\frac{1}{2}\,\frac{1}{2}\,\frac{1}{2})$. Due to the cohesive metal-metal interactions the angle 90-γ is lower (90.9°) than the resulting value from Madelung calculations (93.0°) . The calculated maximum can be shifted to the experimental angle, if only 75 % of the electrons are considered to be localized at the hydrogen atoms. In ζ-NbH the charge at the hydrogen atoms must be reduced to 47 % to get agreement with the experimental angle (fig.3). $T(\frac{1}{2}\,\frac{1}{2}\,\frac{1}{2})$ is the shortest translation between neighboring hydrogen atoms and give rise to the strongest repulsion.

For hydrides with the composition MH_2 the occupation of tetrahedral interstices as shown in fig. 4 give the highest values for bcc and fcc metal lattices. Both can be transformed at constant volume by a decrease of a' and an increase of c'. Hereby the flattened tetrahedra of the bcc lattice get regular with the point symmetry $\bar{4}3m$. In ε-TiH_2, ε-ZrH_2 and ε-HfH_2 the c'/a' value is below the maximum Madelung lattice energy, which might be due to a residual bcc metal-metal cohesive interaction of partly filled t_{2g}-orbitals.

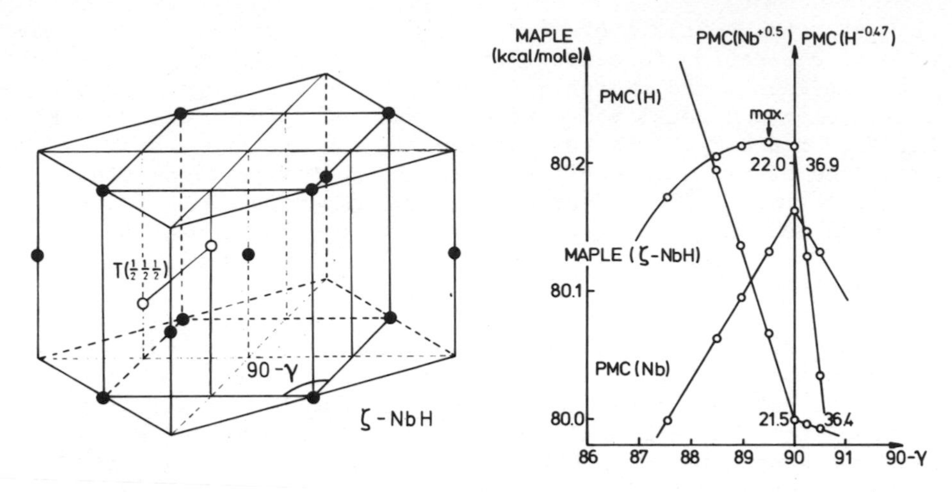

Fig. 3: PMC and MAPLE-values for ζ-NbH-structure $(M_2^{+0.5}H^{-0.47}e_{0.53}$, $b \cdot c = 23.4784$ Å, $a = 3.421$ Å)

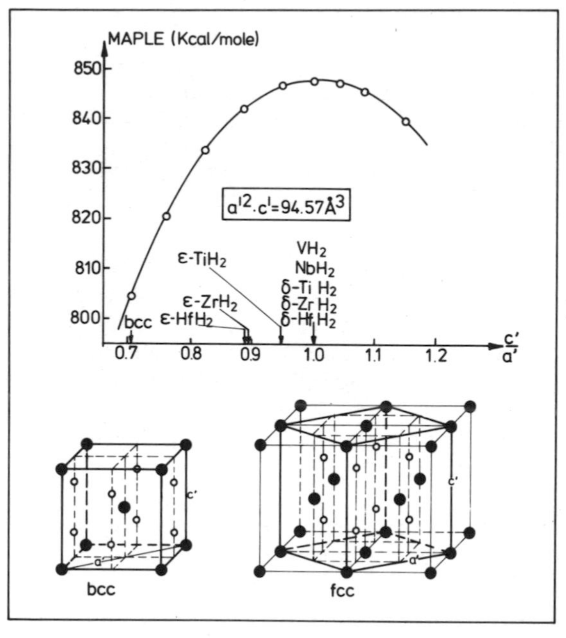

Fig. 4: MAPLE-values for MH_2-transition from bcc to fcc

REFERENCES

1 K. Mackay, Hydrogen compounds of the metallic elements, E.F.Ston, London (1966)
2 A.R. Miedema, Philips Techn. Rundschau 33, 157 (1974)
3 J. Hauck and H.J. Schenk, in preparation
4 J. Hauck, phys. stat. sol.
5 V. Heine and D. Weaire, Solid State Phys. 24, 249 (1970)

NEUTRON SCATTERING FROM PARAMAGNETIC CERIUM

B.D. Rainford

Physics Department, Imperial College, London SW7

ABSTRACT

Mechanisms governing the linewidths and line shapes in inelastic neutron scattering experiments are briefly discussed. The scattering from γ-Ce is shown to be consistent with rapid Korringa relaxation of the f electrons by the conduction electrons. For α-Ce no magnetic scattering is observable. In $CeAl_2$, while linewidths are large, indicating rapid relaxation, their temperature dependence is anomalous.

Inelastic neutron scattering has been widely used over the last eight years for studying the crystal field splittings in paramagnetic rare earth intermetallic compounds. But besides the purely spectroscopic information derived in these measurements, there is a great deal of information on the spin dynamics contained in the linewidths and line-shapes. This aspect has not received much attention until recently. There is probably most interest in results on the dynamic susceptibility for elements close to a valence instability and for intermediate valence materials. In this paper, after discussing the origins of linewidths in neutron results, data are presented for three cases : a) γ-Ce, which has a well defined magnetic configuration $4f^1$ b) α-Ce which is almost non magnetic and c) $CeAl_2$ which is on the verge of being non-magnetic.

For rare earth compounds with well defined $4f^n$ configurations the neutron cross section consists of a number of peaks corresponding to transitions within and between crystal field eigenstates. Thus for cubic Ce compounds, where there are only two eigenstates, a Γ_7 doublet and a Γ_8 quartet, we expect three peaks centred at zero energy transfer ($\Gamma_7 \to \Gamma_7$ and $\Gamma_8 \to \Gamma_8$ transitions), at $+\Delta(\Gamma_8 \to \Gamma_7)$ and at $-\Delta(\Gamma_7 \to \Gamma_8)$, where Δ is the crystal field splitting. These transitions are broadened by various relaxation mechanisms; spin-lattice relaxation, involving both the phonons and electrons, and spin-spin relaxation due to the dipolar and exchange interactions between the local moments. The various mechanisms may be distinguished by the temperature dependence and wave vector dependence of the widths. Since spin-lattice relaxation involves only a single local moment, the widths are independent of wave vector. Lineshapes are Lorentzian. On the other hand when the local moments are coupled by an exchange interaction $\mathcal{J}_{rr'}$, the broadening varies with wave vector q , as was first shown by de Gennes. He found that the second moment of the scattering function at high temperatures is given by[1]

$$\langle \hbar^2 \omega^2 \rangle_q = \frac{2}{3} (g_J - 1)^2 J(J+1) \sum_{r,r'} \mathcal{J}^2_{r,r'} [1 - e^{i\underline{q}\cdot(\underline{r}-\underline{r}')}] \quad ..(1)$$

This quantity, which is directly related to the individual widths of transitions, is essentially temperature independent at high temperatures. The line shape cannot be calculated analytically, but is close to Gaussian at the Brillouin zone boundaries, and Lorentzian at the zone centres. Widths due to spin lattice relaxation vary with temperature with different power laws depending on the temperature regime and the exact mechanism. In practice it appears that widths are relatively small in insulators and semiconducting materials, but large in metals, often being comparable in magnitude to the crystal field splittings themselves. We conclude that the dominant mechanism for spin lattice relaxation in intermetallic compounds is the Korringa relaxation of the local moment by the conduction electrons via the exchange coupling

$$\mathcal{H}_{sf} = (g_J - 1)\, \mathcal{J}_{sf}\, \underline{J}\cdot\underline{s} \quad ....(2)$$

For this case the linewidth of a transition between crystal field states i and j is given[2,3] by

$$\Gamma_{ij} = \frac{2\pi}{\hbar} (g_J - 1)^2 \rho^2(E_F) \mathcal{J}^2_{sf} P_{ij} \frac{\Delta}{2} \coth\left(\frac{\Delta}{2K_B T}\right) \quad ..(3)$$

where $\rho(E_F)$ is the density of conduction electron states at the Fermi energy and P_{ij} is given by

$$P_{ij} = \sum_{m_s m_s'} |\langle m_s, i | \underline{J}\cdot\underline{s} | m_s', j \rangle|^2 \quad ....(4)$$

For small Δ or for high temperatures the linewidth increases linearly with temperature. But for finite Δ there is a finite linewidth at low temperatures. The situation where both spin lattice and spin spin relaxation are important is complicated and has not been adequately discussed so far. For cerium metal and intermetallic compounds, however, the exchange coupling is relatively weak. Estimates[2] of the second moment based on Eq. 1 and a simple RKKY model for the exchange interaction yield a linewidth of about 1 meV for β-Ce. This should be a good estimate for γ-Ce also since the conduction electron density and interatomic spacings are very similar to those of β-Ce. For $CeAl_2$ this linewidth would be even smaller. In reality (see below) the measured linewidths are larger by an order of magnitude, in fact so large that discrete crystal field transitions cannot be observed at room temperature. We conclude that the spin-spin relaxation can be ignored in these cases compared to the very rapid spin-lattice relaxation.

Cerium Metal Measurements have been made of the inelastic paramagnetic scattering as a function of temperature[2], at atmospheric pressure in the γ phase and as a function of pressure[4] at room temperature across the γ-α transition. The temperature variation for γ-Ce is shown in Fig. 1a. A broad, featureless spectrum is observed extending out to about 70 meV. The width of the spectrum is seen to increase with temperature. The lineshape has been fitted by adding together three Lorentzian curves centred at zero energy transfer and at $\pm\Delta$, with Δ and the individual linewidths as variable parameters. The individual integrated areas of the three peaks were constrained to agree with the calculated neutron cross sections for the $\Gamma_7\Gamma_7$, $\Gamma_8\Gamma_8$, $\Gamma_7\Gamma_8$ and $\Gamma_8\Gamma_7$ transitions. The best fits gave a value of Δ of 10.6 meV, though the uncertainty on this value is of course large. The calculated lineshapes, however, do give a good account of the

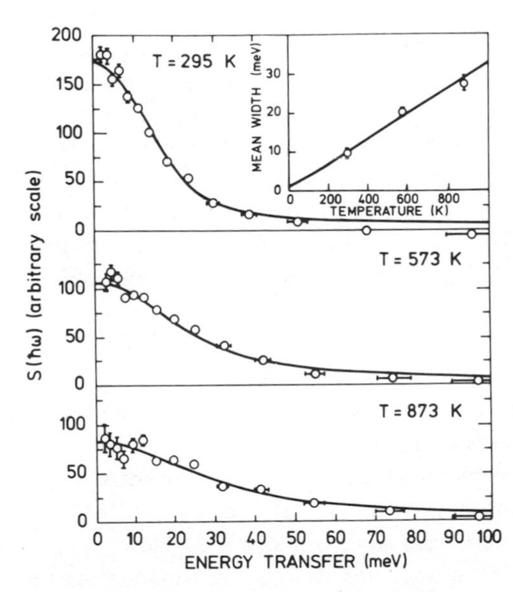

Fig. 1a. Neutron scattering from γ-Ce at atmospheric pressure. Lineshapes are calculated for Δ = 10.6 meV

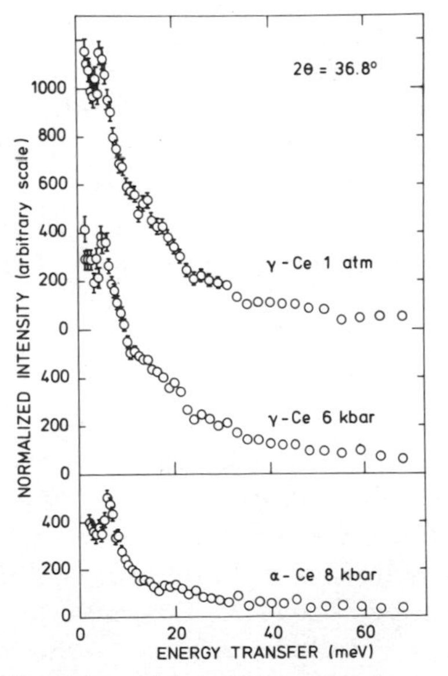

Fig. 1b. Neutron scattering at 300K from Ce metal at 1 atmos., 6 Kbar and 8 Kbar From reference (4).

observations. The mean widths of the Lorentzian peaks are shown inset in the figure. These are seen to increase roughly linearly with temperature. The solid curve in the inset was calculated from Eq. 3 above, with P_{ij} averaged over the initial and final eigenstates for the various transitions, and the product $|\mathcal{J}_{sf}\,\rho(E_F)|$ adjusted to fit the observed gradient at high temperatures. The value of $|\mathcal{J}_{sf}|\,\rho(E_F)$ was found to be 0.23. If we assume a value for $\rho(E_F)$ of 1.6 states per eV atom[5], this yields $|\mathcal{J}_{sf}| = 0.14$ eV, in reasonable agreement with other estimates for Ce. The temperature dependence of the scattering for γ-Ce is thus compatible with the Korringa mechanism of spin lattice relaxation. The phase transition to the non-magnetic α phase takes place at 7.5 Kbar at room temperature. Measurements of the inelastic scattering, made with a triple axis spectrometer[4], at atmospheric pressure, 6 Kbar and 8 Kbar are shown in Fig. 1b. In the γ phase the widths appear to increase somewhat with increasing pressure. Fits of the lineshape to three Lorentzian peaks as discussed above yield linewidths at 6 Kbar which are 40% greater than those at atmospheric pressure. This suggests that $\mathcal{J}_{sf}$ increases with pressure by about 20% on going to 6 Kbar. This effect is predicted by Coqblin's theory of the phase transition in Ce. Theoretical estimates[7] predict a somewhat larger increase in $\mathcal{J}_{sf}$ than that observed. However the data lend qualitative support to Coqblin's model. The data for 8 Kbar show that a dramatic reduction in intensity occurs on passing through the $\gamma \rightarrow \alpha$ phase transition. This corresponds to the decrease, or total loss, of the paramagnetic scattering as the 4f electron becomes delocalised. Careful comparison with the spectrum of La measured under identical conditions has led us to conclude that there is no residual magnetic scattering, within experimental uncertainties, in the α phase. The scattering seen in Fig. 1b is attributed to phonon and multiphonon scattering. From the absence of magnetic scattering above the phase transition we conclude that either there is zero occupancy of the 4f states in α Ce, or that if there is partial occupation of the 4f states, the dynamical response of these electrons is spread over such a wide energy spectrum that it is unobservable in the present experiment.

$\underline{CeAl_2}$ This compound, while appearing to have a normal magnetic behaviour, consistent with a $4f^1$ configuration, at high temperatures, displays many anomalous properties at low temperature[3,8]. The inelastic paramagnetic scattering[2] is shown in Fig. 2. The spectra are quite similar to those for cerium and fits to the line shapes as described above yield a value for Δ of 9 meV approximately. However the temperature variation of the linewidths does not fit the predictions of the Korringa relaxation. The linewidth extrapolates to a large finite value at low temperatures. This anomalous behaviour is obviously closely related to the anomalous behaviour of the resistivity, which displays a broad maximum near 80 K[8]. The resistance anomaly has been attributed to a Kondo divergence for electrons inelastically scattered between crystal field states ("Kondo-sidebands").

However the shape of ρ(T) is similar to that observed in many nearly magnetic actinide intermetallics. Jullien et al.[10] have attributed the anomalies in the latter case to the dynamical response of the f electrons which are assumed to occupy a narrow f band. The same mechanism might be appropriate to $CeAl_2$. In this case the linewidths observed in the neutron spectra would be directly related to the spin fluctuation frequency ($1/T_{sf}$) of the f electrons.

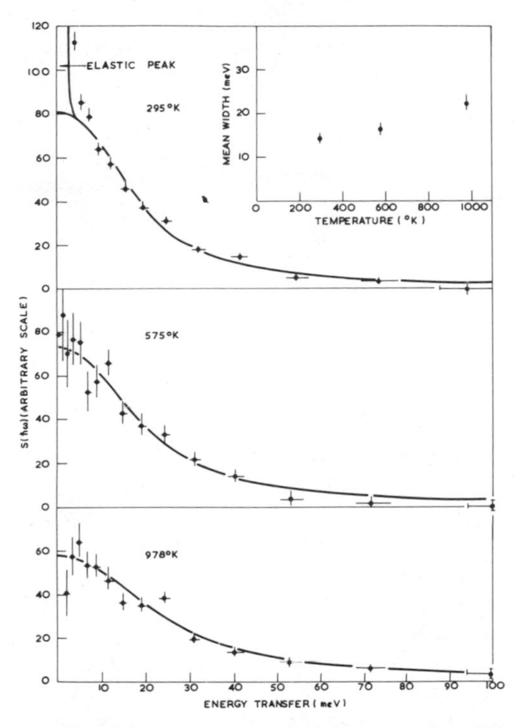

Fig. 2 Inelastic neutron scattering from $CeAl_2$
The lineshapes are calculated for Δ = 9 meV

REFERENCES

1. P.G. de Gennes, J. Phys. Chem. Solids, 4, 223 (1958)
2. B.D. Rainford, D.Phil.Thesis Oxford University (1969) unpublished
3. P. Fulde and I. Peschel, Adv. in Phys. 21, 1 (1972)
4. R.W. Hill and J.M. Machado da Silva, Phys.Lett. 30A, 13 (1969)
5. B.D. Rainford, B.Buras and B. Lebech (to be published)
6. Estimated from the electronic specific heat of La, allowing for electron-phonon enhancement
7. B.Coqblin and A.Blandin, Adv. in Physics, 17, 281 (1968)
8. C.F. Ratto et al. Adv. in Physics, 18, 489 (1969)
9. K.H. Buschow and H.J. Van Daal, Solid State Comm. 8, 364 (1970)
10. F.E. Maranzana, Phys. Rev. Lett., 25, 239 (1970)
11. R. Jullien, M.T.Beal-Monod and B.Coqblin, Phys.Rev. B9, 1441(1974)

PRESSURE DEPENDENCE OF THE MAGNETIC PROPERTIES OF SOME RARE EARTH INTERMETALLIC SYSTEMS*

R.P. Guertin

Physics Dept., Tufts University

Medford, Mass., 02155, U.S.A.

ABSTRACT

The magnetization and susceptibility of several rare earth intermetallic compounds have been measured under hydrostatic pressure, P, up to 10 kbar and in magnetic fields, H, up to 60 kG. For all the materials studied the magnetic properties are strongly influenced by crystalline electric field (CEF) effects. For the Pr-monochalcogenides and $TmAl_3$ the low temperature van Vleck-like susceptibility, χ, decreases with increasing P. This is consistent with an increase of the CEF interaction as the lattice constant is reduced. This trend is also observed for the CEF level structure of dilute Tb in superconducting $LaAl_2$ (where the impurity pairbreaking is reduced with increasing P). However, for the Pr- and Tm- monopnictides χ increases with increasing P. The pressure dependence of the magnetic ordering temperatures for singlet ground state TbSb, Pr_3Tl and TmS is also reported. The effect of P is very strong on the low temperature magnetic anomalies of TmSe; the transition fields increase with increasing P at a rate of +1.7 kG/kbar, and these results are discussed in terms of the mixed-valence nature of TmSe.

INTRODUCTION

Although a great deal of experimental data has been reported showing various crystalline electric field (CEF) effects in metallic materials, comparatively little has been done showing the influence of hydrostatic pressure on these materials. In this paper

*Supported by National Science Foundation, Grant # DMR-75-09494.

we present the results of experiments which probe the response to hydrostatic pressure of the magnetic properties of several rare earth intermetallic systems. The magnetic properties of these systems are strongly influenced by CEF effects and many of the materials are familiar to those working in the field of CEF effects in metals and alloys. The types of systems discussed are paramagnetic singlet ground state systems, magnetically ordered singlet ground state systems and mixed-valence TmSe. The general features of the experimental procedures have been described elsewhere.[1]

PARAMAGNETIC SINGLET GROUND STATE SYSTEMS

In this section we consider the pressure dependence of the low temperature magnetic susceptibility, χ, of several paramagnetic singlet ground state rare earth intermetallic compounds.[2,3] Also discussed are the results of a related experiment involving a superconductor doped with a rare earth impurity having a non-magnetic CEF ground state.[4]

The compounds investigated were several of the Pr- and Tm-monopnictides and Pr-monochalcogenides[2] and $TmAl_3$.[3] In all cases the CEF of the neighboring ions partially lifts the 2J+1 degeneracy of the spin-orbit ground state given by Hund's rules, and a (non-magnetic)singlet CEF ground state results. The two lowest lying CEF levels are Γ_1 (singlet) and Γ_4 (triplet) for the monopnictides and monochalcogenides[5] and Γ_2 (singlet) and $\Gamma_5^{(2)}$(triplet) for $TmAl_3$.[6] The splitting, Δ, between the two lowest lying levels is determined from neutron inelastic scattering[5,6] and is $25<\Delta<157$ K for the materials discussed. Since the experiments reported here were all carried out at 4.2 K, only the singlet state is thermally populated and the low field CEF-only magnetic susceptibility, χ_c, is van Vleck-like and given by[7]

$$\chi_c = \frac{2Ng_J^2\mu_B^2|\langle\Gamma_4|J_z|\Gamma_1\rangle|^2}{k_B\ \Delta} \qquad (1)$$

for the monopnictides and monochalcogenides. For $TmAl_3$ χ_c is given by a similar expression but with two terms, connecting Γ_2 with $\Gamma_5^{(2)}$ and with $\Gamma_5^{(1)}$.

We might expect the measured susceptibility to <u>decrease</u> with increasing P because Δ (which is proportional to the overall CEF interaction strength) should increase as the neighboring ligands move closer to the rare earth ion. The splitting, Δ, is proportional to R^{-5} (where R is the rare earth-ligand distance) for cubic systems in which the fourth order term in the CEF Hamiltonian dominates. Thus a 1% reduction in lattice constant (corresponding to $5<P<30$ kbar for the materials studied) would cause a 5% decrease in χ_c. The pressure dependence of χ may be written[8]

$$(1/\chi_c)(d\chi_c/dP) = -\kappa\gamma \qquad (2)$$

where $\kappa = -(1/V)(dV/dP)$ is the compressibility and $\gamma = (1/\Delta)(d\Delta/dP)$ is the Gruneisen constant for the lowest lying CEF level. Using the notation of Lea, Leask and Wolf[9] (LLW) we derive an expression for γ which is appropriate to the CEF level structure of the materials discussed. This may be written

$$\gamma = 5/3 + \{2\Delta_o(1+x)\}/3\Delta \qquad (3)$$

where $|x|<1$ (restricted in this case to negative x) measures the relative strength of the fourth and sixth order terms in the LLW Hamiltonian and Δ_o is the value of Δ for x=0. The second term in Eq. (3) is quite small for the cases considered. An equivalent expression to Eq. (3) has been derived by Umlauf *et al.*[10]

Pr-Monochalcogenides

The Pr-monochalcogenides studied were PrS, PrSe and PrTe, and in all cases χ decreased with increasing P, consistent with Eqs. (2) and (3). The results for PrS, however, were rather unusual: χ_c decreased at a rate of 1%/kbar (in accordance with $\gamma \simeq 5/3$) but then increased for P>5 kbar.[11] The reason for this non-monotonic behavior is not understood. The pressure dependence of χ for all the paramagnetic singlet ground state compounds studied is listed in Table I.

Table I: Summary of the pressure-dependence of the susceptibility of singlet ground state systems. $(1/\chi_c)(d\chi_c/dP)$ is derived from Eqs. (1)-(3). $(1/\chi)(d\chi/dP)$ is measured.

Sample	Δ (K)	$\frac{1}{\chi_c}\frac{d\chi_c}{dP}$ (10^{-3} kbar^{-1})	$\frac{1}{\chi}\frac{d\chi}{dP}$ (10^{-3} kbar^{-1})
PrS	157	-10.1	non-monotonic
PrSe	121	-2.2	-2
PrTe	78	-5.6	-1
PrAs	133	-5.1	+10
PrSb	72	-3.5	+7
PrBi	68	-3.9	+9
TmAs	31	-4.1	+6
TmSb	25	-2.6	+3
$TmAl_3$	45	-6	-6

Pr- and Tm-Monopnictides

The monopnictides studied were PrAs, PrSb, PrBi, TmAs and TmSb; and in contrast to the results described above, χ increased with increasing P. An example of the data for the monopnictides is shown in Fig. 1 where we plot $[\chi(P)-\chi(0)]/\chi(0)$ vs P for PrBi. The dashed line in Fig. 1 shows the decrease predicted by Eqs. (2) and (3).

On the other hand, the positive P-dependence of χ is consistent with measurements of the Knight shift under pressure of ^{141}Pr and ^{169}Tm in some monopnictides.[12] Furthermore, Ott and Luthi[8] have shown that the thermal expansion of TmSb has a CEF-related Schottky-like anomaly at low temperatures. Their data are fit well with $\gamma \simeq -1.5$ rather than $\gamma=+1.71$ predicted by Eq. (3). Also, using $\gamma \simeq -1.5$, the pressure dependence of χ from Eq. (2) is close to that found experimentally for TmSb. Thus Δ apparently decreases with increasing P for TmSb and the observed increase in χ cannot be attributed to P-dependence of the exchange, as originally suggested.[2] The results for the monopnictides are particularly surprising because the CEF levels derived from neutron inelastic scattering are close to those calculated from a point charge model when a (realistic) charge $Z \simeq -2e$ is assigned to the pnictide ligands. Furthermore, previous experimental data on the monopnictides[13] has been found to be consistent with the predictions of the point charge model.

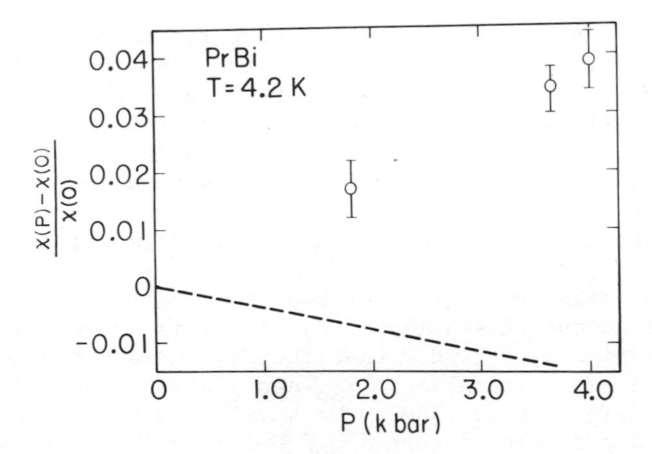

Fig. 1: Relative change of the susceptibility of PrBi at 4.2 K for three hydrostatic pressures. The dashed line shows the behavior predicted by Eqs. (2) and (3).

$TmAl_3$

$TmAl_3$ has the Cu_3Au structure and the Γ_2-$\Gamma_5^{(2)}$ splitting Δ is 45 K, as determined by neutron inelastic scattering.[6] Unlike the monopnictides and monochalcogenides above, all of which have the NaCl structure, the CEF levels for $TmAl_3$ cannot be fit by a point charge model using realistic ligand charges. The fit to the measured LLW[9] parameters x and W by point charges on the neighboring Al and Tm ions gives Z_{Al}=+16e and Z_{Tm}=+50e. Evidently, screening and covalency strongly influence the CEF interaction in $TmAl_3$. Furthermore, the measured susceptibility at $T<<\Delta$ is 35% less than the value calculated from Eq. (1), whereas for the monopnictides and monochalcogenides, χ is slightly enhanced. We find $(1/\chi)(d\chi/dP) = -6 \times 10^{-3}$ $kbar^{-1}$ in close agreement with the predictions of Eqs. (2) and (3). These results show that $TmAl_3$ is in the same category with the Pr-monochalcogenides, where Δ apparently increases with increasing P.

Pressure Dependence of T_S of $La_{1-x}Tb_xAl_2$

For superconductors containing magnetic impurities, the transition temperature, T_S, is depressed by the spin-dependent exchange scattering. When CEF effects are included, this scattering is a function of the relative thermal populations of the CEF levels as well as virtual scattering processes where the intermediate state is a higher CEF level. The superconducting-normal phase boundary for a system containing impurities with a non-magnetic CEF ground state may be written[14]

$$\ln(T_S/T_{So}) + \psi\{\tfrac{1}{2} + (\pi^2 T_S \tau_n)^{-1} y(\Delta/2T_S)\} - \psi(\tfrac{1}{2}) = 0 \quad (4)$$

where T_{So} is the transistion temperature for the undoped sample and $\psi(x)$ is the digamma function. The scattering time τ_n is inversely proportional to the number of impurities and depends on the conduction electron-localized moment exchange constant, and the function y(x) is monotonically decreasing. According to Eq. (4) the depression of T_S is attenuated for an impurity with a non-magnetic CEF ground state compared to an impurity with comparable exchange but where CEF effects are not important, e.g. Gd impurities.

We have measured dT_S/dP for two Tb doped $LaAl_2$ samples and dT_{So}/dP for undoped $LaAl_2$ (T_{So}=3.24 K). This system was selected because experiments showed[15] that the Γ_3(non-magnetic doublet) -Γ_5 splitting, Δ, is comparable to T_{So} and y(x) is expected to be sensitive to $\Delta/2T_S$ in this range. We have found that dT_S/dP is less for the Tb doped samples than dT_{So}/dP for $LaAl_2$ and that the depression is weaker for increasing concentration of Tb. Based on analysis using Eq. (4) we conclude that the effect of P is to increase Δ. This comes about through a further attenuation with increasing P of the pair-breaking term of Eq. (4). Similar results have been obtained by Umlauf *et al.*,[10] who include in their analysis a small P-dependence of the exchange constant, found by measuring dT_S/dP for Gd doped $LaAl_2$. Although quantitative comparison is difficult, we can relate these results to those tabulated in Table I above by noting that Δ increases with P as for the Pr-monochalcogenides and $TmAl_3$. Umlauf *et al.*[10] have shown that the results are rather insensitive to the particular CEF level scheme.

MAGNETICALLY ORDERED SINGLET GROUND STATE SYSTEMS

If the interion exchange in singlet ground state systems is sufficiently strong, magnetic ordering can result; this "induced" ordering phenomenon has been discussed by many authors.[16] In this section we discuss briefly the results of the pressure dependence of the ordering temperature for three intermetallics; Pr_3Tl, TbSb and TmS, where magnetic order is induced.

Pr_3Tl orders feromagnetically at $T_C \simeq 12$ K and has been extensively studied with neutron scattering and other techniques.[17,18] Within the framework of the molecular field model[18] the parameter η (which is proportional to J/Δ, where J is the interion exchange) determines whether the system will order magnetically: if $\eta<1$ the system will remain paramagnetic down to the lowest temperatures; for $\eta>1$ the system should order magnetically. It has been shown[19] that $dT_C/dP=-1.0$ K/kbar for Pr_3Tl. This very strong reduction was attributed to an increase of Δ with increasing P since it was found that $dT_N/dP=0$ for Nd_3Tl, where the ordering does not depend on Δ.

For Pr_3Tl, which has a Γ_1 ground state and $\Delta \simeq 80$ K, the parameter η is 1.014, i.e., close to the critical value for induced ordering.[19] In contrast, η is about 3.5 for TbSb,[20] which also has a Γ_1 ground state but with Δ close to the Néel temperature, ($T_N=15$ K). Our measurements show $dT_N/dP=+0.02\pm.03$ K/kbar for TbSb. Thus it seems that when η is much larger than the critical value, any effect on the pressure dependence of Δ (which in this case might be expected to decrease with P, in analogy with TmSb) is presumably of minor importance in determining the pressure dependence of the ordering temperature.

TmS appears to order antiferromagnetically at $T_N=5.2$ K based on specific heat and magnetic susceptibility measurements.[21] We have found $dT_N/dP=+0.15$ K/kbar for TmS, and it is tempting to ascribe this increase to a reduction of Δ. However, this conjecture is probably premature until other pressure-related experiments are performed on TmS. The results reported in this section are illustrated in Fig. 3.

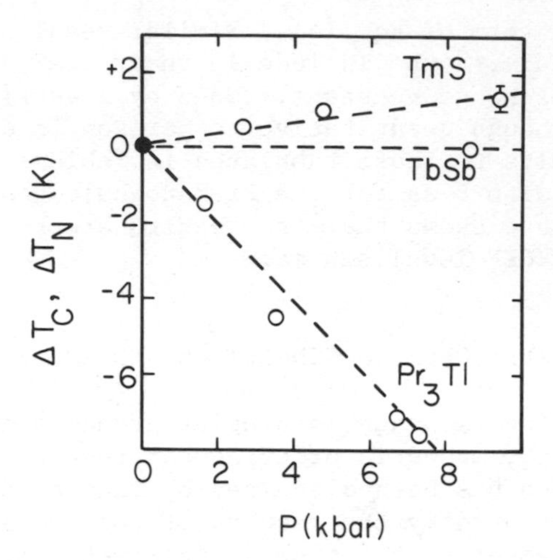

Fig. 2: Relative change of the magnetic ordering temperature vs pressure for Pr_3Tl (T_C=12 K), TbSb (T_N=15 K) and TmS (T_N=5.2 K).

"MIXED-VALENCE" TmSe

The mixed-valence phenomenon found for certain rare earth intermetallics has attracted a great deal of attention recently.[22] For TmSe several experiments have found evidence for mixing between the states associated with the $4f^{12}$ and $4f^{13}$ configurations (Tm^{3+} and Tm^{2+}, respectively). For example, the lattice constant for TmSe lies between that expected for $Tm^{2+}Se$ and $Tm^{3+}Se$ (the latter being the smaller of the two) and the Curie-Weiss moment (6.32 μ_B/Tm) is between that expected for free ion Tm^{3+} and Tm^{2+}. Recently, Ott et al[23] reported the existence of several low temperature magnetic transitions in the H-T plane using magnetostriction, thermal expansion and ac susceptibility techniques. To our knowledge neutron scattering has so far failed to show long range magnetic order at low temperatures.[23]

We have used magnetization, σ, vs field, H, measurements to investigate the low temperature magnetic anomalies.[24] Our measurements were carried out at various pressures up to 10.4 kbar, in fields up to 60 kG and for $1.4<T<20$ K. The sample used was a piece of the same sample used both by Ott et al[23] and in x-ray photoemission studies.[25] The σ vs H $\|$[100] data for P=0 shows the following: for T<3 K σ increases sharply at H≃6 kG and for T>3 K σ vs H is more S-shaped. The maxima of $d\sigma/dH$, determined by numerical differentiation of the data, were chosen as the transition fields, H_T. The H_T-T phase diagram obtained from the σ vs H data at P=0 is in good agreement with the data of Ref. 23.

In Fig. 3 is shown σ vs H at T=1.4 K for several pressurizations of the sample. Note that as P increases, H_T increases (at a rate +1.7 kG/kbar), the step in σ at H_T decreases, and the low field slope decreases and approaches a constant value. (The initial decrease of σ with P is about an order of magnitude larger than observed for any of the singlet ground state systems discussed above). A very reasonable extrapolation of the data of Fig. 3 shows that for P≃20 kbar, the step in σ at 1.4 K is expected to disappear.

The H_T-T phase diagrams of TmSe for P=0 and for five other pressures are shown in Fig. 4. The various P=0 transitions are labelled I-IV after Ref. 23. For T<3 K and low H, σ becomes temperature independent (this is indicated as boundary I). At II a small hysteretic (with H) change in σ is observed. The transitions at III are presumably first order while at IV they are of higher order;[26] thus the point at the intersection of III and IV is a critical point. The locus of these points (one for each P) forms a "critical line", L_1, on the H-T-P surface. (It is interesting to extrapolate L_1 to P≃20 kbar, where the step in σ (see Fig. 3) is expected to vanish. We speculate that another critical line, L_2, exists for P≃20 kbar, H≃40 kG and T<3 K; if L_1 and L_2 intersect nontangentially they would form yet another critical point of some kind.)

The expected effect of P on the valence of TmSe can be deduced from considerations of the lattice constant: increasing P favors $Tm^{3+}Se$, which has the smaller ionic volume. A 4% reduction in volume (corresponding to P≃20 kbar) should lead to complete conversion to $Tm^{3+}Se$. Although this argument is independent of our data, we note that it is consistent with the extrapolation of the data of Fig. 3: at P≃20 kbar the step in σ (labelled III in Fig. 4) is expected to vanish. Although the actual mechanism driving the transitions at III and IV is not clear, we note that the volume increases ($\Delta V/V \simeq 10^{-4}$) discontinuously at III and continuously at IV.[26] Since valence and ionic volume are related, a field-induced valence change

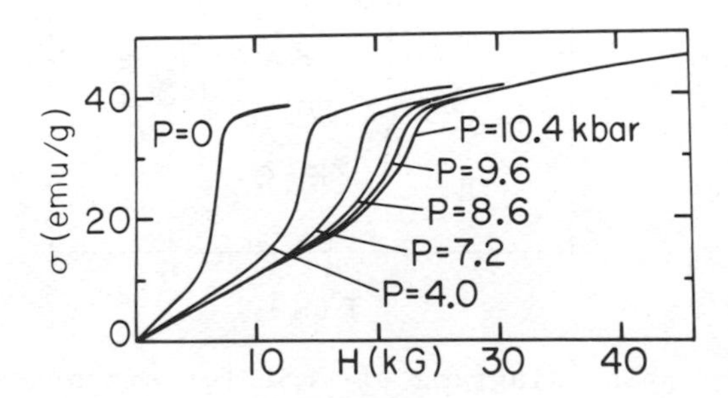

Fig. 3: Magnetization, σ, vs H of TmSe for six pressurizations T = 1.4 K.

may take place at these transitions. A volume increase favors $Tm^{2+}Se$, but the increase is far too small for complete conversion. The transitions at III and IV may also be associated with a field-induced lifting of the configuration degeneracy.

A detailed analysis of the complex magnetic properties of TmSe (and their pressure dependence) must take into account the CEF effects. The two Tm configurations should have quite different CEF-related magnetic properties: Tm^{2+} is a Kramers ion (J=7/2) and would have a magnetic ground state, whereas Tm^{3+} (J=4) would probably have a singlet (Γ_1) ground state with $\Delta \simeq 20$ K, in analogy with the Tm-monopnictides and Pr-monochalcogenides discussed above. We defer discussion of the nature of long range magnetic ordering (if any) in TmSe until neutron scattering experiments have been performed on samples showing the same magnetic transitions described here.

ACKNOWLEDGMENTS

The author wishes to express his gratitude to the Francis Bitter National Magnet Laboratory, M.I.T., where he has been a Visiting Scientist for the last seven years, and in particular to Dr. Simon Foner of that laboratory with whom the author has collaborated on much of the work reported here.

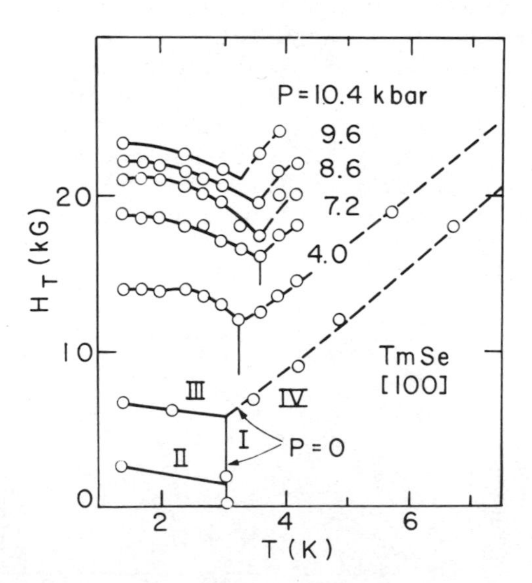

Fig. 4: H_T vs T phase diagrams of TmSe for six pressurizations. Numerals I-IV label the phase boundaries at P=0, using the notation of Ref. 23.

REFERENCES

1. R.P. Guertin and S. Foner, Rev. Sci. Instrum. 45, 863 (1974).
2. R.P. Guertin, J.E. Crow, L.D. Longinotti, E. Bucher, L. Kupferberg and S. Foner, Phys. Rev. B12, 1005 (1975).
3. J.E. Crow, R.P. Guertin and S. Foner, Phys. Lett. 54A, 324 (1975).
4. R.P. Guertin, W. Boivin, J.E. Crow, A.R. Sweedler and M.B. Maple, Solid State Commun. 13, 1889 (1973).
5. R.J. Birgeneau, E. Bucher, J.P. Maita, L. Passell and K.C. Turberfield, Phys. Rev. B8, 5345 (1973) and references therein.
6. H.L. Davis, H.A. Mook and E. Bucher, Bull. Am. Phys. Soc. 19, 206 (1974).
7. B.R. Cooper and O. Vogt, Phys. Rev. B1, 1211 (1974).
8. H.R. Ott and B. Luthi, Phys. Rev. Lett. 36, 600 (1976).
9. K.R. Lea, M.J.M. Leask and W.P. Wolf, J. Phys. Chem. Solids 23, 1381 (1962).
10. E. Umlauf, P. Holzer, J. Keller, M. Dietrich, W. Gey and R. Meier, Z. Physik 271, 305 (1974).
11. See Fig. 8 of Ref. 2.
12. H.T. Weaver and J.E. Schirber, AIP Conf. Proc. 24, 49 (1975).
13. See, for example M.E. Mullen, B. Luthi, P.S. Wang, E. Bucher, L. D. Longinotti, J.P. Maita and H.R. Ott, Phys. Rev. B10, 186 (1974).
14. P. Fulde, L.L. Hirst and A. Luther, Z. Physik 230, 155 (1970).
15. E. Umlauf, G. Pepperl and A. Meyer, Phys. Rev. Lett. 30, 1173 (1973).
16. See, for example, B.R. Cooper, Phys. Rev. 163, 444 (1967).
17. R.J. Birgeneau, J. Als-Nielsen and E. Bucher, Phys. Rev. B6, 2724 (1972).
18. K. Andres, E. Bucher, S. Darack and J.P. Maita, ibid., 2716.
19. E. Bucher, J.P. Maita, G.W. Hull, Jr., J. Sierro, C.W. Chu and B. Luthi, Proceedings of the First Conference on Crystalline Electric Field Effects in Metals and Alloys, Univ. of Montreal, June, 1974. P. 221.
20. T.M. Holden and W.J.L. Buyers, ibid.. P. 32.
21. E. Bucher, K. Andres, F.J. diSalvo, J.P. Maita, A.C. Gossard, A.S. Cooper and G.W. Hull, Jr., Phys. Rev. B11, 500 (1975).
22. For a review see C.M. Varma, Rev. Mod. Phys. 48, 219 (1976).
23. H.R. Ott, K. Andres and E. Bucher, AIP Conf. Proc. 24, 40 (1975).
24. R.P. Guertin, S. Foner and F.P. Missell. Submitted for publication.
25. M. Campagna, E. Bucher, G.K. Wertheim, D.N.E. Buchanan and L. D. Longinotti, Phys. Rev. Lett. 32, 885 (1974).
26. Y. Shapira, private communication.

STUDY OF HYPERFINE INTERACTIONS IN SOLIDS BY INELASTIC SPIN FLIP SCATTERING OF NEUTRONS

A. HEIDEMANN[x] and D. RICHTER[+]

[x]Institut Laue-Langevin, 156X Centre de Tri, 38042 GRENOBLE CEDEX, France
[+]Institut für Festkörperforschung, KFA Jülich, 517 JULICH, West Germany.

ABSTRACT

Inelastic spin flip scattering of neutrons is presented as a new experimental technique to study hyperfine interactions in solids. A short outline of the theory is given. The advantages and limitations of the method are discussed by comparison with other techniques such as Mössbauer spectroscopy and NMR. A number of nuclei are quoted which are suited for the method. The application of the new method is demonstrated by a number of examples where the hyperfine field in compounds containing hydrogen, cobalt and vanadium was investigated.

I. INTRODUCTION

In the last decade a considerable improvement of the energy resolution of neutron spectrometers has been achieved. Energy changes as low as $3 \cdot 10^{-8}$ eV or 8MHz can be detected with the backscattering technique[1,2]. The neutron spin echo principle[3] should shift this resolution limit again by one order of magnitude to lower frequencies. Therefore a good overlap exists now between inelastic neutron scattering and other techniques such as NMR, Mössbauer effect, Raman and Infrared spectroscopy. One application of the neutron spectroscopy with very high energy resolution is the study of hyperfine interactions. In this paper we discuss this subject in some detail. In section II we outline the theory of inelastic spin flip scattering of neutrons and compare the method with other techniques. In section III some results of the investigation of hyperfine fields in compounds containing hydrogen, cobalt and vanadium are presented.

II. EXPERIMENTAL METHOD

The interaction potential between nuclei and a neutron is given by a sum of a spin independent and a spin dependent part. The latter enables us to measure spin dependent energy splittings of nuclear ground states such as Zeeman or quadrupol splitting by inelastic neutron scattering. We want to discuss the case of Zeeman splitting. The energy difference ΔE between two nuclear Zeeman levels is given by

$$\Delta E = \frac{\mu\mu_k}{I} H_{hf} \qquad (1)$$

where μ is the number of nuclear magnetons μ_k, I the nuclear spin and H_{hf} the magnetic field. For such a system the double differential scattering cross-section has the following form[4]:

$$\frac{d^2\sigma}{d\Omega d\omega} = \frac{k}{k_o} e^{-2W} \frac{1}{3} I(I+1)\alpha'^2 \left[\delta(\hbar\omega)+\delta(\hbar\omega+\Delta E)+\delta(\hbar\omega-\Delta E)\right] \qquad (2)$$

where k_o, k are the wave vectors of the incident and scattered neutron, $\hbar\omega$ is the energy transfer, e^{-2W} the Debye-Waller factor and α' the incoherent scattering length. Eq.(2) has been derived assuming no correlation between the nuclear spin system and the lattice as well as no nuclear spin-spin interactions. Zero phonon approximation has been applied. The first delta function in Eq.(2) centered at $\hbar\omega = 0$ results from neutrons scattered without spin flip; the last two delta functions centered at $\pm \Delta E$ are due to neutrons scattered with spin flip. Therefore one can measure hyperfine interactions with neutrons provided that :

1. The energy resolution is equal or better than the hyperfine splitting.
2. The incoherent scattering length α' is sufficiently large to get a signal.

The required energy resolution is achieved by the backscattering technique which has been realised at the FRJ reactor Jülich and with the IN10 at the Grenoble HFR[2]. Table I shows a number of nuclei which fulfil the second condition. In Fig.1a we give a characteristic example of an energy spectrum of neutrons scattered incoherently in ferromagnetic cobalt. In the following we compare the neutron method with Mössbauer and NMR techniques:

1. <u>Mössbauer Effect</u>. All nuclei in Table I are not Mössbauer isotopes. Therefore a new class of substances is accessible with the neutron method. Though the energy width of Fe^{57} is $0.9\cdot 10^{-8}$eV compared to $30\cdot 10^{-8}$eV in the neutron case, the sensitivity to internal magnetic fields of a Mössbauer experiment in iron is roughly the same as that of a neutron experiment in cobalt due to the higher nuclear magnetic moment of cobalt. The neutron energy spectrum with its three lines is simpler than the general rather complex

Nucleus	Spin	$4\pi\alpha'^2 I(I+1)$ [barn]	ΔE [μeV]
H^1	1/2	80	1.77
Li^7	3/2	0.6	0.69
Na^{23}	3/2	1.4	0.47
Sc^{45}	7/2	4.4	0.43
V^{51}	7/2	5.1	0.46
Co^{59}	7/2	6.4	0.42
Zr^{91}	5/2	0.7	0.17
La^{139}	7/2	1.7	0.25

Table 1: Nuclei suited for inelastic spin flip scattering. ΔE is calculated for a magnetic field of 100 kOe.

Mössbauer spectra. This is advantageous for investigations of substances with more than one hyperfine field or with field distributions. In the neutron case the Debye Waller factor is near to one even at room temperature. Therefore no intensity problems arise at elevated temperatures.
2. NMR has the advantage of a very high resolution. However, from this a number of difficulties arises, which do not exist in the neutron experiment. It is often rather tedious to find an unknown resonance. This is especially true in antiferromagnetic samples where no enhancement factor exists. Further on short relaxation times make the signal to noise ratio small. Investigations of mixed systems in the high concentration range are often impossible. Magnetically hard materials are not easily accessible. The sensitivity of the neutron experiment does not depend on relaxation times and enhancement factors. Therefore all nuclei in a substance are contributing equally to the signal. Always the bulk properties are measured. NMR experiments are performed at a transferred momentum of $q \sim 0$ Å^{-1} whereas the neutron experiments can cover a q-range: $0.1 < q < 2$ Å^{-1}. Only in the case of nuclear spin-spin correlations which have been neglected in the derivation of Eq.(2) would different results be obtained by NMR and neutrons.

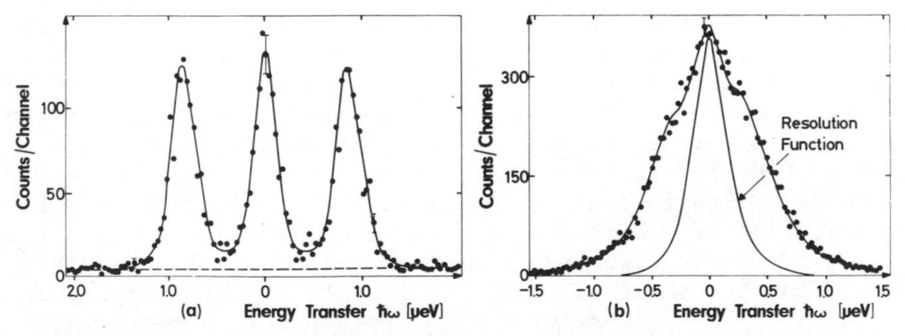

Fig. 1: Energy spectrum of neutrons scattered in a) cobalt at 300 K, b) $TbFe_2H_4$ at 4.2 K.

III. APPLICATIONS

Until now only three of the eight nuclei shown in Table I have been used to measure hyperfine interactions with neutrons: Vanadium[4,5] cobalt[2,6] and hydrogen[7]. We will present for each of these nuclei one example: Hyperfine interaction in the rare earth hydride $TbFe_2H_4$, in CoPt alloys with different degrees of atomic order and V_3O_7.

1. $\underline{TbFe_2H_4}$. No NMR data are available on the hyperfine fields at protons in magnetically ordered substances. The first successful attempt to determine this type of interaction was achieved at $TbH_{1.9}$ by neutron scattering[7]. Here we present a similar study on ferromagnetic $TbFe_2H_4$ (T_c = 326 K) with a cubic C_{15} structure where the iron moments are antiparallely ordered with respect to the terbium moments. The neutron scattering experiments were performed at 4.2 and 70 K. The resulting spectra could be fitted satisfactorily with a convolution of the resolution function with a scattering law consisting of an elastic δ -function $\delta(\hbar\omega)$ and two shifted (ΔE) and broadened (δE) Lorentzians (see Fig. 1b). The results are given in Table II. The broadening of the inelastic peaks indicates that there are proton sites which are nonequivalent with respect to their magnetic surroundings. The hyperfine field at the proton is a sum of mainly two components: the dipole field due to the localized moments of the Tb and Fe atoms and the Fermi contact field. With the knowledge of the magnetic structure and the positions of the protons the dipole field can be calculated. The remaining contact field is directly related to the conduction electron polarization (CEP) without corrections concerning the core polarization. Therefore the CEP-value can be obtained quantitatively as soon as reliable calculations of the electron density at the protons are available.

2. CoPt. The ferromagnetic alloy CoPt can exist in various stages of disorder. Going from the ordered to disordered phase the magnetic moment μ which is mainly localized on the cobalt site[8] increases by about 80%. Due to the simultaneous presence of magnetic hardness[9] and disorder, NMR experiments are very difficult. We investigated the influence of the disorder on the internal magnetic field at the Co nucleus by neutron scattering. The position and the width of the field distribution were determined for three different degrees of order. (see Table III). The decrease of the hyperfine field per Co moment with increasing order as well as the width of the field distribution can be explained by the influence of the CEP[10].

3. $\underline{V_3O_7}$ belongs to the group of vanadium oxides with the

T [K]	ΔE [kOe]	δE [kOe]
4.2	18.5	15
70	14.9	10

Table II: Size ΔE and width δE of the hyperfine-splitting of the proton in $TbFe_2H_4$

Table III: Size ΔE and width δE of the hyperfine field at the cobalt nucleus in CoPt at 4.2K. S is the short range order parameter (obtained by X ray diffraction).

S	ΔE [kOe]	δE [kOe]	$\Delta E/\mu_{Co}$ [kOe]
0	186	42	113
0.6	151	30	106
1	117	0	94

general formula V_nO_{2n+1}. From magnetisation and susceptibility measurements[11] it was estimated that V_3O_7 undergoes a phase transition at T_c = 18 K from a paramagnetic to a ferromagnetic state, where 1/3 of the vanadium atoms occupy magnetic sites with a magnetic moment of about $1\mu_B$ leading to an estimated hyperfine field of about 100 kOe. NMR experiments at 4.2 K failed to find the internal field. Fig.2 shows the results obtained from the neutron experiments. The internal field extrapolated to OK is 204 kOe which is until now the largest value ever found in vanadium oxides. It approaches zero at T_c = 5.2 K. The reason for the failure of the NMR experiment may be twofold. The measured field is two times larger than expected. Possibly the relaxation times were already too short to observe any transient signal. The T_c value disagrees with the older one. This can be due to the fact that the former experiments were performed in strong external fields up to 50 kOe.

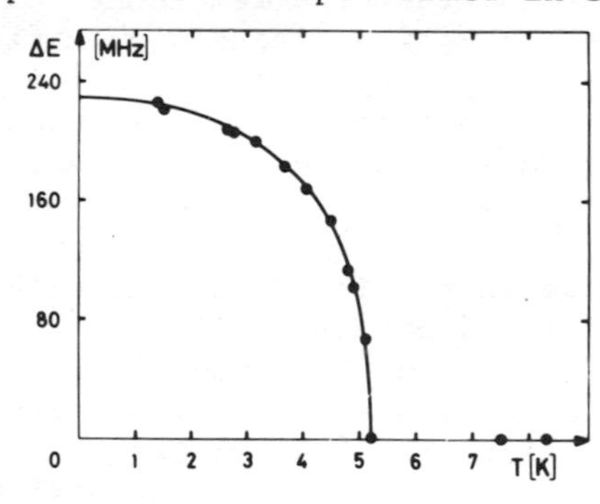

Fig.2: Temperature dependence of the hyperfine field at the vanadium nucleus in V_3O_7

REFERENCES

1. B. Alefeld, Kerntechnik 14, 15 (1972)
2. A. Heidemann, Z. Physik B20, 385 (1975)
3. F. Mezei, Z. Physik, 255, 146 (1972)
4. A. Heidemann, Z. Physik, 238, 208 (1970)
5. A. Heidemann, K. Kosuge, S. Kachi, Phys. stat. sol (a)35, 481 (1976)
6. A. Heidemann, D. Richter, K. Buschow, Z. Physik, B22, 367 (1975)
7. D. Richter, J. Magnetism and Magnetic Materials 2, 109 (1976)
8. B. Van Laar, Journal de Physique 25, 600 (1964)
9. D.C. Martin, J. Phys. F. Metal Phys., 5, 1031 (1975)
10. D. Richter, K.H.J. Buschow, A. Heidemann, to be published
11. M. Bayard, J. Grenier, M. Pouchard, P. Hagenmüller, Mat. Res. Bull. 9, 1137 (1974)

A METHOD FOR CALCULATING THE ELECTROSTATIC POTENTIAL IN CRYSTAL LATTICES

S. Matthies

Joint Institute for Nuclear Research, Dubna
P.O. Box 79, 101000 Moscow, USSR

ABSTRACT

A method for calculating the electrostatic potential in multipole lattices is discussed. The method formulae employing a form-factor type expression of the basis and special sums over the simple Bravais lattices are given.

A typical problem in solid state theory is the determination of the electrostatic potential, connected with a given density $\rho^{\ell,m}(\vec{r})$ of multipole charges in the unit cell of a crystal lattice. The purpose of this report is to give a brief account of the basic idea and the main results of a method for calculating the electrostatic potential in crystal lattices with high accuracy, which was lately described in detail by the author in papers (1) and (2).

For more visuality we consider in the following the potential in a lattice of point charges
$(\ell=0,\rho(\vec{r}) = \sum_{j=1}^{\infty} Z_j\delta(\vec{r}-\vec{r}_j))$

$$V_o(\vec{r}) = \sum_{j=1}^{\infty} Z_j \ / \ |\vec{r}_j-\vec{r}| \quad , \tag{1}$$

where the sum extends over all charges Z_j in the knots $\vec{r}_j$. The difficulties of the calculation of the potential (1) are connected with the fact that the sum over j converges very slowly and moreover conditionally only, i.e. the result of a "direct" evaluation depends on the order of summation.

To obtain a more rapid convergence of sums of the form (1), various methods have been developed which may be classified by two types:

- direct summation methods developed by Evjen (3) and recently by Hajj (4) and
- Ewald-type methods (5).

In the crystal-field theory the following representation of (1) is used ($\vec{r}_1=0$; $r<\min r_j$; $j>1$):

$$V_o(\vec{r}) = \sum_{L=0}^{\infty} \sum_{M=-L}^{L} A_{L,M}\, r^L\, Y_{L,M}(\Omega_{\vec{r}}) + Z_1/r \quad , \quad \text{where} \tag{2}$$

$$A_{L,M} = 4\pi(-1)^M/(2L+1) \sum_{j=2}^{\infty} Z_j Y_{L,M}(\Omega_{\vec{r}_j})/r_j^{L+1} \equiv \frac{4\pi(-1)^M}{2L+1} W_{L,-M} \quad . \tag{3}$$

In order to calculate the "crystal-field parameters" $A_{L,M}$, not only direct methods but also Ewald-type methods may be used, which have been done by de Wette (6) for the sums

$$W_{L,M} = \sum_{j=2}^{\infty} Z_j Y_{L,M}(\Omega_{\vec{r}_j})/r_j^{L+1} \quad . \tag{4}$$

However, these methods cannot separate the sums over the charges of the basis of the unit cell from sums over simple Bravais lattices which may be tabulated*. This leads automatically to a strong raise of computer time, if we have to consider the case of a continuous charge distribution $\rho(\vec{r})$ in the unit cell instead of isolated point charges. A continuous charge distribution has to be introduced in order to account for the conduction electrons.

The starting point of our method is the division of the sum (4) into three parts: the contribution of the basis ($i=1,2,\ldots,N$), of the neighbouring cells, and of the infinite rest of the lattice

---

* Kanamori et al. (7) succeed in separating a basis structur factor, but their approach works only for isolated point charges.

$$W_{L,M} = W^{B}_{L,M} + W^{(n_o-1)}_{L,M} + W^{(n_o)}_{L,M} ,\qquad \text{where}$$

$$W^{B}_{L,M} = \sum_{i=2}^{N} Z_i Y_{L,M}(\Omega_{\vec{r}_i})/r_i^{L+1} ,$$

$$W^{(n_o-1)}_{L,M} = \sum_{n=1}^{n_o-1} \sum_{i=1}^{N} Z_i Y_{L,M}(\Omega_{\vec{r}_i+\vec{r}_{\vec{n}}})/|\vec{r}_i+\vec{r}_{\vec{n}}|^{L+1} , \tag{5}$$

$$W^{(n_o)}_{L,M} = \sum_{n=n_o}^{\infty} \sum_{i=1}^{N} Z_i Y_{L,M}(\Omega_{\vec{r}_i+\vec{r}_{\vec{n}}})/|\vec{r}_i+\vec{r}_{\vec{n}}|^{L+1}$$

and $\vec{r}_j=\vec{r}_i+\vec{r}_{\vec{n}}$. $\vec{r}_{\vec{n}}$ is the vector of a knot of the simple Bravais lattice, lying on the surface, which has the number n. There are $24n^2+2$ knots:

$$\vec{r}_{\vec{n}} = \vec{a}\, N^n_a + \vec{b}\, N^n_b + \vec{c}\, N^n_c ,$$

$$\left\{N^n_a = -n,n \quad N^n_b = -n,-(n-1),\ldots,n \quad N^n_c = -n,-(n-1),\ldots,n\right\} ,$$

$$\left\{N^n_a, N^n_b = -(n-1),-(n-2),\ldots,(n-1) \quad N^n_c = -n,n\right\} ,$$

$$\left\{N^n_a = -(n-1),(n-2),\ldots,(n-1) \quad N^n_b = -n,n \quad N^n_c = -n,-(n-1),\ldots,n\right\} .$$

For any simple Bravais lattice a number $\bar{n}_o$ may be found for which $d<|\vec{r}_{\vec{n}}|$ $(n\geq\bar{n}_o)$, where d is the length of the largest diagonal of the unit cell. For $r_1<r_2$

$$\frac{Y_{\ell,m}(\Omega_{\vec{r}_1+\vec{r}_2})}{|\vec{r}_1+\vec{r}_2|^{\ell+1}} = \sum_{L=o}^{\infty} \sum_{M=-L}^{L} C_{\ell,m;L,M}\, r_1^L\, Y_{L,M}(\Omega_{\vec{r}_1}) \frac{Y_{\ell+L,m-M}(\Omega_{\vec{r}_2})}{r_2^{\ell+L+1}} ,$$

where

$$C_{\ell,m;L,M} = (-1)^{L+M} \left(\frac{4\pi(2\ell+1)(\ell+L-m+M)!(\ell+L+m-M)!}{(2(L+\ell)+1)(2L+1)(\ell+m)!(\ell-m)!(L+M)!(L-M)!}\right)^{1/2} ,$$

$W^{(n_o)}_{L,M}$ may be represented for $n_o\geq\bar{n}_o$ by

$$W_{L,M}^{(n_o)} = \frac{2L+1}{4\pi}(-1)^M A_{L,-M}^{(n_o)} = \sum_{L'=o}^{\infty} \sum_{M'=-L'}^{L'} C_{L,M;L',M'} D_{L',M'} \sigma_{L+L',M-M'}^{(n_o)} . \qquad (6)$$

In this way one succeeds in introducing into $W_{L,M}^{(n_o)}$ an expression of form factor type of the basis

$$D_{L',M'} = \sum_{i=1}^{N} Z_i r_i^{L'} Y_{L',M'}(\Omega_{\vec{r}_i}) = \int_{\text{unit cell}} \rho_o(\vec{r}) r^{L'} Y_{L',M'}(\Omega_{\vec{r}}) d\vec{r}$$

and in separating universal sums for the appropriate Bravais lattice

$$\sigma_{\lambda,\mu}^{(n_o)} = \sum_{n=n_o}^{\infty} Y_{\lambda,\mu}(\Omega_{\vec{r}_{\vec{n}}}) / r_{\vec{n}}^{\lambda+1} . \qquad (7)$$

These sums may be tabulated by means of which the calculation of the parameters $A_{L,M}$ is considerably simplified. The series (6) converge according to the estimation of paper [1] more quickly than

$$\sum_{L'} (L'+L)^{L+1} / n_o^{L'} .$$

The greater n_o the more rapid the series (6) converge, but with increasing n_o the calculation size for the sums $W_{L,M}^{[n_o-1]}$ is rising. E.g. for the NaCl lattice it was shown in paper [2] that for the evaluation of the parameters $A_{L,M}$ in cubic lattices with a precision of ten decimal places it is sufficient to choose n_o=3 and to break off the series (6) at L'=14. Because of the crystal-field symmetry in the space point under consideration as well as the symmetry of the given Bravais lattice a series of terms of the sum (6) may be zero. This additionally simplifies the calculations.

The main quantities of the given calculating method are the sums $\sigma_{\lambda,\mu}^{(n_o)}$. The optimum ways of their calculation are described in [2]. In the same paper numerical values of these sums for the simple cubic Bravais lattice (n_o=1,3,5 and λ=4,6,...,22) and formulae for the case of multipole lattices are given. The contribution of the infinite rest of the multipole lattice to the potential value contains the same sums $\sigma_{\lambda,\mu}^{(n_o)}$ as in the case of a monopole lattice. Therefore it seems to be useful to tabulate these sums for the most typical Bravais lattices of metals.

We have estimated [2] the ratio of expenditure, which is needed to calculate with a precision of ten decimal places the potential (1) in cubic lattices if the Ewald-method or our method are used.

This ratio for the evaluation of $V_O(\vec{r})$ at M points in the unit cell (n_O=3) may be represented in a rough approximation by (N,M>1)

$$\frac{122\ M + 275\ N\ M}{528\ M + 12.5\ N\ M + 529\ (N+1)} \quad . \tag{8}$$

From (8) follows that our method gives a considerable gain of time for large M and N, which is especially valuable if we have to calculate the potential at many points or if we have a complicated basis as in the case of conduction electrons (N>>1). The ratio (8) favours even more the present method if the charge distributions have a high point symmetry.

REFERENCES

1. S. Matthies, phys. stat. sol. (b) 74, 69 (1976)
2. S. Matthies, phys. stat. sol. (b) 74, 531 (1976)
3. H.M. Evjen, Phys. Rev. 39, 675 (1932)
4. F.Y. Hajj, J. Phys. C7, 1069 (1974) and earlier works
5. M.P. Tosi, Solid State Phys. 16, 1 (1964)
   S. Adler, Physica 27, 1193 (1961)
   I. Birman, J. Phys. Chem. Solids 6, 65 (1958)
6. F.W. de Wette, Physica 25, 1225 (1959) and earlier works
7. J. Kanamori, T. Moriga, K. Mitizyki, T. Nagamiya, J. Phys. Soc. Japan 10, 93 (1955)

NEUTRON INELASTIC SCATTERING STUDY OF THE COMPOUND $CeAl_3$

A.P. Murani[+], K. Knorr[*] and K.H.J. Buschow[§]

[+] Institut Laue-Langevin, 38042 Grenoble, France
[*] Institut für Physik, Uni. Mainz, 65 Mainz, B.R.D.
[§] Philips Research Lab., Eindhoven, The Netherlands

ABSTRACT

We have carried out time-of-flight inelastic scattering measurements on a polycrystalline $CeAl_3$ sample in the temperature range between 5K and 100K with neutrons of incident energy of 18.3 meV at a scattering angle of 7.5° (q_{el} = 0.4Å^{-1}). Our measurements at 5K reveal a broad double-humped spectrum on the neutron energy loss side which appears as if due to two broad upward transitions from the ground state to two higher-lying states at energies close to around 5.2 and 7.6 meV. With increasing temperature the double-humped structure is apparently replaced by a "single" broad peak with its centre of gravity shifting progressively towards lower energy transfers. A simple interpretation of the results in terms of transitions between normal crystal field split energy levels is not possible. However, allowing for the role of the strong s-f interaction, which could significantly modify the crystal field states, there are indications in the data that the ground state is composed principally of the $|\pm 3/2>$ state with the other states having the character of the $|\pm 5/2>$ and the $|\pm 1/2>$ doublet states lying higher up in energy, in that order.

INTRODUCTION

The intermetallic compound $CeAl_3$ (hexagonal, Ni_3Sn structure) exhibits some very interesting physical properties[1,2]. The unusual electrical resistivity behaviour-increase of ρ with decreasing temperature, passing a maximum at about 35K and then decreasing again at lower temperatures- being its most striking feature which has brought the compound wide attention.

The importance of the role of the strong s-f interaction together with the splitting of the 6-fold degenerate f-state of the Ce^{3+} ion into 3 doublets by the crystalline electric field of the lattice has been recognised in the resistivity behaviour as, for example, in the Kondo-side-band theory of Maranzana[3] and also in a more generalised s-f interaction theory of Cornut and Coqblin[4]. In the latter, however, a Kondo minimum is always predicted at low temperatures whatever the magnetic quantum state of the ground doublet, contrary to what is now clearly established to be the case[2]. The specific heat measurements on $CeAl_3$ by Mahoney et al[5] show that the total magnetic entropy of the system (extrapolated to zero temperature) is in excess of that expected from the simple splitting of the 4f level into three doublet states, but is less than the value expected for the complete lifting of the degeneracy of the 4f level. Also, in their neutron diffraction measurements on the compound the authors find no evidence of any magnetic ordering down to low temperatures, suggesting therefore that the s-f interaction is primarily responsible for the observed departures from the simple crystal field behaviour. The recent measurements by Andres et al[2] of the low temperature electrical resistivity, magnetic susceptibility and the specific heat of the compound have helped to further illuminate the role of the s-f exchange interaction and provided additional proof for the absence of magnetic ordering in the compound. An alternative explanation of the electrical resistivity of the compound has been offered by the authors[2] in terms of resonant scattering of conduction electrons passing a maximum when kT is approximately equal to the overall CEF splitting.

In the following we report time-of-flight neutron inelastic scattering measurements on the compound, carried out with a view to investigate the possible CEF level scheme of the 4f ions, a knowledge of which is useful for a proper understanding of the various physical properties of the system, and for which satisfactory experimental information has so far not been available.

EXPERIMENT

The measurements were made on polycrystalline $CeAl_3$ prepared by arc-melting and vacuum annealing at 900°C for several weeks. X-ray diffraction showed no detectable trace of any other phase. The inelastic neutron scattering experiments were performed on the modified IN7 time-of-flight spectrometer at the Institut Laue-Langevin. Neutrons of incident energy 18.3 meV were obtained in the 2nd order Bragg reflection from graphite double monochromators and pulsed by means of a Fermi chopper. The sample was placed inside a variable temperature aluminium cryostat a short distance from the chopper. A group of six He^3 counters were mounted at the end of a helium-filled flight tube providing a path-length of 2.35 m between the sample and the detectors. The low scattering angle of 7.5° (mean) was chosen so as to keep the momentum transfer as small as possible over the energy range of interest. The elastic energy resolution of the spectrometer was 7 % under these conditions.

Fig. 1 shows the observed spectra at several temperatures where the intensity of the scattered neutrons is plotted vs. the energy transfer on the time-of-flight scale. Each spectrum was obtained over a roughly constant counting period of 3 days. Because of the temporary arrangement of the spectrometer with its insufficient shielding, the time-independent neutron background was relatively high, a significant contribution to which was from the "off" background of the chopper due to its partial transparency, especially to higher order neutrons in the beam.

ANALYSIS AND DISCUSSION

The observed spectra are in general rather broad, such that their analysis in terms of the simple well defined transitions between crystal field split states of the 4f ion does not seem possible. Indeed, such a simple interpretation of the spectra for a system with strong s-f interaction may itself be questionable, especially in view of the specific heat results[5] which show significant departures from the behaviour (including entropy considerations) expected for a simple crystal field system. On the other hand the role of the crystalline field is shown to be important in the interpretation of the electrical resistivity behaviour of the compound. Hence it appears that a proper interpretation of the neutron scattering results should include transitions between crystal field states, but also allow for their possible modifications due to the strong s-f interaction.

Among the observed spectra the low temperature spectrum at 5K shows a definite double-humped inelastic structure reminiscent of two upward transitions from the ground state. Confidence in the reality of the structure was gained by measurements repeated on another time-of-flight spectrometer, although in the direct measurement of $S(q,\omega)$ with a triple axis spectrometer (in the constant q and constant k_f mode) only a single broad peak at around $\sim$ 7.5 meV was found. This observation, however, does not necessarily contradict the time-of-flight spectra where the scattered neutron intensity in equal time channels is weighted towards lower energy transfers thus favouring the observation of the additional hump at the lower energy.

Results of our approximate attempts to fit the 5K spectrum to two inelastic Lorentzian lines of half-width $\sim$ 2 meV centred about energy transfers of $\sim$ 5.2 and $\sim$ 7.6 meV are shown in Fig. 1. Some allowance is also made for the central quasi-elastic line whose

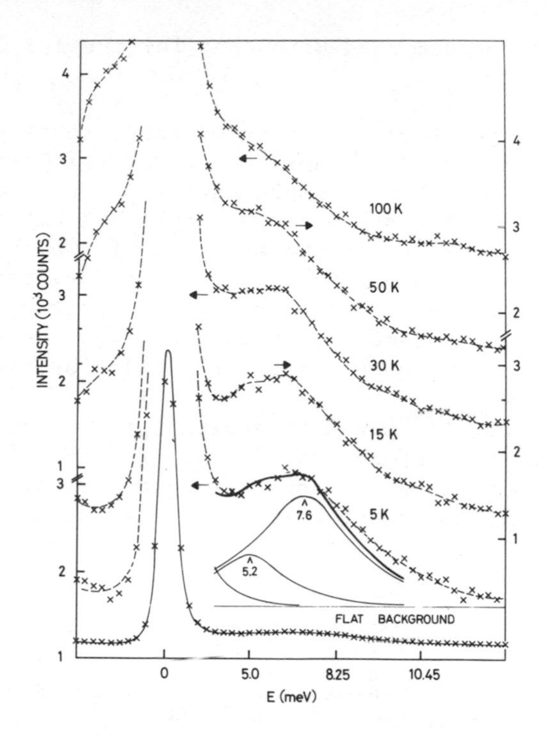

Fig. 1: Scattered intensity of the TOF spectra vs. energy transfer at various temperatures for polycrystalline $CeAl_3$. The thick continuous cruve shows the result of the fit constructed from various components indicated by the fainter lines.

presence is strongly evident in the spectrum (and has been verified in a supplementary experiment with lower incident energy). Observation of two upward transitions from the ground state suggests that it has the character of the $|\pm 3/2>$ doublet which is in accordance with the interpretation of the resistivity data[3]. However, the relative intensity of the two fitted Lorentzians is in the ratio $\sim$ 1:3. If they were simply related to the transitions $|\pm 3/2> \rightarrow |\pm 5/2>$ (at $\sim$ 5.2 meV) and $|\pm 3/2> \rightarrow |\pm 1/2>$ (at $\sim$ 7.6 meV), the intensities should be in the ratio 1 : 1.6, suggesting that as a result of the s-f interaction (which appears responsible for the observed additional entropy[5] interpreted by those authors in terms of lifting of the degeneracy of the ground state) the intensity of the $|\pm 3/2> \rightarrow |\pm 5/2>$ transition is reduced by about 50 %, or that the intensities of both the transitions are affected but the reduction is greater, in proportion, for the above transition compared with that for the $| \pm 3/2 > \rightarrow | \pm 1/2 >$ transition.

From the temperature variation of the intensity it is clear that the observed spectra can only be of magnetic origin. Furthermore measurements on a $LaAl_3$ sample have shown no evidence of any similar structure in the same energy range. The higher temperature spectra if interpreted in the above scheme of two transitions from the ground state suggest not only that the transitions broaden further with increasing temperature but it appears also that their relative intensities become more balanced, i.e. the ratio of the intensities seem to approach the value 1 : 1.6 as required.

In conclusion, the present results appear consistent with the interpretation of many of the physical properties of the compound $CeAl_3$. Clearly further neutron scattering measurements are needed for a fuller understanding of this very interesting system.

REFERENCES

1. K.H.J. Buschow, H.J. van Daal, F.E. Maranzana and P.B. Van Aken, Phys. Rev. B3, 1662 (1971)

2. K. Andres, J.E. Graebner and H.R. Ott, Phys. Rev. Lett. 35, 1779 (1975).

3. F.E. Maranzana, Phys. Rev. Lett. 25, 239 (1970)

4. B. Cornut and B. Coqblin, Phys. Rev. B5, 4541 (1972)

5. J.V. Mahoney, V.U.S. Rao, W.E. Wallace and R.S. Craig, Physical Rev. B9, 154 (1974).

CRYSTAL ELECTRIC FIELDS IN RARE-EARTH Al_2 COMPOUNDS

H. Happel[+], P. v. Blanckenhagen[x], K. Knorr[o], A. Murani["]

+ Physikalisches Institut der Universität Frankfurt
x Kernforschungszentrum Karlsruhe, IAK
o Institut für Physik der Universität Mainz
" Institut Laue-Langevin, Grenoble

ABSTRACT

Neutron time-of-flight measurements have been performed on $REAl_2$ compounds (RE = Pr, Ho, Er, Tm) in the paramagnetic region. Resolved crystal field transitions are observed in $ErAl_2$ and $TmAl_2$. We deduce crystal field parameters (x = 0.16, W = -0.030 meV) and (x = -0.28, W = +0.040 meV) for $ErAl_2$ and $TmAl_2$ respectively.

In the $REAl_2$ (RE = rare earth) intermetallic compounds the RE^{3+}-ions are subject to a crystalline electric field (CEF) of cubic symmetry. The cubic CEF is described by two parameters A_4 and A_6 or x and W [1]. Though there are a large number of measurements on the concentrated $REAl_2$ series([2] and references therein), no spectroscopic measurements of the CEF-splittings have been carried out up to now as it is possible by inelastic neutron scattering in the paramagnetic region. Instead of this, the CEF-parameters published were either deduced from magnetic measurements or from neutron scattering results on the dilute systems $Y_{0.85}Er_{0.15}Al_2$ [3] and $Y_{0.75}Tm_{0.25}Al_2$ [4]. In order to determine the CEF-parameters directly in the $REAl_2$ series we have carried out inelastic neutron scattering experiments on the four compounds which have the lowest Curie temperatures (RE = Tm, Er, Ho, Pr).

Polycrystalline samples were prepared by arc-melting stoichiometric amounts of 99.9 % RE and 99.999 % Al (both obtained from Metals Research) in a high purity argon atmosphere followed by levitation melting for further homogenization. Neutron diffraction did not give any indication for the existence of a second phase.

Neutron time-of-flight (TOF) measurements were performed on the spectrometer IN7 at the ILL, Grenoble, and on the reactor FR2 in Karlsruhe between 4.2 K and room temperature. Varying the energy of the incident neutrons between 5 meV and 50 meV the momentum transfer for elastic scattering covered the range from 0.1 to 2.5 $Å^{-1}$. Some of the TOF spectra for $ErAl_2$, $TmAl_2$, and $HoAl_2$ above the Curie temperatures are given in Figs. 1-3.

The information on the CEF-transitions yielded by the TOF-spectra was of different quality for the four systems investigated. The situation was most clear for $ErAl_2$. Here we observed one transition at 7.0 meV on the neutron energy loss side at 15 K. Raising the temperature this transition became weaker in intensity and a second line at 5.2 meV started growing. At 80 K the second one was finally equal to the former one (Fig. 1). At elevated temperatures the corresponding lines could be seen on the neutron energy gain side of the spectrum though with worse resolution. The magnetic origin of these transitions was verified by their Q-dependence and by a comparison with TOF-results on $LaAl_2$ [5]. From these observations we conclude on the following conditions:

1. In $ErAl_2$ there is a transition from the ground to an excited state at 7.0 meV.
2. The 5.2 meV transition starts from an excited state. It must have a larger matrix element than the 7.0 meV transition.

For $TmAl_2$ we observed one resolved CEF-transition at 4.9 meV. In addition the elastic line was considerably broadened (2.5 meV FWHM). At 12 K which is just above the onset of magnetic order the elastic line adapted an asymmetric shape with a shoulder at about 1.3 meV. We therefore propose to interprete these results by a first excited CEF-state at 1.3 meV and a second one at 4.9 meV.

For $HoAl_2$ a broad quasielastic line and a shoulder at 6.3 meV were detected. Finally for $PrAl_2$ we observed a large line broadening without any structure in the inelastic spectra. The information on the last two compounds was considered insufficient to try a

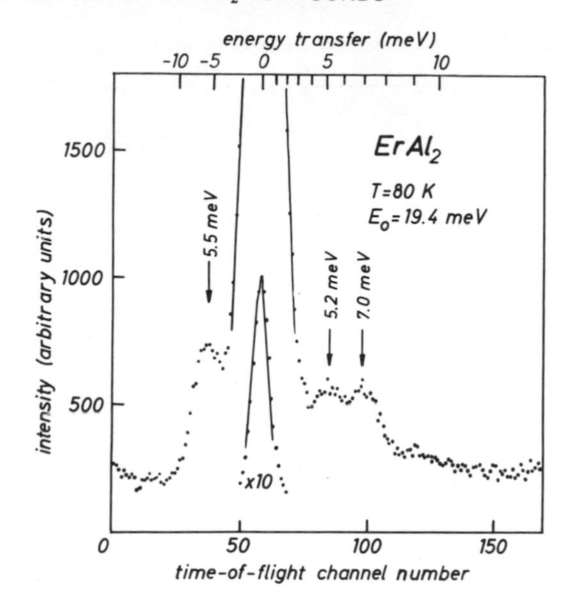

Fig. 1 Neutron time-of-flight spectrum for $ErAl_2$ at 80 K

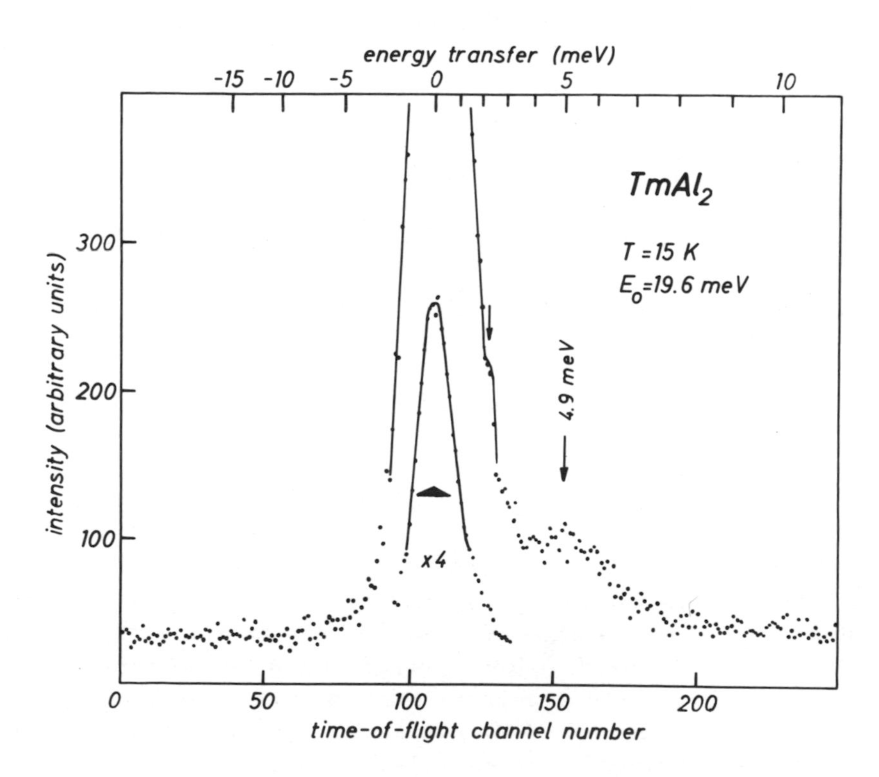

Fig. 2 Neutron time of flight spectrum of $TmAl_2$ at 15 K

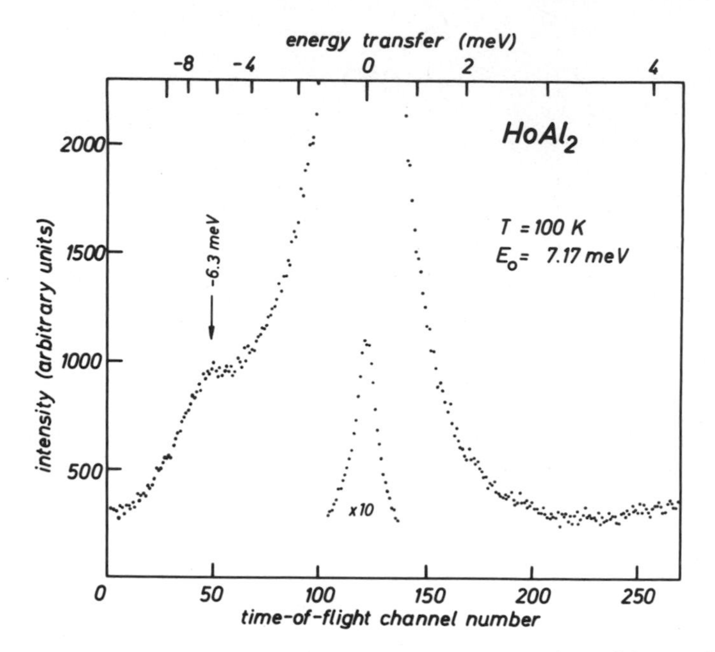

Fig. 3 Neutron time of flight spectrum of $HoAl_2$ at 100 K

determination of CEF-parameters. One has to conclude that in these samples the exchange interaction is of considerable strength. For $PrAl_2$ it might even exceed the CEF-interaction. This idea is supported by our investigation of the line width ΔE of the quasi-elastic scattering as a function of momentum transfer Q and of temperature. The Q-dependence of ΔE for $PrAl_2$ looks similar to those one expects for exchange-coupled S-state ions in the para-magnetic region [6].

The determination of the CEF-parameters of $ErAl_2$ is straight-forward. Analyzing the level scheme given in [1] and the matrix elements calculated by Birgeneau [7], the conditions given above are fulfilled only for CEF-parameters $x = +0.16 \pm 0.02$ and $W = -0.030 \pm 0.002$ meV or $A_4a^5 = -58 \cdot 10^4$ meV $\cdot$ Å and $A_6a^7 = -17 \cdot 10^6$ meV$\cdot$Å (a=lattice parameter). A_4a^5 and A_6a^7 should be constant for the $REAl_2$-series within a point-charge model (PCM-extrapolation). It now turns out that a shift of these figures to $A_4a^5 = -39 \cdot 10^4$ meV$\cdot$Å and $A_6a^7 = -15 \cdot 10^6$ meV$\cdot$Å (or $x = -0.28$ and $W = +0.040$ meV) gives a good description of the experimental results on $TmAl_2$. The modification of the reduced CEF-parameters A_4a^5 and A_6a^7 are probably due to the uncertainty of the nonrelativistic matrix elements $\langle r^n \rangle$ used in the calculation.

Applying PCM-extrapolation from $ErAl_2$ and $TmAl_2$ to $HoAl_2$, we obtain x=0.2 and W=+0.02 meV. In the corresponding level scheme there are a number of transitions of comparable strength and with energies ranging from 1.8 to 7.5 meV which could explain the observed large quasielastic scattering. More statements concerning the CEF-parameters for $HoAl_2$, however, are not possible up-to-now.

It is interesting to note that the present neutron data on $ErAl_2$ are quite similar to results on the related pseudobinary system $Y_{0.85}Er_{0.15}Al_2$ [3]. The neutron energy gain side of our neutron spectrum on $ErAl_2$ at 80 K can be considered identical to that on (Y, Er) Al_2. The additional information we obtained at low temperatures on the neutron energy loss side forced us, however, to abandon the CEF-splitting scheme given in ref. 3).

The CEF-parameters of $ErAl_2$ which were derived recently by a molecular field calculation from magnetization measurements [8] fail to explain our neutron results. On the other hand, our CEF-parameters do not explain the easy direction of magnetization in ferromagnetic $ErAl_2$. Discussing the magnetic anisotropy, one might wonder if one really can neglect the effect of anisotropic exchange due to 5d-conduction electrons. That such contributions exist in the $REAl_2$ compounds was shown by NMR measurements [9].

REFERENCES

1) K.R.Lea, M.J.M.Leask, W.P.Wolf, J.Phys.Chem.Solids 23, 1381 (1962)
2) P.Bak, Excitations and Magnetic Properties of Rare-Earth Al_2 Compounds, Risö Report No. 132 (1974)
3) H. Heer, A.Furrer, E. Walker, A. Treyvaud, H.-G.Purwins, J. Kjems, J. Phys. C 7, 1207 (1974)
4) A. Furrer, W. Bührer, H. Heer, H.-G.Purwins, E.Walker, Int. J. Magnetism 4, 63 (1973)
5) N. Nücker et al., to be published
6) P.G.de Gennes, J. Phys. Chem. Solids 4, 223 (1958)
7) R.J. Birgeneau, J. Phys. Chem. Solids 33, 59 (1972)
8) H.-G.Purwins, E. Walker, B. Barbara, M.F.Rossignol, A. Furrer, J. Phys. C 9, 1025 (1976)
9) N. Kaplan, E. Dormann, K.H.J. Buschow, D. Lebenbaum, Phys. Rev. B7, 40 (1973)

CRYSTAL-FIELD SPLITTING AND RELAXATION EFFECTS IN CeAs

H. Heer, A. Furrer, and W. Hälg, Institut für Reaktortechnik ETHZ, EIR, CH-5303 Würenlingen, Switzerland

O. Vogt, Laboratorium für Festkörperphysik, ETH, 8093 Zürich, Switzerland

ABSTRACT

Neutron inelastic scattering experiments have been performed in paramagnetic CeAs in order to determine the crystal-field splitting within the $^2F_{5/2}$ ground state multiplet. The crystal-field splitting increases with decreasing temperature. The point-charge model fails to explain the temperature dependence as well as the size of the crystal-field parameters. The line widths are discussed in terms of various line broadening effects.

INTRODUCTION

The neutron inelastic scattering technique is now well established to examine crystal-field effects in metallic rare earth systems. Whereas measurements of the bulk properties cannot reflect details of the crystal field because of its integral nature, neutron spectroscopy yields detailed information on the crystal-field energy levels through the peak positions in the observed energy spectra as well as on the crystal-field wave functions through the peak intensities and on relaxation phenomena through the peak widths. In the present paper we report neutron inelastic scattering experiments for polycrystalline CeAs in the paramagnetic state. This compound crystallizes in the rocksalt structure (a=6.072 Å) and shows antiferromagnetic ordering below T_N=7.5 K. The crystal-field Hamiltonian may be written in the form

$$\mathcal{H} = B_4 \, (O_4^0 + 5 \, O_4^4) \quad , \tag{1}$$

where B_4 is the crystal-field parameter, and the O_n^m are the Stevens operator equivalents[1]. Under the action of $\mathcal{H}$ the free ion $^2F_{5/2}$ ground state multiplet splits into a doublet Γ_7 and a quartet Γ_8. The observed crystal-field transition peaks show three main features: A much smaller energy transfer than expected from a point-charge model, a strong temperature dependence of the crystal-field splitting Δ (Δ=360 B_4), and a rather large and temperature-dependent line width. The line widths are discussed in terms of various line broadening mechanisms.

EXPERIMENTAL RESULTS

The inelastic neutron scattering experiments were performed on a triple-axis spectrometer at the reactor Diorit, Würenlingen. The measurements have been carried out in the neutron energy loss configuration with the analyser energy kept fixed at 14.96 meV. The neutron scattering vector was held constant at Q=2.3 $Å^{-1}$. Measurements at additional neutron momentum transfers $\hbar Q$ have been performed in order to confirm the crystal-field origin of the observed inelastic transition peak. The neutron spectra have been analysed by a least-squares fitting procedure, in which the crystal-field transition peaks have been approximated by Gaussians. The results are listed in Table I. The widths of the crystal-field transition lines (full width at half maximum) have been obtained by taking account of the instrumental resolution. In contrast to other light rare earth monopnictides[2,3], the crystal-field splittings and the line widths are strongly temperature-dependent. Moreover the line widths are much larger than those for the praseodymium and neodymium monopnictides. This feature will be discussed in the following section.

Table I: Observed energies, Δ_{obs}, and line widths, γ_{obs}, and calculated mean-squared displacement, $\langle u^2 \rangle$, exchange broadening, γ_{ss}, and Stark broadening, γ_{sl}, of the crystal-field transition peak of CeAs.

T (K)	Δ_{obs} (meV)	γ_{obs} (meV)	γ_{ss} (meV)	γ_{sl} (meV)	$\langle u^2 \rangle$ ($Å^2$)
12	13.7±0.1	2.3±0.2	0.1	0.5	0.006
77	13.0±0.2	3.6±0.2	0.2	0.5	0.008
160	12.1±0.2	3.9±0.3	0.3	0.7	0.015
293	11.9±0.2	4.1±0.4	0.4	1.0	0.026

LINE BROADENING EFFECTS

The energy widths of crystal-field levels result from a number of effects, but relaxation effects are generally the most important broadening mechanisms.

a) Spin-Spin Relaxation

The importance of including spin-spin relaxation effects in the dynamic effective field approximation in the Hamiltonian of a rare earth system has been demonstrated by Furrer and Heer[4]. In CeAs as well as in CeP[5], however, the crystal field dominates the exchange interaction, and the exchange broadening, γ_{SS}, does not exceed 0.4 meV at room temperature (Table I).

b) Spin-Lattice Relaxation

In a phenomenological picture the spin-lattice relaxation mechanism can be described by the modulation of the crystalline electric field due to thermal motion of the ions. The microscopic origin of this effect is the strain field associated with the phonon modes. We call this oscillation of the crystal field around its equilibrium value the dynamical Stark effect. Based on results of general fluctuation theory[6] the probability distribution of the mean displacement $\langle u \rangle$ of the ions from their equilibrium position is a Gaussian. We have calculated the crystal-field splitting for cerium monopnictides for a large number of configurations of the six nearest-neighbouring ligand ions determined by a Monte Carlo procedure, and correlations between the ionic positions have been neglected. The results have been obtained in the form of histograms as shown in Fig. 1. Fig. 2 shows the calculated line widths for two different crystal-field splittings Δ. Within the given range, the line width turns out to be linearly dependent on $\langle u \rangle$ as well as on the crystal-field splitting Δ. The shape of the crystal-field transition line due to the spin-lattice relaxation can be approximated by a Gaussian, centered at the position of the static crystalline electric field.

In order to determine the line broadening due to spin-lattice relaxation, we have to estimate the mean displacement of the ligand ions. Detailed information about lattice vibrations can be obtained from measurements of the phonon dispersion curves. However, only a few measurements for rare earth metals have been reported and only one for the rare earth monopnictides (NdSb[7]). The mean-squared displacements $\langle u^2 \rangle$ of the ions can be estimated from the Debye-Waller

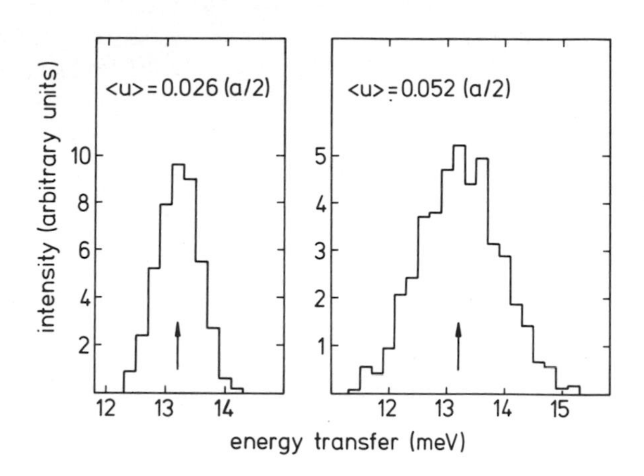

Fig. 1: Histograms of a crystal-field transition peak calculated from a Monte Carlo procedure as described in the text. The arrow denotes the peak position for a static crystal field (Δ=13.2 meV)

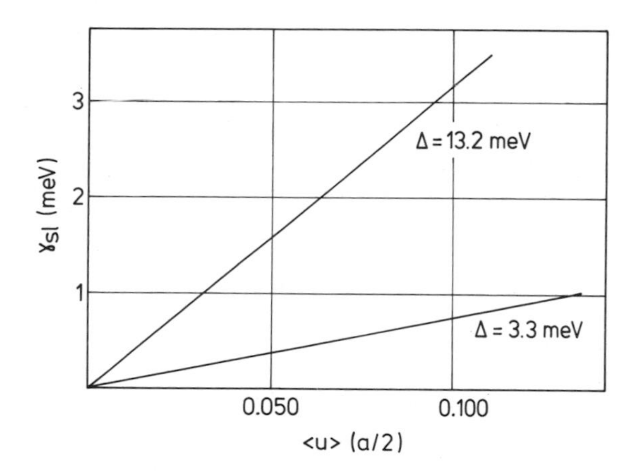

Fig. 2: Line broadening γ_{sl} due to the dynamical Stark effect versus mean ionic displacement $\langle\mu\rangle$ for two crystal-field splittings Δ.

factor. Table I lists the mean-squared displacements $<u^2>$ calculated from an estimated Debye temperature[8] θ_D=260 K. The values are about 30 % larger than the values for NdSb[7] determined from phonon dispersion curves, but follow the same temperature variation. Under the assumption that the probability distribution of the mean ionic displacement is a Gaussian, we obtain $<u^2>=\pi<u>^2$. The corresponding line broadening due to spin-lattice relaxation, γ_{sl}, can be obtained from Fig. 2 and is listed in Table I. Comparing γ_{sl} with the experimentally determined line widths shows a rather minor influence of this relaxation effect to the total intrinsic line width.

CONCLUSION

In contrast to the praseodymium and neodymium monopnictides[2,3], additional line broadening mechanisms are important for CeAs. It is most likely that spin-electron relaxation effects give the major line-broadening contribution. Following Korringa[9] we expect $\gamma(T)\sim T$. This is in contradiction to the temperature dependence of the remaining line width. It is most likely that the deviation can be attributed to the Kondo effect, due to the proximity of the single 4f-electron to the Fermi surface. Indeed, measurements of the electrical resistivity[10] show an anomalous temperature behaviour below 150 K. In the same temperature region we observe a strong shift of the crystal field transition peak, which similarly might be due to the Kondo effect.

We may conclude that for the cerium monopnictides the generally accepted picture of configuration interaction causing the external electrostatic field to be screened is oversimplified, and overlap, charge penetration and correlation seem to be important[2]. In addition one has the complication of possible shielding and 5d virtual bound state effects arising from the conduction electrons[11]. Application of band structure calculations to the problem of crystal-field splittings in these compounds would be of great interest.

REFERENCES

1. K.W.H. Stevens, Proc. Phys. Soc. A65, 209 (1952)
2. K.C. Turberfield, L. Passell, R.J. Birgeneau, and E. Bucher, J. Appl. Phys. 42, 1746 (1971)
3. A. Furrer, J. Kjems, and O. Vogt, J. Phys. C5, 2246 (1972)
4. A. Furrer and H. Heer, Phys. Rev. Letters 31, 1350 (1973)

5. H. Heer, A. Furrer, and W. Hälg, J. Magn. and Magn. Materials 3, 55 (1976)

6. A. Münster, in "Fluctuation Phenomena in Solids", ed. by R.E. Burgess (Academic Press, New York, 1965), p. 179

7. N. Wakabayashi and A. Furrer, Phys. Rev. B13, 4343 (1976)

8. R.J. Birgeneau, E. Bucher, J.P. Maita, L. Passell, and K.C. Turberfield, Phys. Rev. B8, 5345 (1973)

9. J. Korringa, Physica 16, 601 (1950)

10. T. Tsuchida, M. Kawai, and Y. Nakamura, J. Phys. Soc. Japan 28, 528 (1970)

11. G. Williams and L.L. Hirst, Phys. Rev. 185, 407 (1969)

LINE WIDTH OF CRYSTAL-FIELD EXCITATIONS DUE TO ELECTRONS AND PHONONS

Klaus W. Becker[+] and Peter Fulde

Max-Planck-Institut für Festkörperforschung
D-7 Stuttgart 80, Germany
[+]Permanent address: Theoretische Festkörperphysik, Technische Hochschule, D-61 Darmstadt

ABSTRACT

A discussion is given of the line width of CEF excitations in Rare-Earth (RE)-systems due to electrons and phonons. The temperature and wave-number dependence of the line width is discussed for both cases. A connection is established to calculations of the line width resulting from the interaction of magnetic excitations.

I. INTRODUCTION

During the last years large progress has been made both experimental and theoretical, in our understanding of the excitations of RE-systems. The crystalline electric field (CEF)-splitting of the RE-ions leads thereby to a variety of effects in the temperature and wave number dependence of the excitations. The bulk of the experimental informations is provided by inelastic neutron scattering and the measured dispersion curves are rather well understood at present. The experimental advances have made it possible to study not only the dispersion of the excitations but also their line widths. It is therefore of interest to look into the theoretical aspects of this problem. We do this by limiting ourselves to the interactions of CEF split RE-ions with conduction electrons and phonons. The excitations of the coupled RE-ion system are obtained from the poles of a susceptibility or response function. The imaginary part of such a pole gives the line width of the excitation. Obviously it is of importance which susceptibility one is considering. For example the neutron differential cross section is connected with the magnetic susceptibility. On the other hand the quadrupolar susceptibility is required if the perturbation on the system has quadrupolar character (lattice deformations). Thus different methods of measurement can yield different line widths of the excitations of the RE-system.

II. INTERACTION WITH CONDUCTION ELECTRONS

Conduction electrons and RE-ions interact with each other via the general k-f interaction H_{k-f}. Due to Hirst[1] it can be written as a sum of terms $H^{(\Lambda\Sigma)}$. Each of these terms consists of a product of an operator acting on the RE-ion and an operator describing the scattering of the conduction electrons. Λ ranges from 0 to 6 and marks that the operators transform under space rotations like a total angular momentum state Λ. Σ is 0 or 1 and indicates that the operators transform under spin rotations like Σ. The commonly used isotropic exchange interaction is characterized by $(\Lambda,\Sigma)=(0,1)$. It is usually written as $H_{ex}=-(g-1)J_{ex}\underset{\sim}{\sigma}\underset{\sim}{J}$ and shows the transformation properties directly.

We calculate in the following the frequency and wave number dependent magnetic susceptibility: first for a single RE-ion and second for a lattice of RE-ions.

1. Single RE-Ion

In the absence of H_{k-f} the susceptibility of a single ion is given up to a factor $\mu_B^2 g^2$ by

$$U(\omega,T)=\sum_{ij}^{2J+1}|M_{ij}|^2\frac{\varepsilon_{ij}}{\varepsilon_{ij}^2-(\omega+i\eta)^2}(n_j-n_i) \quad (1)$$

Here $\varepsilon_{ij}=\delta_i-\delta_j$ where the δ_u are the energies of the CEF-states. Furthermore $M_{ij}=\langle i|J|j\rangle$ and $n_i=\exp(-\delta_i/T)(\sum\exp(-\delta_j/T))^{-1}$. For the special case of two singlets seperated by an energy δ, Eq. (1) reduces to

$$U(\omega,T)=-\frac{2M^2\delta\tanh\delta/2T}{\omega^2-\delta^2} \quad (2)$$

In the presence of H_{k-f} this equation is changed into

$$U(\omega)=-\frac{2M^2\delta\tanh\delta/2T\,[1-\delta^{-2}(\Delta^2-i\omega\gamma/2)]}{\omega^2-\delta^2-\Delta^2+i\gamma\omega} \quad (3)$$

Here $\Delta^2+\omega \mathrm{Im}\gamma$ determine the shift in the level seperation and $i\omega \mathrm{Re}\gamma$ describes the damping. Since we are mainly interested in the latter we will neglect the level shift in the following. If there is only $(\Lambda,\Sigma)=(0,1)$ present then to lowest order in the coupling $\mathrm{Re}\gamma$ is given by [2,3].

$$\mathrm{Re}\gamma=4\pi M^2(N(0)J_{ex})^2(g-1)^2\delta\coth^{\delta}/_{2T} \tag{4}$$

The presence of other interactions (Λ,Σ) leads simply to a modification of the prefactors. Different interactions contribute additively to Δ^2 and γ. This requires a replacement of

$$|M|^2J^2_{ex}(g-1)^2 \text{ by } |M_{tot}|^2J^2_{tot}$$

The positive pole of $U(\omega)$ is given by $\omega\simeq\delta-i^{\gamma}/_2$. For $T>>\delta$ one finds the Korringa broadening $\gamma\sim T$ while for $T\lesssim\delta$ the damping γ approaches a constant[4].

One can generalize Eq. (3) to a CEF multi-level scheme by using a projector formalism. In that case one finds

$$U(\omega,T)=U_{st}(T)-\omega\sum_{\substack{jk\\il}} M_{ij}M^*_{kl}\Omega^{-1}_{klij}\ \frac{n_j-n_i}{\varepsilon_{ij}} \tag{5}$$

where $U_{st}(T)=U(\omega=0,T)$ (see Eq. (1)). Furthermore up to second order perturbation theory

$$\Omega_{klij}=(\omega+\varepsilon_{ij})\delta_{ik}\delta_{j\ell}+\frac{1}{T}\ \frac{\varepsilon_{ij}}{(n_j-n_i)}\ (kl|L_{k-f}$$
$$\times\ (1-\frac{1}{L_o+\omega+i\eta}\ L_{k-f})|ij) \tag{6}$$

Hereby $|ij)=|i\rangle\langle j|$ denotes the transition operators which connect the CEF-states $|i\rangle$, $|j\rangle$. The Liouville operators L_o, L_{k-f} and L correspond to the unperturbed Hamiltonian H_o (sum of the CEF and free electron Hamiltonian), to H_{k-f} and to $H=H_o+H_{k-f}$. They have the property $L_o|A)=|[H_o,A])$ etc. Finally the scalar product $(A|B)$ is defined by [5]

$$(A|B)=T\int_0^{1/T} d\lambda\langle e^{\lambda H}A^+e^{-\lambda H}B\rangle_{\text{therm. av.}} \tag{7}$$

With the help of Eqs. (5-7) one can calculate $U(\omega,T)$ to second order in H_{k-f} for an arbitrary CEF-level scheme. For two singlets one obtains again Eq. (3). If all CEF-levels are non-degenerate then to a good approximation Ω is diagonal and can be trivially inverted. In the more general case the inversion has to be done numerically.

2. Lattice of RE-Ions

The interaction between the RE-ions is mediated by the conduction electrons. The magnetic susceptibility $\chi(q,\omega)$ is then given (up to a factor $g^2\mu_B^2$) by

$$\chi(q,\omega) = \frac{U(\omega)}{1-J_{eff}(q,\omega)U(\omega)} \tag{8}$$

For a two level system $U(\omega)$ is given by Eq. (3). Furthermore

$$J_{eff}(q,\omega) = 2(g-1)^2 J_{ex}^2 \{\chi_e(q,\omega) + A(q,\omega)\} \tag{9}$$

Here $\chi_e(q,\omega)$ denotes the Lindhard function. The function $A(q,\omega)$ varies slowly in q and ω. It results from a self-energy correction and a sum over reciprocal lattice vectors. The magnetic excitations are determined from the poles of $\chi(q,\omega)$. Their dispersion is indicated in Fig. 1. For a two level system there are two branches for small values of q due to the fact that we are dealing with a coupled conduction electron-RE ion system. For the lower branch it is always $\omega \ll qV_F$.

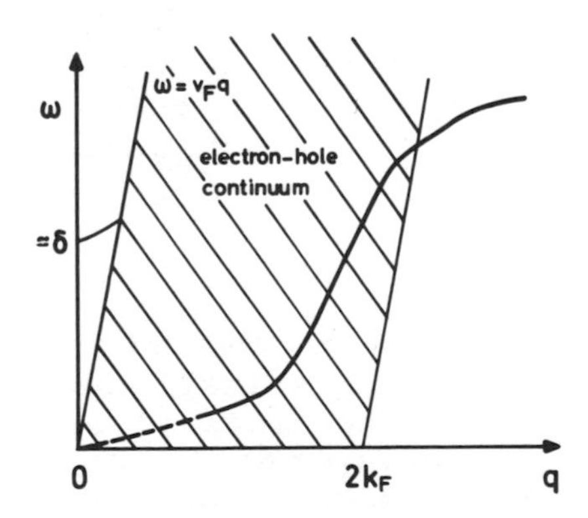

Fig. 1: Excitation spectrum of the coupled ion-conduction electron system. The dashed line indicates the paramagnon pole (from Ref. (2)).

In that case $\mathrm{Im}\chi_e(q,\omega) = \frac{i\pi}{2}N(O)\,\omega/qv_F$. It is useful to introduce $\lambda^2 = 4N(O)J_{ex}^2(g-1)^2|M|^2\delta\tanh\delta/2T$. There are three distinct regions in q-space for which we shall discuss the excitations corresponding to the lower branch.

a) $qv_F<\lambda$: The pole is purely imaginary (paramagnon pole). It appears only close to T_c where one finds $\omega = -iv_p|q|$. The paramagnon velocity $v_P \simeq v_F\left(\frac{T-T_c}{T_c}\right)$.

b) $\lambda < qv_F < \lambda\left(1+\frac{\lambda^2\pi^2}{16}(\delta^2-\lambda^2)^{-1}\right)^{1/2}$:

Also in this regime the pole is purely imaginary. Its dispersion joins smoothly to the regime (a) but near the upper limit of the present regime the position of the pole is nearly independent of $|q|$. Near T_c it is $(\delta^2-\lambda^2)/\delta^2 \simeq \delta/T_c \,\sinh^{-1}\delta/T_c\,(T-T_c)/T_c$ and the regime can become large.

c) $qv_F > \lambda\left(1+\frac{\lambda^2\pi^2}{16}(\delta^2-\lambda^2)^{-1}\right)^{1/2}$: Here one finds

$$\omega = \left(\delta^2 - \frac{\lambda^2}{N(O)}(\chi_e(q,O)+A(q,O))\right)^{1/2} - \frac{i}{2}\Gamma \tag{10}$$

In evaluating Γ one has to keep in mind that the different interactions (Λ,Σ) contribute differently to $i\omega\gamma$ in the single ion susceptibility $U(\omega)$ and to $J_{eff}(q,\omega)$. While all interactions (Λ,Σ) contribute to γ only a limited number of them contributes to $J_{eff}(q,\omega)$. A full discussion of the problem will be given elsewhere (6). Here we note merely that if only the interaction $(\Lambda,\Sigma)=(O,1)$ is present one obtains

$$\frac{\Gamma}{2} = 2\pi N^2(O)J_{ex}^2(g-1)^2|M|^2\left\{\delta\coth\frac{\delta}{2T} + \left(\frac{1}{2N(O)qv_F}-1\right)\delta\tanh\delta/2T\right\} \tag{11}$$

This demonstrates clearly the two contributions to the line width of the magnetic excitations. One contribution (first term within the bracket) arises from the one ion problem. It increases with increasing temperature. The other contribution results from the decay of the CEF excitations into electron-hole excitations while they are transfered from one ion to the next. This contribution decreases with increasing temperature since the hopping probability decreases (the band of excitations becomes smaller as T increases).

Finally we comment on the second branch which appears at low q-values. There the damping is the same as for a single ion before the branch merges with the electron-hole continuum where it becomes heavily damped. The link between $\chi(\underset{\sim}{q},\omega)$ and the neutron scattering data is provided by the relation

$$\frac{d^2\sigma}{d\Omega d\omega} \sim \frac{1}{1-\exp(-\omega/T)} \operatorname{Im}\chi(\underset{\sim}{q},\omega) \tag{12}$$

In passing we note that there should be Kohn anomalies in the magnetic excitation branches in the same way as they occur for phonons.

Finally we remark that the above considerations can be generalized to a lattice of ions with a full CEF level scheme. This has been done but will be reported elsewhere.

III. INTERACTION WITH PHONONS

We consider now the line width of the CEF-excitations due to the interaction with the phonons. The magnetoelastic interaction has been discussed in detail in Ref. (7). Here we want to consider only the linear strain interaction, i.e. we neglect the rotational as well as the non-linear strain terms. Thus

$$H_{me}=\sum_{m\underset{\sim}{q}\mu} V_1(\underset{\sim}{J}^m;\underset{\sim}{q}\mu)\exp(i\underset{\sim}{q}\underset{\sim}{R}_m)(a_{\underset{\sim}{q}\mu}+a^+_{-\underset{\sim}{q}\mu}) \tag{13}$$

where $\underset{\sim}{R}_m$ denotes the positions of the RE-ions. The $a_{\underset{\sim}{q}\mu}$ are phonon destruction operators of wave number $\underset{\sim}{q}$ and polarization μ. V_1 contains the coupling constant g, an operator $O(\underset{\sim}{J}^m)$ bilinear in J_μ which has quadrupole character, a factor $Q(\underset{\sim}{q}\mu)=2^{-1}(e_\alpha(\underset{\sim}{q}\mu)q_\beta+e_\beta(\underset{\sim}{q}\mu)q_\alpha)$ where $\underset{\sim}{e}(\underset{\sim}{q}\mu)$ is the polarization vector and an appropriate normalization factor. The quadrupolar CEF-excitations are found from the poles of the quadrupole susceptibility. We calculate this susceptibility for a single ion and for a lattice of ions.

1.) Single Ion

The quadrupole susceptibility $U_Q(\omega,T)$ in the absence of H_{me} is of the form of Eq. (1) but with M_{ij} denoting the quadrupole-matrix element $M_{ij}=\langle i|O(\underset{\sim}{J})|j\rangle$.

We consider again the simple case of two singlets with $M_{12}\neq 0$ but $M_{11}=M_{22}=0$. In the presence of H_{me} one finds that $U_Q(\omega)$ is of the same form as Eq. (3). In the Debye approximation the damping is given by

$$\mathrm{Re}\gamma=3\pi\alpha|M|^2 g^2 \frac{\omega^2}{M_\alpha v_o^2\ \omega_D{}^3}\ \delta\coth\frac{\delta}{2T} \tag{14}$$

for $|\omega|\leq\omega_D$ and zero otherwise. α is a constant of order one, M_α is the averaged atomic mass and v_o is the sound velocity. One notices that the temperature dependence of γ is the same as found before when the damping was caused by the conduction electrons. The poles of $U_Q(\omega,T)$ are of the form $\omega=\delta-i\mathrm{Re}\gamma/2$ as found before.

It should be mentioned at this point that the phonons contribute also to the damping in the single ion magnetic susceptibility. The contribution is the same as for the quadrupole susceptibility. This situation does no longer hold for a lattice of RE-ions.
The above considerations can be extended to the full CEF-level scheme in close analogy to Eqs. (5-7).

2. Lattice of RE-Ions.

The phonons mediate a quadrupolar interaction between different ions. The quadrupole susceptibility $\chi_Q(\underset{\sim}{q},\omega)$ is given by an expression equivalent to Eq. (8), with U replaced by U_Q. For a two level system one finds

$$J_{eff}(\underset{\sim}{q},\omega)=\frac{g^2}{2M_\alpha}\left\{\frac{Q^2(\underset{\sim}{q})}{\omega_{\underset{\sim}{q}}}D(\underset{\sim}{q},\omega)-\frac{1}{N}\sum_{\underset{\sim}{k}}\frac{Q^2(\underset{\sim}{k})}{\omega_{\underset{\sim}{k}}}D(\underset{\sim}{k},\omega)\right\} \tag{15}$$

$D(\underset{\sim}{q},\omega)$ denotes the phonon propagator $D(\underset{\sim}{q},\omega)=-2\omega_{\underset{\sim}{q}}/(\omega^2-\omega_{\underset{\sim}{q}}{}^2)$. The second term is a selfenergy correction the real part of which can be also neglected if Δ^2 and $\mathrm{Im}\gamma(\omega)$ are neglected as we did.

It is instructive to compare Eq. (15) with Eq. (9) where the selfenergy correction is contained in $A(\underset{\sim}{q},\omega)$. The poles of $\chi_Q(\underset{\sim}{q},\omega)$ yield the coupled phonon and CEF-excitation modes including their damping (Fig. 2). Without going into the details which will be presented elsewhere we list the following results. Except in the vicinity of the cross-over point of the coupled excitations one finds for the CEF-like excitations

$$\omega \simeq \delta - ig^2 3\pi |M|^2 \alpha \frac{\delta^3}{\omega_D{}^3 M_\alpha v_o{}^2} \sinh^{-1} \delta/_T \tag{16}$$

For the phonon like excitations one obtains

$$\omega \simeq \omega_q - ig^4 3\pi |M|^4 \alpha \frac{\delta^2 \omega_q{}^4}{(\omega_q{}^2 - \delta^2)^2 \omega_D{}^3 (M_\alpha v_o{}^2)^3} \cosh^{-2} \frac{\delta}{2T} \tag{17}$$

Thus the damping is of order g^4. The temperature dependence is in both cases very simple.

IV. COMPARISON WITH OTHER MODEL CALCULATIONS

Very often the Hamiltonian $H_{int} = \sum_{ij} J_{ij} \underset{\sim}{J}_i \underset{\sim}{J}_j$ is used to describe the interacting RE-ions on a lattice. Here the conduction electrons (or phonons) which can absorb energy are already eliminated. Thus within this model a damping of the excitations can only arise from their mutual interaction (8). It is evident that it is therefor at most of order $(J_{ij})^2$. In a metal $J_{ij} \sim J_{ex}{}^2$. Therefore the interaction of excitations should be of secondary importance as compared with the excitation of electron-hole pairs. In insulators a similar argumentation holds with respect to phonons. Only if the mutual interaction of RE-ions is a result of direct overlap it is likely that the damping is dominated by the interaction of the CEF-excitations.

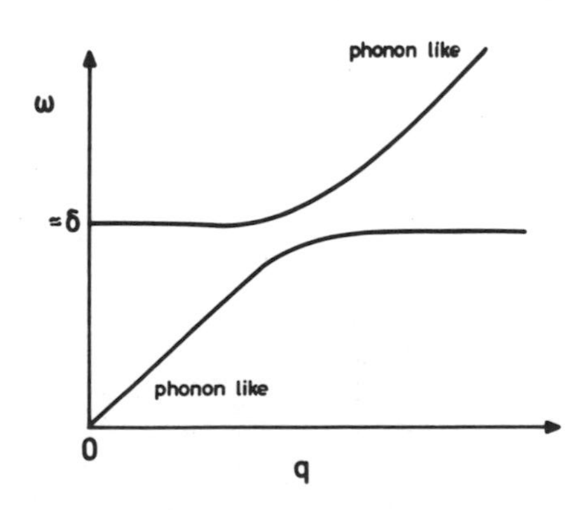

Fig. 2: Excitation spectrum of the coupled ion-phonon system.

REFERENCES

+ This work was partly performed within the programme "Sonderforschungsbereich 65, Festkörperspektroskopie Darmstadt - Frankfurt".

1. L. L. Hirst, Solid State Commun. 5, 751 (1967)
2. P. Fulde and I. Peschel, Adv. Phys. 21, 1 (1972)
3. K.-D. Schotte, phys. stat. sol. (b) 71, 221 (1975)
4. B. Rainford, Thesis Oxford University (1969)
5. H. Mori, Progr. theor. Phys. 34, 399 (1965)
6. K. W. Becker and P. Fulde, to be published
7. V. Dohm and P. Fulde, Z. Phys. B 21, 369 (1975)
8. P. Bak, Phys. Rev. B 12, 5203 (1975)

INELASTIC COULOMB SCATTERING OF CONDUCTION ELECTRONS IN $(\underline{La}Pr)Sn_3$

J. Keller and P. Holzer

Fachbereich Physik, Universität Regensburg

8400 Regensburg, Germany

ABSTRACT

The influence of inelastic Coulomb scattering of conduction electrons on the superconducting properties of the impurity system $(\underline{La}Pr)Sn_3$ is investigated.

In a recent series of papers[1,2] the superconducting properties of the nonmagnetic ground state-system $(\underline{La}Pr)Sn_3$ have been investigated. So far the theoretical analysis has concentrated on the pairbreaking effect of the elastic and inelastic conduction-electron-impurity spin-exchange interaction. In this communication we investigate the additional effect of inelastic charge scattering from the aspherical part of the impurity 4f-shell. Because of the time reversal symmetry of this interaction it acts as an additional pairing mechanism for conduction electrons.[3,4] The level scheme of Pr^{3+} ions $(\Gamma_1(0.K)-\Gamma_5(8.4K)-\Gamma_4(14.6K)-\Gamma_3(25.K)$ established from seperate measurements[1] offers a good possibility to distinguish both effects by their different temperature dependence.

Following the method outlined in Ref. 4 we may write the Hamiltonian for the scattering of conduction electrons with momentum $\underset{\sim}{k}$ from the multipole moments Q_L of the 4f-shell with total angular momentum J as

$$H_{CL} = \frac{1}{N} \sum_{kk'\alpha} \sum_{M=-L}^{+L} f_L^M(\underset{\sim}{k}'\underset{\sim}{k}) Q_L Y_L^M(\underset{\sim}{J}) C^+_{\underset{\sim}{k}'\alpha} C_{\underset{\sim}{k}\alpha}$$

Here the $Y_L^M(J)$ are Steven's operators. The term with L=0 leads to the usual elastic charge scattering, the L>0 part allows inelastic transitions. In the following we will consider only the contribution of the quadrupole moment (L=2). The operators Y_2^M are given in Ref. 4; in order to facilitate the comparison with the exchange interaction we chose the normalization such that

$$\sum_M \mathrm{Tr}\,\{(Y_2^M)^+(Y_2^M)\} = J(J+1)(2J+1)$$

The charge and exchange scattering rates of conduction electrons $\tau_{c,ij}^{-1}$ and $\tau_{ex,ij}^{-1}$ corresponding to transitions between impurity states i and j, which are used in the electronic selfenergy,[5] can then be written as

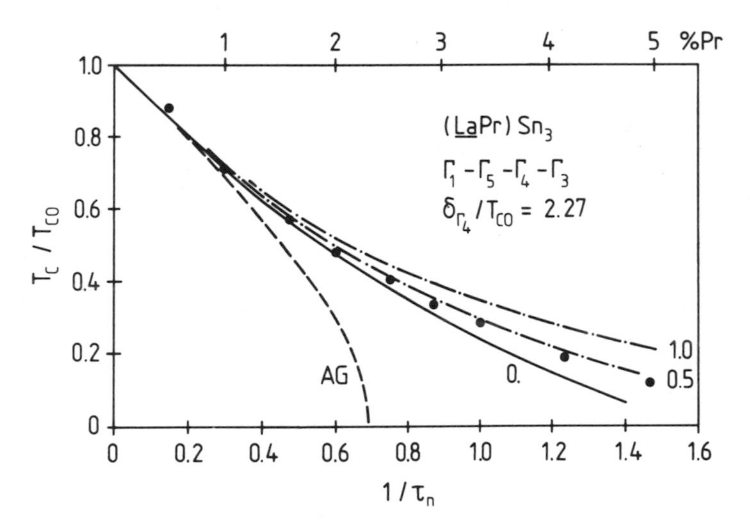

Fig. 1: Reduced transition temperature T_c/T_{co} vs. Pr concentration. The solid line is calculated from the indicated level scheme with exchange scattering only. The numbers at the dot-dashed lines indicate the ratio $\tau_c^{-1}/\tau_{ex}^{-1}$ of charge and exchange scattering. Solid circles, UCSD data;[1] T_{co}=6.4K. The result of the Abrikosov-Gorkov theory (dashed line) is shown for comparison. For all curves $(dTc/dn_I)n_I$=o has been used as adjustable parameter.

$$\tau_{c,i,j}^{-1} = \tau_c^{-1} \sum_{M=-2}^{+2} |\langle i|Y_2^M|j\rangle|^2$$

$$\tau_{ex,ij}^{-1} = \tau_{ex}^{-1} \sum_{k=1}^{3} |\langle i|J^k|j\rangle|^2$$

The quantities τ_c^{-1}, τ_{ex}^{-1} are proportional to the impurity concentration n_I. The matrix elements are easily calculated from the eigenstates of the impurities in a cubic crystalline field tabulated by Lea et al.[6]

The effect of inelastic transitions on the superconducting condensate and the transition temperature can be characterized by a pairing and a pairbreaking strength which depends on the ratio δ/T, where δ is the transition energy.[3] The pairing strength vanishes for $\delta=0$ and reaches a maximum for $\delta/T\simeq 10$. The pairbreaking effect is strongest for degenerate energy levels and decreases with increasing δ/T.

For Pr^{3+} ions in $LaSn_3$ the ground state is nonmagnetic. The

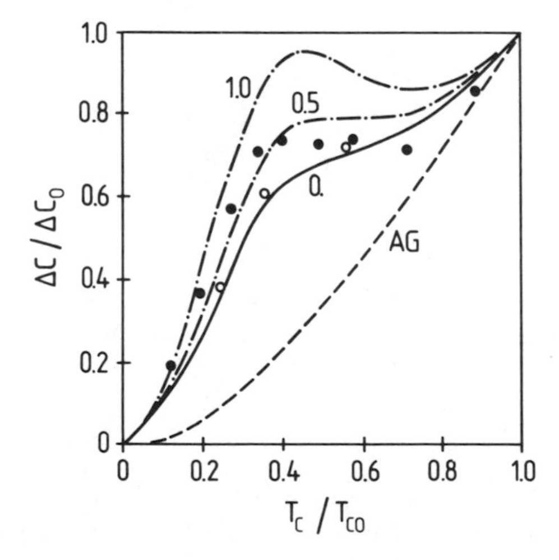

Fig. 2: Reduced specific heat jump $\Delta C/\Delta C_o$ vs. reduced transition temperature T_c/T_{co} for the (LaPr)Sn_3 system. The notation of the curves is the same as in Fig. 1. Solid circles UCSD data;[1] open circles, BL data.[1]

first excited level Γ_5 is coupled to the ground state only via the charge scattering, the following Γ_4 level is coupled to the ground state via the exchange scattering. The influence on the superconducting transition temperature therefore is as follows: At small concentrations (high temperature) the pairbreaking effect from Γ_1-Γ_4, Γ_5-Γ_5, Γ_5-Γ_4, Γ_4-Γ_4 transitions dominates. It gradually decreases with increasing impurity concentration causing a positive curvature in T_c vs. n_I. The additional pairing effect from Γ_1-Γ_5 transitions is negligible at high temperatures but becomes sizable at low temperatures when the ratio $\delta(\Gamma_1-\Gamma_5)/T$ reaches its optimum value. This causes an additional bending of the T_c curve, which is shown in Fig. 1 for different values of the relative strength $\tau_c^{-1}/\tau_{ex}^{-1}$ of the charge and exchange scattering.

In the jump of the specific heat at the phase transition shown in Fig. 2 the effect of additional pairing interactions is much more pronounced, and leads to characteristic deviations which can not be produced by exchange scattering alone. In this case, however, there is also a discrepancy between the data from different

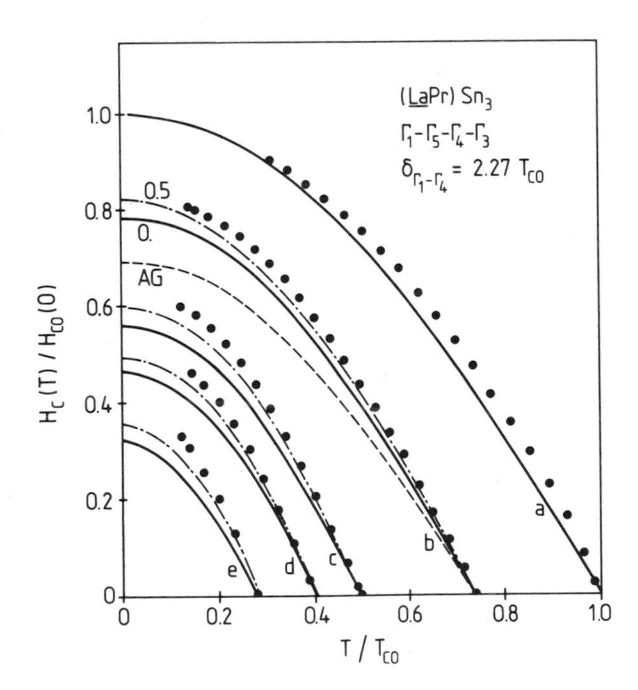

Fig. 3: Reduced thermodynamical critical field H_c/H_{co} vs. reduced temperature T/T_{co} for samples of $\underline{\text{La}}\text{Pr})\text{Sn}_3$ with different Pr^{3+} concentrations, a (0. at %), b (1. at %), c (2. at %), d (2.6 at %), e (3.4 at %); solid circles UCSD data.[2] The notation of the curves is the same as in Fig. 1.

experimental groups making a conclusion about the role of the inelastic charge scattering difficult.

Let us finally discuss the thermodynamical critical field H_c. The theoretical curves shown in Fig. 3 are obtained from a calculation of the free energy. The experimental points are obtained from heat capacity measurements.[2] Also in this case a decrease of the pairbreaking strength towards low temperatures is observed, and an additional pairing effect gives close agreement between theory and experiment. The remaining deviations may be due to strong coupling effects and are also visible in the H_c curve of the pure sample.

Our calculations have shown the influence of an additional pairing mechanism associated with impurity scattering on the thermodynamic superconducting properties. From a comparison with experimental results a value of 0.3-0.5 for the relative strength $\tau_c^{-1}:\tau_{ex}^{-1}$ of inelastic charge and exchange scattering can be estimated.

REFERENCES

1. R.W. McCallum, W.A. Fertig, C.A. Luengo, M.B. Maple, E. Bucher, J.P. Maita, A.R. Sweedler, L. Mattix, P. Fulde, and J. Keller, Phys.Rev.Letters 34, 1620 (1975)
2. R.W. McCallum, C.A. Luengo, M.B. Maple, A.R. Sweedler, Proc. LT 14, Vol. 2, 537 (1975) and J. Keller, P. Holzer, R.W. Mc Callum, C.A. Luengo, M.B. Maple, A.R. Sweedler, to be published
3. P. Fulde, L.L. Hirst, and A. Luther, Z.Physik 230, 155 (1970)
4. P. Fulde and I. Peschel, Advan.Phys. 21,1 (1972)
5. J. Keller and P. Fulde, J. Low Temp. Phys. 4, 289 (1971)
6. K.R. Lea, M.J. Leask, and W.P. Wolf, J.Phys.Chem.Solids 23, 1381 (1962)

CRYSTAL FIELD AND PHYSICAL PROPERTIES OF RX_3 COMPOUNDS

P. Lethuillier

Laboratoire de Magnétisme, C.N.R.S.,

166X, 38042-Grenoble-Cedex, France

ABSTRACT

The sign change of the CEF parameters of RX_3 compounds (R = Ce, Pr, Nd and X = Sn, Pb, In) is shown to arise from the large variation of the exchange contribution between cerium and neodymium compounds. The influence of the crystal field on physical properties of these compounds is examined.

I - INTRODUCTION

The recently determined crystalline electric field (CEF) parameters of RX_3 compounds (R = Ce, Pr, Nd and X = Sn, Pb, In) which have the cubic $AuCu_3$-type structure exhibit puzzling sign changes[1], the origin of which we shall briefly discuss in section II. This anomalous variation of the CEF parameters will allow us to understand some physical properties influenced by the crystal field : this will be section III. We shall determine, in section IV, the exchange integrals in some $La_{1-n}R_nSn_3$ superconducting compounds by the measurement of T_c for small rare earth (RE) concentrations.

II - SIGN CHANGES OF THE CEF PARAMETERS OF RX_3 COMPOUNDS (R = Ce, Pr, Nd and X = Sn, Pb, In)

The CEF parameters of these compounds have been recently determined[1-4]. In these RX_3 compounds, as there is a large number of electrons, the point charges of the ligand ions are screened and their contribution to the crystal field is expected to be weak. In order to explain the observed sign changes of the CEF parameters (figures 1 and 2), we shall discuss the contribution of the conduction electrons surrounding the rare earth ion which is expected to

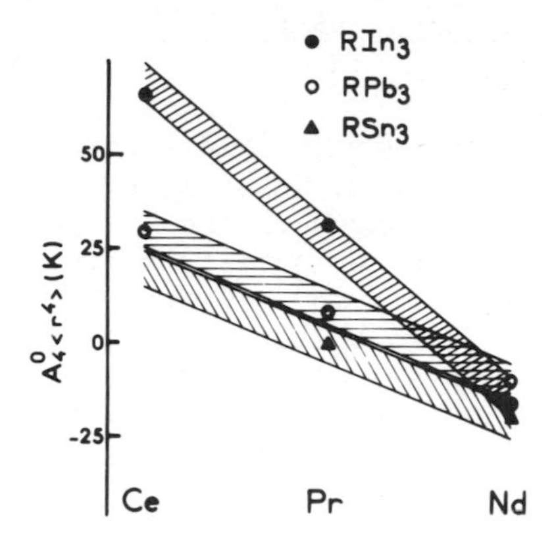

Fig. 1 : Variation of the fourth order CEF parameters of the RX_3 compounds with the RE ion

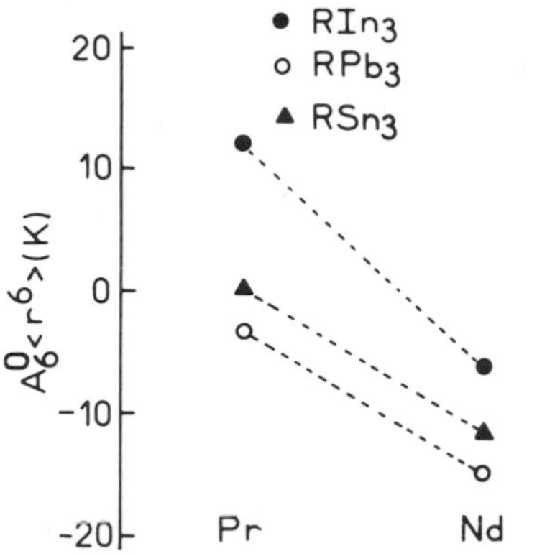

Fig. 2 : Variation of the sixth order CEF parameters of the RX_3 compounds with the RE ion.

be predominant. It consists of the Coulomb and exchange contributions. The crystal field Hamiltonian is : $H_c = B_4 O_4 + B_6 O_6$. From selection rules, valid for the conduction electrons localized around the rare earth site, it is possible to show that those which have a d-like character can contribute only to the fourth order CEF parameter and those which have an f-like character can contribute to the fourth and sixth order CEF parameters. We may suppose that the Coulomb-contribution from d-electrons to $A_4^0\langle r^4\rangle$ is weak because the d-like bands which predominate at the Fermi level[5-7] have both e_g and t_{2g} characters and these two types of electrons give opposite contributions[8]. On the contrary, the exchange contributions to $A_4^0\langle r^4\rangle$ of these conduction electrons keep the same sign[9]. When going from neodymium to cerium compounds, the 4f-5d overlap increases and the d-like exchange term may become larger than the corresponding Coulomb term. We may also infer that the exchange contribution of the f-band electrons to $A_4^0\langle r^4\rangle$ and $A_6^0\langle r^6\rangle$ varies very rapidly between neodymium and cerium compounds owing to the larger f character (large hybridization) of the band in cerium compounds.

For these reasons, it is logical to conclude that the sign change of the CEF parameters between neodymium and cerium compounds (figures 1 and 2) is related to the large variation of the exchange contribution, which is thus positive for both parameters. On the contrary, the Coulomb terms are likely negative.

III - CRYSTAL FIELD EFFECT ON PHYSICAL PROPERTIES OF SOME RX_3 COMPOUNDS

Shenoy et al[10] have determined an electric field gradient (efg) positive in $PrSn_3$ and negative in $NdSn_3$. This efg sign change seems to be related to the similar sign changes of the CEF parameters in these RX_3 compounds. In $PrSn_3$, a resistivity minimum, interpreted as a Kondo effect, has been observed[11,12], which is directly related to the small crystal field splitting ($\Delta \leq 20$ K). Indeed, above 20 K, the ground state appears as 2J+1 = 9 times degenerate and thus the decrease of the logarithmical resistivity term may overcome the relatively weak spin disorder increase, leading to the observed resistivity minimum.

The theory of Peschel and Fulde[13] predicts a thermopower (TEP) maximum at $\Delta/3$ for a two-level system separated by an energy Δ. In satisfactory agreement with this theory, Bucher et al[3] have observed TEP maxima at about $\Delta/2$ for some Pr or Tm compounds (where Δ is the total crystal field splitting). In $CePb_3$[14] and $CeIn_3$[15], thermopower maxima have been observed respectively at 40 and 75 K whereas the crystal field splitting is $\Delta = 66$ K in $CePb_3$[1] and $\Delta = 150$ K in $CeIn_3$[2] confirming the precedent results. $CePb_3$ has a very low Kondo temperature[16] : $T_K << 1$ K while the Kondo temperature of $CeIn_3$, calculated in the Anderson model is $T_K = 1.7$ K[17]. In both cases, the Kondo temperature is quite lower than the crystal field splitting and the general result observed by Bucher et al remains valid. However, the compounds $CeSn_3$ and $CeBe_{13}$ have high Kondo temperatures[16] and their thermopower maxima[14] occur at temperatures close to their Kondo temperatures.

IV - SUPERCONDUCTING PROPERTIES OF SOME $La_{1-n}R_nSn_3$ COMPOUNDS

Fulde et al[18] have taken into account the crystal field effect on the transition temperature of a matrix containing RE impurities. The initial depression of T_c with the concentration n is then given by the expression :

$$\left(\frac{dT_c}{dn}\right)_{T_{c_o}} = - \frac{\Pi^2}{8k_B} N(E_F)\ \Gamma^2 (g-1)^2\ F(T_{c_o}) \qquad (1)$$

with the usual notations ; $F(T_{c_o})$ is defined in reference 19. The $La_{1-n}R_nSn_3$ samples have been prepared in an induction furnace under argon atmosphere and T_c was measured by the appearance of diamagnetism at the transition. The experimental results are given in figure 3. They are in approximate agreement with the results of Schmid et al[20]. The curve of T_c versus n exhibits a positive curvature in the (La:Ce)Sn_3 system. In order to explain the experimental result, we have taken into account simultaneously the crystal field effect and the influence of the Kondo effect, calculated by Müller-Hartmann et al[21]. The dashed curve has been calculated using the

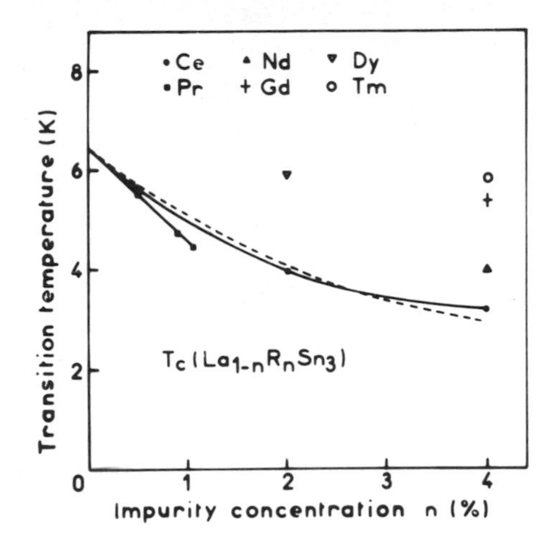

Fig. 3 : Superconducting transition temperature of some $La_{1-n}R_nSn_3$ compounds versus the concentration n. The dashed curve has been calculated with equations 2 and 3. Error bars on T_c : ± 0.2 K.

equations : $\mathrm{Ln}(T_c/T_{c_o}) + \Psi(1/2 + \rho) - \Psi(1/2) = 0$ (2)

with $$\rho = \frac{\Pi^2 n(g-1)^2\, F(T_c)}{4T_c\, N(E_F)\, \left[\mathrm{Ln}^2(T_c/T_K) + \Pi^2(g-1)^2\, F(T_c)\right]} \qquad (3)$$

where Ψ is the digamma function and with $N(E_F) = 1$ state/eV and $T_K = 140$ K. We have supposed that the crystal field splitting is $\Delta = 45$ K, the doublet being the lowest level. The agreement with the experiment is satisfactory. For the system $(La:Pr)Sn_3$, the application of the theory of Müller-Hartmann et al, neglecting the crystal field effects, would lead to a Kondo temperature T_K : 1.2×10^{-5} K with a density of states $N(E_F) = 0.33$ states/eV. Since T_K is very low, the theory of Fulde et al, valid in the first Born approximation, may give a good order of magnitude of the exchange integral. Using thus equation (1), we have determined in the systems $(La:Pr)Sn_3$ and $(La:Nd)Sn_3$ respectively $|\Gamma| = 0.22$ and $|\Gamma| = 0.19$ eV with a value of the density of states $N(E_F) = 0.33$ states/eV/at. These values are in very good agreement with the determination of Γ in $PrSn_3$ and $NdSn_3$ by N.M.R. measurements[22] : $\Gamma = -0.2$ eV. In the $(La:Gd)Sn_3$ system, we find $|\Gamma| = 0.02$ eV. The 4f shell appears thus more localized in this system than in the Pr or Nd systems.

REFERENCES

1. P. Lethuillier and J. Chaussy, J. Physique, 37, 123 (1976).

2. A.M. Van Diepen, R.S. Craig, and W.E. Wallace, J. Phys. Chem. Solids, 32, 1867 (1971).

3. E. Bucher, J.P. Maita, G.W. Hull, J. Sierro, C.W. Chu and B. Lüthi, Conference of Montreal (1974).

4. P. Lethuillier, J. Pierre, K. Knorr and W. Drexel, J. Physique, 36, 329 (1975).

5. A.J. Freeman and D.D. Koelling, J. Physique, C3, 33, 57 (1972).

6. D.M. Gray and L.V. Meisel, Phys. Rev. B, 5, 1299 (1972).

7. W.D. Grobman, J. Phys. Chem. Solids, 33, 1307 (1972).

8. G. Williams and L.L. Hirst, Phys. Rev. B, 185, 407 (1969).

9. D.K. Ray, Private communication.

10. G.K. Shenoy, B.D. Dunlap, G.M. Kalvius, A.M. Toxen and R.J. Gambino, J. Appl. Phys., 41, 1317 (1970).

11. A.I. Abou Aly, S. Bakanowski, N.F. Berk, J.E. Crow and T. Mihalisin, Phys. Rev. Letters, 35, 1387 (1975).

12. P. Lethuillier and P. Haen, Phys. Rev. Letters, 35, 1391 (1975).

13. I. Peschel and P. Fulde, Z. Phys., 238, 99 (1970).

14. J.R. Cooper, C. Rizzuto and G. Olcese, J. Physique, Colloque C1, 32, 1136 (1971).

15. R.J. Gambino, W.D. Grobman and A.M. Toxen, Appl. Phys. Letters, 22, 506 (1973).

16. P. Lethuillier and C. Lacroix - Lyon-Caen, to be published.

17. C. Lacroix - Lyon-Caen and P. Lethuillier, to be published.

18. P. Fulde and I. Peschel, Adv. Phys., 21, 1 (1972).

19. P. Lethuillier, Phys. Rev. B, 12, 4836 (1975).

20. W. Schmid and E. Umlauf, Communications on Physics, 1, 67 (1976) and references therein.

21. E. Müller-Hartmann and J. Zittartz, Phys. Rev. Letters, 26, 428 (1971).

22. F. Borsa, R.G. Barnes and R.A. Reese, Phys. Stat. Sol., 19, 359 (1967).

INFLUENCE OF THE CRYSTALLINE FIELD ON THE KONDO EFFECT : CERIUM AND YTTERBIUM IMPURITIES

A. K. Bhattacharjee[+], B. Cornut[*] and B. Coqblin[†]

[+]Groupe de Physique des Solides de l'ENS, Université Paris VII, place Jussieu,75221 Paris Cedex 05, France
[*]CRTBT, C.N.R.S., 166X, 38042 Grenoble Cedex, France
[†]Physique des Solides, Université Paris-Sud, Centre d'Orsay, 91405 Orsay, France

I - THE EFFECTIVE HAMILTONIAN

The presence of a 4f level close to the Fermi level is responsible for the Kondo effect occurring in alloys containing anomalous rare-earth impurities such as LaCe, YCe[1], AuYb[2] or also in compounds such as $CeAl_2$ and $CeAl_3$[3]. The crystalline field effect is known to be important in such alloys or compounds and thus it affects deeply the properties due to the Kondo effect.

The purpose of the present paper is to present a model describing both the Kondo effect and the crystalline field (CF) effect in Cerium and Ytterbium alloys and then to discuss briefly the different properties : resistivity, thermopower and relaxation rate. The detailed calculations are given in ref. 4, 5, 6,7 and 8.

The derivation of the effective Hamiltonian appropriate for Ce or Yb impurities has been previously described[4,5]. Let us recall the main points : If M denotes a CF eigenstate, C_M^* the creation operator for a 4f electron in the corresponding M substate and C_{kM}^* the creation operator for a conduction electron partial wavefunction of wavenumber k, total angular momentum j (5/2 for Ce and 7/2 for Yb) with a magnetic z-component $M=\langle j_z \rangle$, the effective Hamiltonian derived by the Schrieffer-Wolff transformation for the $4f^1$ (or $4f^{13}$) configuration can be written as :

$$\mathcal{H} = -\sum_{\substack{k,k' \\ M,M'}} J_{MM'} C_{k'M'}^* C_{kM} \left(C_M^* C_{M'} - \delta_{MM'} \langle n_M \rangle \right) + \mathcal{V} \sum_{\substack{k,k' \\ M}} C_{k'M}^* C_{kM} \qquad (1)$$

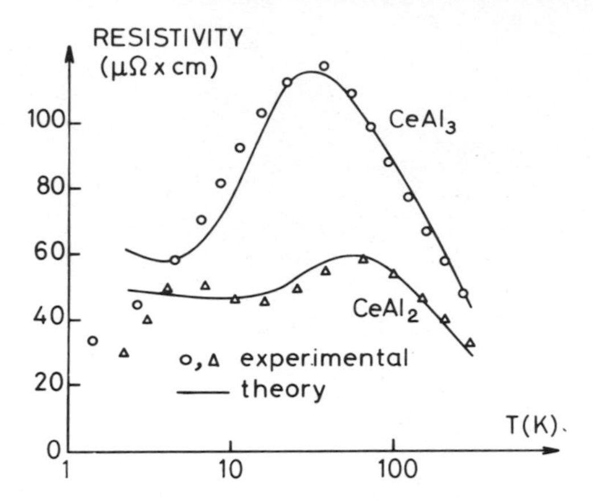

Fig. 1 : The magnetic resistivities of $CeAl_2$ (Δ) and $CeAl_3$ (o)[11], the theoretical curves are computed using the parameters given in ref. 5.

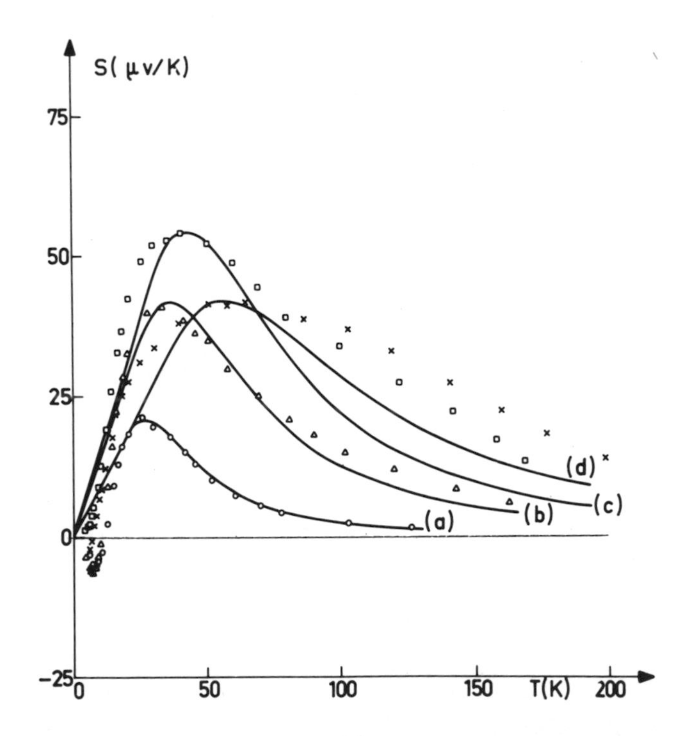

Fig. 2 : The experimental points[12] give the differences between the Seebeck coefficients of $Ce_{1-x}La_xAl_3$ and of $LaAl_3$ for x = 0.99 (o), 0.90 (Δ), 0.50 (◘), 0(x) and the full lines represent the corresponding theoretical curves obtained with the parameters given in ref. 6.

where the exchange integrals $J_{MM'}$ are given by :

$$J_{MM'} = \frac{|V_{kf}|^2}{2} \left(\frac{1}{E_M} + \frac{1}{E_{M'}} \right)$$

V_{kf} is the mixing parameter, E_M (< 0) the energy compared to the Fermi level for the CF level of M value. The different notations are explained in ref. 5.

II - THE ELECTRICAL RESISTIVITY

The third-order perturbation theory resistivity ρ has been computed exactly for any configuration of levels split by the CF[5] and the detailed calculations can be found in ref. 5. In the case of two levels separated by an energy Δ, the resistivity presents a peak at a temperature around Δ and, at high and low temperatures compared to Δ, ρ behaves proportionally to Log T. If we take all the $J_{MM'}$ equal to J, the coefficient of the Logarithmic slope of the low temperature or high temperature resistivity is proportional to $(\lambda_i^2 - 1)\, J^3$, where λ_i is the degeneracy λ_o of the ground state at low temperatures and the total degeneracy $\lambda_\infty = 2j + 1$ of the multiplet at high temperatures. Figure 1 shows a typical fit in logarithmic scale of the magnetic resistivities of $CeAl_2$ and

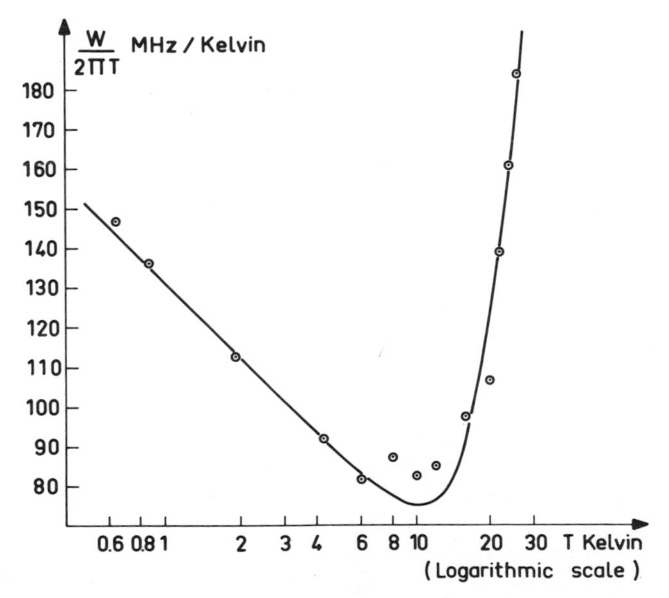

Fig. 3 : Plot of the relaxation rate W divided by 2 T versus Log T[7] for AuYb alloys. The points are the experimental results and the full line is the theoretical fit obtained with the parameters given in ref. 7.

$CeAl_3$, i. e. the differences between their resistivities and the resistivities of respectively $LaAl_2$ and $LaAl_3$. A good agreement with experiment is found between 10 and 300 K, especially for the high temperature Logarithmic behaviour ; the fits are obtained by taking CF splittings of order 100 K in $CeAl_2$ and 50 K in $CeAl_3$ and the 4f level lying of order 0.05 eV below the Fermi level. On the contrary, the very low temperature is not fitted by the present model, especially in $CeAl_3$ where the resistivity is continuously decreasing with decreasing temperature. This behaviour is probably connected to a non magnetic behaviour of Cerium in such compounds at very low temperatures[9].

III - THE THERMOELECTRIC POWER

The third-order perturbation-theory thermoelectric power has been also computed with the Hamiltonian (1) in the case of two levels splits by the CF effect[6]. When the direct scattering term is zero, a very large negative thermopower is obtained, while for a sufficiently large and negative v value, a positive thermopower is obtained. The thermopower curve presents a peak at a temperature corresponding to 1/3 to 1/6 of the total CF splitting Δ. Figure 2 shows the fits obtained by the model for the $Ce_{1-x}La_xAl_3$ alloys. A good fit is obtained for x = 0.99 and x = 0.9, while the fit is worse for x = 0.5 and x = 0. The fits are obtained by taking the 4f level lying some hundredths of eV. below the Fermi level, relatively large and negative v values and a Δ value increasing from 130 K to 255 K for x varying from 0.99 to 0. Although the values of Δ as deduced by resistivity or thermopower experiments are different in $CeAl_3$, the properties of Ce compounds appear to be well fitted by the present model.

IV - THE RELAXATION RATE

The relaxation rate of the doublet ground state Γ_7 in cubic AuYb alloys has been computed with the Hamiltonian (1) without magnetic field[7]. The third-order perturbation relaxation rate on T behaves as Log T below roughly $\Delta/15$, then goes through a minimum and rises finally very quickly for a temperature larger than $\Delta/10$. The calculation has been applied to the relaxation rate of the ground state Γ_7 of AuYb alloys as deduced by Mossbaüer experiments and figure 3 shows an excellent fit obtained by taking Δ of order 100 K, in agreement with magnetic susceptibility measurements.

Finally, the effect of a magnetic field on the relaxation rate of the ground state of Ytterbium impurity in gold has been computed up to third order in the exchange integrals within the framework of the Hamiltonian (1). It is found that the zero-field TLogT behaviour of the relaxation rate is no more clearly visible for a 5kG

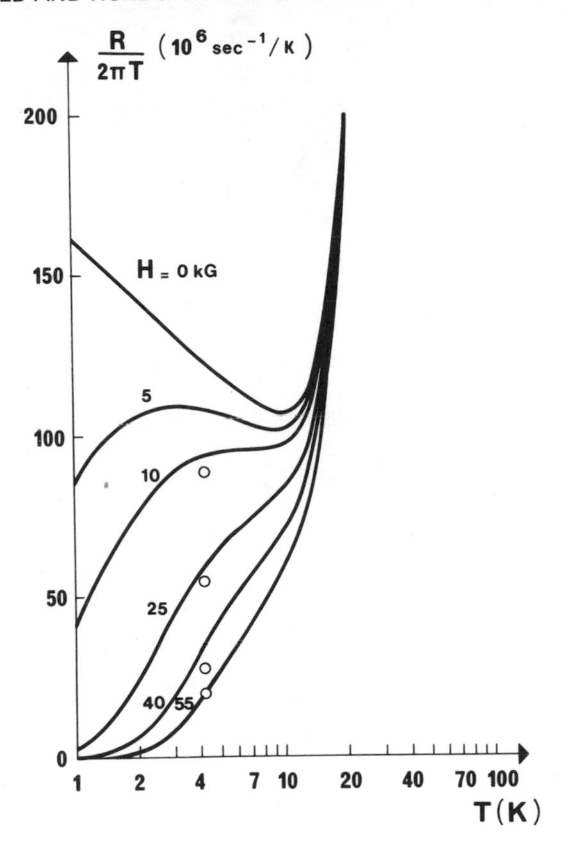

Fig. 4 : Plot of the ground state relaxation rate R divided by $2\pi T$ versus Log T for AuYb alloys. The circles are the experimental points of Stöhr[10] for H = 10, 25, 40 and 55 kG and the full lines are the theoretical curves obtained with the parameters given in ref. 8 for respectively H = 0, 5, 10, 25, 40 and 55 kG.

field in AuYb alloys and has completely disappeared for a 10 kG field, as it is shown on figure 4, where the theory is compared to the 4.2 K results of Stöhr[10]. So the low temperature TLogT behaviour of the relaxation rate, which has been found in a zero field Mossbaüer experiment, cannot be observed in a magnetic resonance experiment where a magnetic field is always present.

So, the present model which describes the CF and Kondo effects can account for different properties of Cerium and Ytterbium alloys and compounds, except at very low temperatures in compounds such as $CeAl_3$ where Cerium appears to undergo a transition to a nonmagnetic state[9].

REFERENCES

(1) T. Sugawara, R. Soga and I. Yamase, J. Phys. Soc. Japan 19, 780 (1964)

(2) A. P. Murani, J. Phys. C Suppl. 3, S 153 (1970); Solid State Comm. 12, 295 (1973)

(3) H. J. Van Daal, F. E. Maranzana and K. H. J. Buschow, J. de Physique, suppl. 32, C 1 - 424 (1971)

(4) B. Coqblin and J. R. Schrieffer, Phys. Rev. 185, 847 (1969)

(5) B. Cornut and B. Coqblin, Phys. Rev. B 5, 4541 (1972)

(6) A. K. Bhattacharjee and B. Coqblin, Phys. Rev. B 13, 3441 (1976)

(7) F. Gonzalez-Jimenez, B. Cornut and B. Coqblin, Phys. Rev. B11, 4674 (1975)

(8) A. K. Bhattacharjee and B. Coqblin, Solid State Comm. 18, 1587 (1976)

(9) B. Coqblin, A. K. Bhattacharjee, B. Cornut, F. Gonzalez-Jimenez, J. R. Iglesias-Sicardi and R. Jullien, J. of Mag. and Mag. Mat. 3, 67 (1976) and references therein.

(10) J. Stöhr, Phys. Rev. B11, 3559 (1975).

(11) K. H. J. Buschow and H. J. Van Daal. Solid State Comm. 8, 363 (1970)

(12) P. B. Van Aken, H. J. Van Daal and K. H. J. Buschow. Proceedings of the 11th Rare-earth Research Conference, Traverse City (USA) October 1974, p. 738

CRYSTAL FIELD EFFECTS ON THE SPECIFIC HEAT OF (LaEr)Al_2

J.G.Sereni

Centro Atómico Bariloche (CNEA-IFB)

8400 - Bariloche, Rio Negro, ARGENTINA

ABSTRACT

Specific heat measurements for the (LaEr)Al_2 system are reported, revealing a Schottky tipe anomaly which corresponds to an energy splitting of δ= 5K between the $\Gamma_8^{(3)}$ ground and $\Gamma_8^{(2)}$ first excited crystalline states of Er^{3+}. The initial slope of $\Delta C/\Delta C_o$ versus T_c/T_{co} is well fitted by the Abrikosov-Gor'kov theory and the depression of T_c with Er impurity concentration is compared with the Fulde, Keller and Peschel theory.

INTRODUCTION

The crystal electric field (CEF) in the cubic matrix $LaAl_2$ lifts the (2J+1) fold degeneracy of the rare earth impurity Hund's rule ground state, producing a Schottky like anomaly in the normal state specific heat (C_m). This anomaly is associated with the thermal population of the excited levels, connecting therefore the temperature (T) of the maximun and its value with the energy splitting (δ) and the relative degeneracy of these levels.

Measurements of the specific heat jump (ΔC) at the superconducting critical temperature (T_c) allow to distinguish between the largely studied magnetic regimes of rare earths in superconducting hosts[1]. There are cases which present deviations of these regimes, such as (LaPr)$_3$Tl [2] and (LaTb)Al_2[3]. These can be explained by including the interaction between the conduction electrons and the CEF-split levels of the impurity. Their superconducting behaviour can be described by including a pair breaking parameter (α), which according with the Fulde, Keller and Peschel theory[4], depends on the possible transitions between the split levels "i" and "j" through $\alpha_{ij}= \alpha|\langle i|J|j\rangle|^2/J(J+1)$.

In this work we report the experimental results of C_m, $\Delta C(T_c)$ and T_c for various impurity concentrations of the system $(\underline{La}Er)Al_2$. Due to the cubic CEF, the sixteen-fold degenerate ground state multiplet of the Er splits into two doublets (Γ_6,Γ_7) and three quartets ($\Gamma_8^{(i)}$) as it is shown in the Lea, Leask and Wolf (LLW) level scheme[5]. Consequently in the absence of a magnetic field, it is expected that this system will have at least a doubly degenerate ground state.

EXPERIMENTAL RESULTS AND ANALYSIS

Arc melted samples, annealed 48 hours at 900°C in vacuum were prepared, in which the Er concentration in La has been varied from 0.36 to 3.5 at.%. The measurements were carried out in a semiadiabatic calorimeter described elsewere[6], and the temperature range covered the whole anomaly region (between 0.4 K and 7 K). The experimental values obtained from the specific heat data of the cubic laves phase $LaAl_2$, are in good agreement with previous measurements[7] and the initial depression of T_c is 0.48 K/at.% Er, which also agrees with previous reported work[8].

C_m was obtained by subtracting the electron and phonon contributions to the total specific heat, and the results are shown in Fig.1. The maximum was found at T_m= 2.1 K and its position was independent of the concentration. It is also possible to observe a contribution due to the exchange coupling of magnetic ions, as the impurity concentration is increased. Therefore a concentration linear extrapolation was made to get the single impurity value. These points are well fitted by a calculated curve, which thermodynamically corresponds to two equally-degenerate levels split by $\delta=(5.0\pm0.1)$ K. The reduced $\Delta C(T_c)/\Delta C_{co}(T_{co})$ versus reduced T_c/T_{co} (ΔC_o and T_{co} are the matrix ΔC and T_c) is fitted in the $T_c\sim T_{co}$ range by the Skalski curve[9] characteristic of a system with a magnetic ground state.

In order to determine the sequence of the CEF split levels, we have compared in Fig.2a and 2b our results for the Schottky peak with the peak height, as calculated from the LLW level scheme[5]. This procedure gives several possibilities for the values of the LLW parameters x and W. These are: $W>0$ with $x= -0.45$ and $W<0$ with $-0.1> x> -1$.

The second Schottky maximun, due to the thermal population of the second excited level, was found at $T_m'= 12\ T_m$.

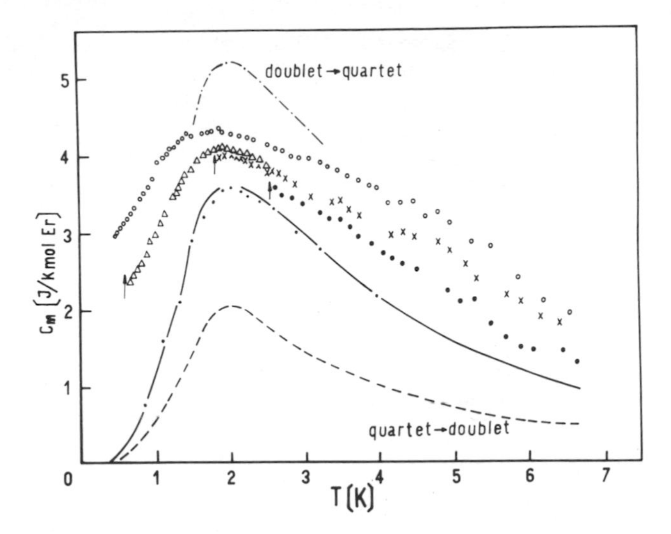

Fig. 1: The Schottky specific heat for $(La_{1-n}Er_n)Al_2$ (n= +0.0108, X 0.0234, Δ 0.025 and o 0.035, • is the extrapolation to n=0). ↑ indicates the transition temperature T_c.

In Fig.2c we show the possible values of the LLW parameter x, assuming that the energy separation of the second excited level (δ') is given by $\delta' = 12\ \delta$. Consequently the possible values of x are those which correspond to a pair of equally-degenerate ground and first excited levels, with a second one separated by twelve times the splitting δ. With this procedure, no additional information can be obtained in the first case (W>0). However if W<0 we get : $-0.3 > x > -0.4$ and $-0.5 > x > -0.6$, which corresponds, in either case, to a system with two quartets $\Gamma_8^{(3)}$ and $\Gamma_8^{(2)}$ as the ground and first excited states respectively.

Unfortunately it is not possible to eliminate the first case with doublets Γ_6 and Γ_7 levels as the lowest states, on the grounds that the pair breaking should be zero due to the fact that $\langle\Gamma_6|J|\Gamma_7\rangle = 0$. On the contrary, the matrix elements $\langle\Gamma_6|J|\Gamma_6\rangle$ and $\langle\Gamma_7|J|\Gamma_7\rangle$ are non zero and consequently the system could show the Abrikosov-Gor'kov (AG)[10] behaviour for $T_c(n)/T_{co}$, similar to that corresponding to impurities with effective spin $S'=1/2$.

In Fig.3, we compare our experimental data with the crystal field adapted AG function for $T_c(n)/T_{co}$. The sample with n=0.025 still exibites superconducting behaviour and it shows a large deviation from the theoretical prediction. For n=0.035 no superconducting transition was observed within the temperature limits of our experiments ($T \sim 0.4$ K).

It is then clear that the observed CEF effects in the (LaEr)Al_2 system are largely due the value of $\delta/T_{CO} = 1.6$, making it an interesting system to be studied. It is important to note that similar values of the LLW parameters were found by neutron spectroscopy from the (Y Er)Al_2 [11] system, which has the same charge symmetry as (LaEr)Al_2.

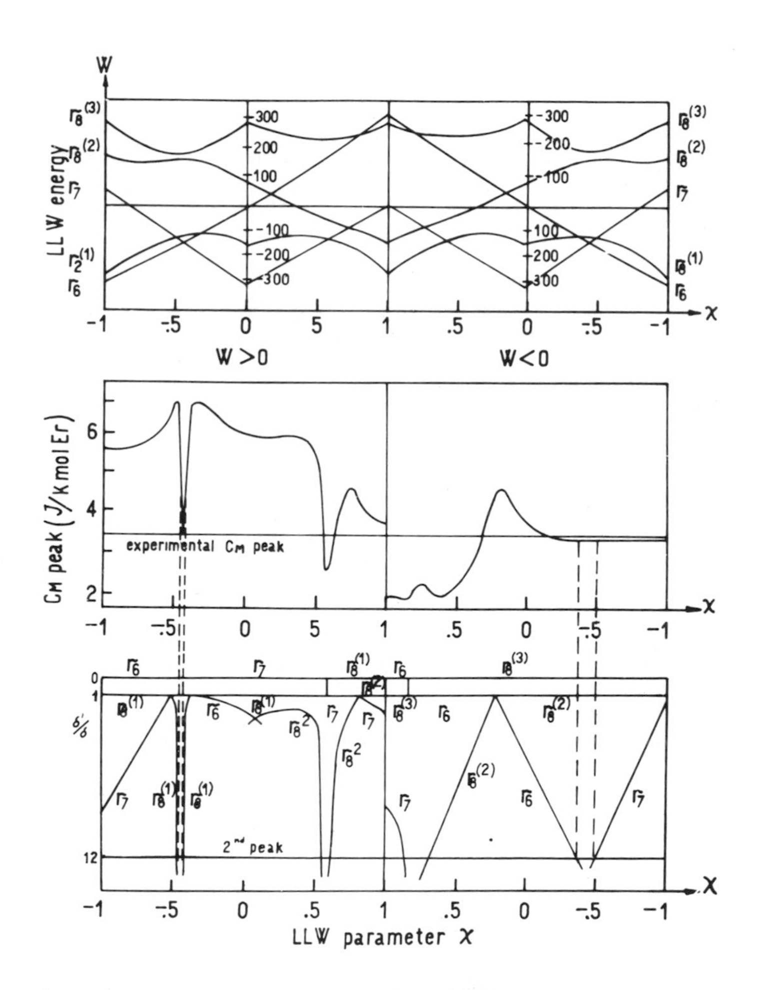

Fig. 2: a) The LLW level scheme of J=15/2 as a function of W and x
b) The peak height calculated from a) versus W and x. The straight line is the (LaEr)Al_2 value.
c) The energy splitting in δ units versus x. The straight line corresponds to the second Schottky peak.

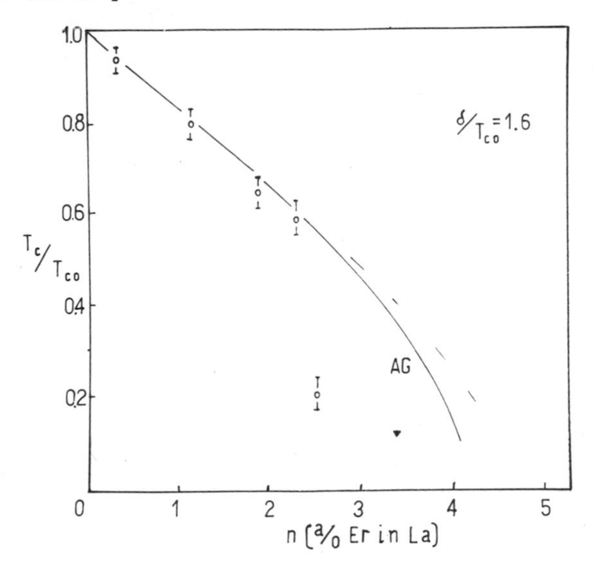

Fig. 3: The reduced critical temperature T_c/T_{co} is plotted versus the impurity concentration n. The dashed line is the $T_c(n)/T_{co}$ function calculated from the Fulde, Keller and Peschel theory with x= -0.6[12]. AG is the Abrikosov-Gor'kov function[10]. ▼ indicates n of the non-superconducting sample and the T_c/T_{co} limit of our experiments.

REFERENCES

1. C.A. Luengo, J.G. Huber and M.B. Maple, J. Low Temp. Phys. 21, 129 (1975) and M.B. Maple, Appl. Phys. 9, 176 (1976)
2. E. Bucher, J. Maita and A.S. Cooper, Phys. Rev. B 6, 2709 (1972)
3. H. Happel and H. Hoenig, Solid State Commun. 13, 1641 (1973)
4. J. Keller and P. Fulde, J. Low Temp. Phys. 4, 289 (1971)
   P. Fulde and I. Peschel, Adv. in Physics 21, 1 (1972)
5. K. Lea, M. Leask and W. Wolf, J. Phys. Chem. Solids 23, 1381 (1962)
6. J.G. Sereni, Ph.D. Thesis, Universidad Nacional de Cuyo, Bariloche, Argentina (1976), (unpublished)
7. R. Hungsberg and K. Gschneidner Jr., J. Phys. Chem. Solids 33, 401 (1972)
8. M.B. Maple, Solid State Commun. 8, 1915 (1970)
9. S. Skalski, O. Betbeder-Martibet and P. Weiss, Phys. Rev. 136, A1500 (1964)
10. A.A. Abrikosov and L.P. Gor'kov, Soviet Phys. JETP 12, 1234 (1961)
11. H. Herr, A. Furrer, F. Volken, A. Treyvaud, H. Purvins and J. Kjems, J. Physsics C 7, 1207 (1974)
12. J. Keller, private communication.

EXPERIMENTAL EVIDENCE FOR THE PRESENCE OF A NON-MAGNETIC 5d VIRTUAL BOUND STATE IN THE 4f-KONDO ALLOY Au Yb

B. Cornut and P. Haen

Centre de Recherches sur les Très Basses Températures,

CNRS, BP 166 Centre de Tri, 38042 Grenoble Cédex, France

ABSTRACT

Low Temperature resistivity and magnetization measurements on Au Yb alloys containing less than 3000 at.ppm of Yb will be briefly recalled. These showed that the Kondo temperature of the Γ_7 groundstate is very low. They also allowed us to determine the magnitude of the crystalline field splitting from the superposed susceptibility observed in the high-field, low temperature part of the magnetization curves. New measurements of the dc-resistivity between 4.2 and 300 K show an anomalous increase which is approximatively logarithmic with temperature. This increase is attributed to the presence of a non-magnetic 5d virtual bound state at the Fermi level. This conclusion is similar to that made on the basis of thermoelectric power measurements.

The first observation of the Kondo effect of Ytterbium impurities diluted in Gold is the Murani's resistivity measurements[1]. The logarithmic decrease of $W/2\pi T$ (W is the relaxation rate) with increasing temperature in Mossbauer experiments[2] is a second proof of Kondo effect in Au Yb alloys. In order to observe the influence of the crystalline field on the Kondo effect we have made magnetization and resistivity measurements on seven Au Yb alloys containing less than 3000 appm. The chemical analysis give the concentrations : 42, 95, 430, 609, 1860 and 2940 ppm. Two samples were activated : they contain only 5 ppm of iron.

MAGNETIZATION

The magnetic measurements (between 0.07 K and 70 K and from 0 to 100 KOe give the following results :

- Above 0.2 K all the magnetization curves can be fitted by the law:

$$M/c = Mo \mathcal{B}_{1/2} \left(\frac{MoH}{KT}\right) + X_{VV} H$$

c is the Ytterbium concentration and $\mathcal{B}_{1/2}$ the Brillouin function for a spin S = 1/2. Mo is equal to 1.69 ± 0.05 μ_B. This value corresponds to the Γ_7 doublet as fundamental level of the 4f hole of Ytterbium. The proportionality of the magnetization to the concentration indicates that the interactions between Ytterbium ions are very weak . The second part of the magnetization which is always proportional to the field is due to the presence of the excited level Γ_8. Thus, the high field - low temperature part of our data permits us to determine quite precisely the distance between Γ_7 and Γ_8 : 102 ± 5 K.

Finally we don't observe above 0.2 K any Kondo behavior. Namely the Kondo temperature of the ground state is very low.

- - Below 0.2 K hyperfine effects are observed. They have been extensively studied in our laboratory by means of various monoisotopic Au Yb alloys[3].

RESISTIVITY

The resistivity between 1.3 K and 300 K was measured on the same samples. Below 4.2 K we observe also the characteristic logarithmic increase due to the Kondo effect of Ytterbium (Fig. 1).

Above 4 K the magnetic resistivity which is the difference between the resistivity of the alloy and the resistivity of the matrix increases roughly logarithmicly with temperature : (Fig. 2).

If we take into account the small errors in the determination of the form factor of our samples (of about 1%), we can claim that apart in a small range of temperature around 40 K the resistivity is proportional to the Ytterbium concentration. Departures are observed around 40 K. They are due to deviations from Matthiessen's rule and are very similar to these observed in Au Cr alloys. We actually omit these deviations in order to consider only the general behaviour of the resistivity.

The use of an effective hamiltonian appropriate for Cerium or Ytterbium impurities is necessary for describing our experiments. This hamiltonian takes into account spin and orbit scattering[4]. Thus we choose J = - 0.37 eV (exchange integral) and V = 0.2 eV (direct scattering potential). This choice is a very good check for the Mossbauer and EPR measurements. It explains also the low temperature logarithmic slope of the resistivity. But this choice

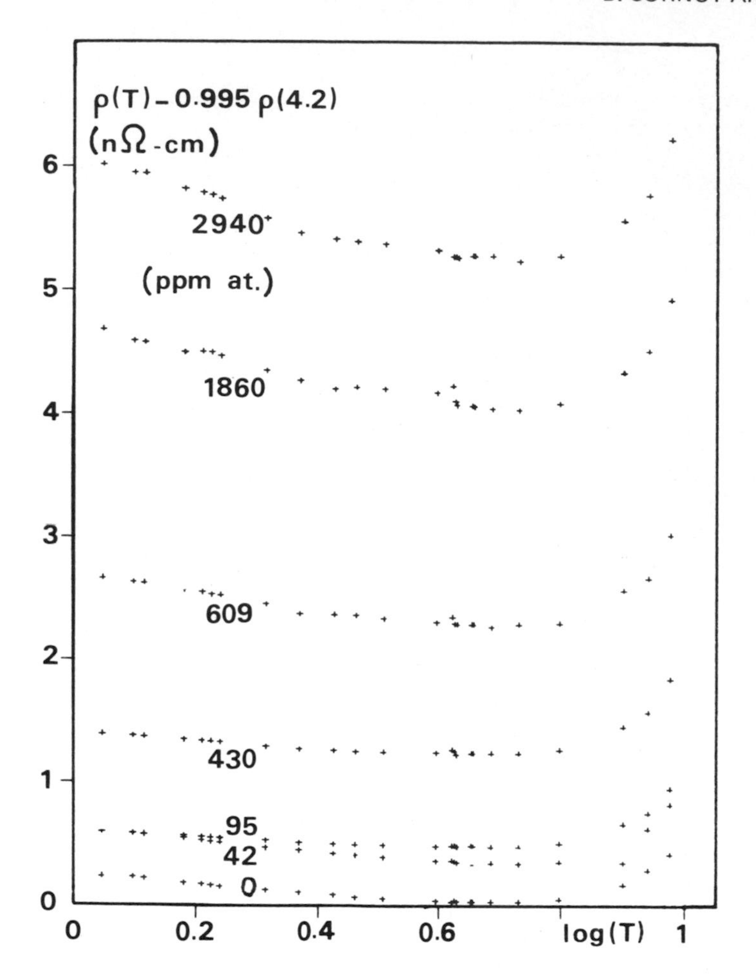

Fig. 1 : Kondo resistivity of Au Yb below 10 K.

implies that the magnetic resistivity is very small at 4.2 K and varies between 10 and 300 K of the same order of magnitude than between 1 and 10 K. So in the scale of Fig. 2 the calculated resistivity appears nearly constant between 1 and 300 K.

Consequently we have to explain the magnetic resistivity and its variations with temperature with an another mechanism of scattering.

Following Bekker and Van Duren[5] we attribute this increase to the presence of a non-magnetic 5d virtual bound state at the Fermi level. Applying the Friedel's sum rule, we deduce from the residual resistivity 1.6 ± 0.2 d-like screening charge which corresponds

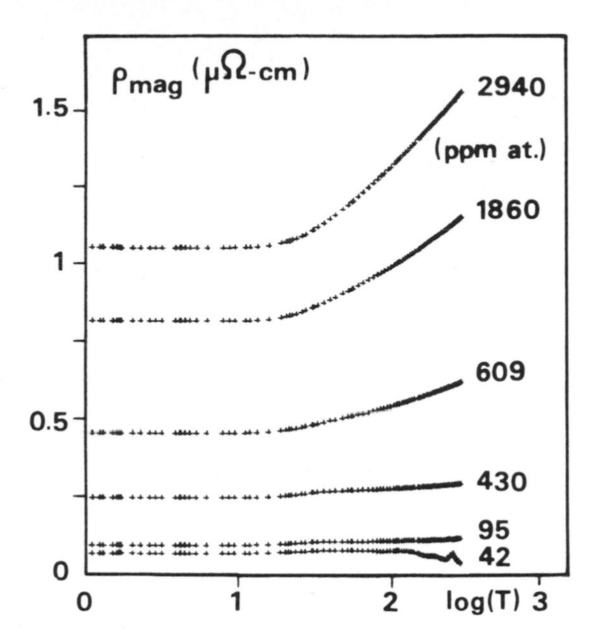

Fig. 2 : Magnetic resistivity of Au Yb alloys between 1 and 300 K.

approximatively to the charge difference of value two between Gold and Ytterbium . Moreover the model of virtual bound state[6] allows us to predict the thermal variation of the resistivity which is in accordance to the experimental one.

In conclusion the Au Yb system is very interesting because of the variety of phenomena which can occur : hyperfine field effects, Kondo effect on 4f hole, crystalline field effects and now virtual bound d.state scattering.

REFERENCES

1. A.P. Murani, Solid State Commun. 12, 295 (1973)

2. F. Gonzalez-Jimenez and P. Imbert, Solid State Commun. 13, 85 (1973)

3. G. Frossati, J.M. Mignot, D. Thoulouze and R. Tournier, Phys. Rev. Lett. 36, 203 (1976)

4. B. Cornut and B. Coqblin, Phys. Rev. B5, 4541 (1972)

5. F.F. Bekker and C.J.A. Van Duren, Proc. of the LT 14th Conference 3, 406 (1975)

6. J. Souletie, J. Phys. F5, 329 (1975).

NMR STUDY OF SINGLET GROUND STATE SYSTEMS

M. A. Teplov

Kazan State University

420008 Kazan, U.S.S.R.

ABSTRACT

A review is presented of theoretical and experimental NMR investigations of rare-earth ions in Van Vleck paramagnets carried out since 1956.

INTRODUCTION

Paramagnetic ions with an even number of electrons in the unfilled shell can be in a singlet ground state in a crystal electric field (CEF). In zero magnetic field the singlet ground state has no permanent magnetic moment, but an applied field admixes the wave functions of excited states into the singlet ground state and produces an "induced" dipole moment. This induced moment in turn produces a hyperfine field at the nucleus which can exceed the applied field by orders of magnitude. It should be noted that this enhancement of the applied magnetic field has proved to be important for practical purposes. Ten years ago Al'tshuler[1] has suggested the use of Van Vleck paramagnets for a magnetic cooling. The new cooling method was presently tested in numerous experiments by Andres and Bucher[2] and found to be very promising. The NMR investigation of singlet ground state ions was first suggested by Zaripov[3] and Elliott[4]. It was shown (see also Ref.5-8) that the NMR study should make it possible to obtain data on CEF symmetry as well as the energy and wave functions of paramagnetic ions in crystals. This report presents a short survey of NMR investigations of singlet ground state systems carried out since 1956 with a special emphasis on the rare-earth compounds[9].

NMR SPECTRA

The magnetic resonance spectra of the nuclei of rare-earth ions in crystals can be described by a spin Hamiltonian whose form depends on the CEF symmetry. Thus for a rhombic symmetry the spin Hamiltonian is

$$\mathcal{H}_I^{rhomb} = -\gamma\hbar \sum_{i=x,y,z} (1+\alpha_i) H_i I_i + D[I_z^2 - \tfrac{1}{3} I(I+1)] + E(I_x^2 - I_y^2)$$

while for an axial and cubic symmetry it has the form

$$\mathcal{H}_I^{ax} = -\gamma\hbar (1+\alpha_\parallel) H_z I_z - \gamma\hbar (1+\alpha_\perp)(H_x I_x + H_y I_y) + D[I_z^2 - \tfrac{1}{3} I(I+1)]$$

and $\mathcal{H}_I^{cub} = -\gamma\hbar(1+\alpha)\bar{H}\bar{I}$

respectively. The parameters α determine the hyperfine magnetic field at the nucleus produced by the polarized electronic 4f-shell and the zero magnetic field splitting parameters D and E apart from the terms due to the quadrupole interaction also contain the so-called "pseudoquadrupole" contributions[10]. The data on the 4f-electronic states contained in experimentally derived spin Hamiltonian parameters can be obtained from the following formulae [3,4,11]:

$$D = D_{hf} + \frac{3e^2qQ}{4I(2I-1)}\ ; \quad E = E_{hf} + \frac{e^2qQ}{4I(2I-1)}\eta\ ;$$

$$D_{hf} = a_J\left(\frac{\Lambda_x + \Lambda_y}{2} - \Lambda_z\right); E_{hf} = a_J\frac{\Lambda_y - \Lambda_x}{2}\ ;$$

$$\alpha_i = 2g_J\mu_B\Lambda_i/\gamma\hbar\ ; \quad \Lambda_i = a_J\sum{}' \frac{|\langle 0|J_i|n\rangle|^2}{\mathcal{E}_n - \mathcal{E}_0}.$$

Here g_J and a_J are the Lande g-factor and the magnetic hyperfine constant for the free ion, $\mathcal{E}_n$ and $|n\rangle$ are the energy and the wave function of the n-th CEF eigenstate of the ion ("0" refers to the singlet ground state). These formulae are valid in the case of low enough temperatures when only the lowest singlet is populated and the value of the hyperfine field H_{hf} at the nucleus is determined by the ground state of the 4f-shell. If the excited states populations differ from zero the nuclear moment is affected by the average field $\langle H_{hf}\rangle$ which decreases with temperature increase, so that the parameter α is always proportional to paramagnetic susceptibility

$$\alpha(T) = a_J\chi^{ion}/g_J\mu_B\gamma\hbar\ ;$$

here χ^{ion} is the magnetic susceptibility per ion. It can be seen that the quantity α in its sense is identical with the paramagnetic part of a chemical shift well familiar in high resolution NMR spectroscopy. Therefore we shall refer to α as "paramagnetic shift". The results of the NMR study of insulators carried out at liquid helium temperatures are given in Table I.

Table I. ^{I4I}Pr and ^{I69}Tm NMR data on dielectric Van Vleck paramagnets.

Sample	Spin Hamiltonian parameters				
	$\alpha_x(0)$	$\alpha_y(0)$	$\alpha_z(0)$	$\lvert D/h \rvert$, MHz	$\lvert E/h \rvert$, MHz
$Pr_2(SO_4)_3\ 8H_2O$ [II]	2.54 ±0.04	I.070 ±0.008	8.43 ±0.I2	5.865 ±0.005	0.730 ±0.00I
$Pr(NO_3)_3\ 6H_2O$ [I2]	4.73 ±0.03	0.664 ±0.008	II.79 ±0.03	6.550 ±0.0I5	I.540 ±0.005
$PrAlO_3$ [I3]	0.539 ±0.020	0.539 ±0.020	6.I8 ±0.50	4.2 ±0.I	
$Pr_2(SeO_4)_3\ 5H_2O$ [I4]				3.84	I.I0
$Pr(CH_3COO)_3\ 2H_2O$ [I4]				4.56	0.4I
PrF_3 [I4]				4.3I	0.30
TmES [I5]	73.35 ±0.05	73.35 ±0.05	0.6 ±0.I		
$Tm_3Ga_5O_{I2}$ [I6]	I3.80	9.85	24.46		
$Tm_3Al_5O_{I2}$ [I7]	5.2I	2.50	95.34		
$LiTmF_4$ [I8]	68.9	68.9	I.765		

In some cases it is possible to derive the first excited state energy of the paramagnetic ion from the temperature dependence of spin Hamiltonian parameters. Thus the energies of Tm^{3+} excited doublets in TmES [I5] and $LiTmF_4$ [I8] have been found.

Table 2.
Observed ^{141}Pr and ^{169}Tm NMR paramagnetic shifts in conducting Van Vleck paramagnets.

Sample	T,K	α_Σ
PrP	4	5.38 ± 0.I2[a]
	27	6.30 ± 0.I2[a]
	76	3.93 ± 0.07[a]
PrAs	4	6.50 ± 0.I2[a]
	27	6.37 ± 0.I2[a]
PrS	4	4.84[c]
PrSe	4	4.4[b]
	4	4.52[c]
PrTe	4	6.78[c]
Pr-metal (cubic)	I.5	37.5 ± 2.5[b]
TmP	4	76.7 ± I.0[a]
	27	4I.2 ± 0.5[a]
TmAs	4	70.9 ± I.0[a]
	27	37.9 ± 0.5[a]
TmSb	4	88.7 ± I.2[a]
	27	39.9 ± 0.5[a]
TmSe	4	66.0[b]
$TmAl_3$	4	39.4[b]

[a]Ref.I9 [b]Ref.20 [c]Ref.2I

The NMR of rare-earth ions in a singlet ground state has also been studied in intermetallic praseodymium and thulium compounds. The CEF symmetry in all the investigated substances is cubic. Therefore in polycrystalline samples a single NMR line characterized by an isotropic paramagnetic shift has been observed. The measurement results are summarized in Table 2. The NMR in metals and intermetallic compounds has some peculiarities due to the presence of conduction electrons. The paramagnetic shift α_Σ is composed of two parts. The first (normally the largest) is a contribution of the crystal field α_{cr} and coincides with a shift in insulators. The second part due to the conduction electrons can in its turn be represented as a sum of two terms $\alpha_{cont} + \alpha_{exc.polar}$. The shift α_{cont} appears as a result of a contact interaction between a nuclear spin and conduction electrons affected by a strong "exchange field" of the rare-earth ions. In order to estimate α_{cont} in cubic intermetallic compounds Jones[19,20] resorted to the uniform conduction electron spin polarization model and obtained the expression

$$\alpha_{cont} = K_0[1+\Gamma(g_J-1)\chi^{ion}/2g_J\mu_B^2],$$

where K_0 is the Knight shift, Γ is the s-f exchange interaction energy. Although Γ in intermetallic praseodymium and thulium compounds is fairly large ($\Gamma \sim$ 0.I-I.0 eV) and the second term in brackets is of the order of I00, the resulting shift α_{cont} proves to be small, only a few per cent of α_{cr}.

Tsarevskii[22] has considered an additional shift $\alpha_{exc.polar}$ appearing due to the polarization of an ionic 4f-shell by the "exchange field" of conduction electrons and has shown that this shift exceeds considerably α_{cont} and for high concentrations of the conduction electrons may even occur to be larger than α_{cr}. The expression for the shift due to the "exchange polarization" of a rare-earth ion's magnetic shell in compounds with cubic CEF symmetry may be written as

$$\alpha_{exc.polar} = \alpha_{cont}\, a_J \Gamma(g_J-1)\chi^{ion}/2a_s g_J \mu_B^2,$$

where a_s is the constant of a contact interaction between conduction electrons and the nuclear moment of the rare-earth ion. Assuming $a_s \sim a_J$ one can find that, for instance, in Pr- and Tm-pnictides the shift $\alpha_{exc.polar}$ may be one or two order of magnitude larger than α_{cont}.

At low temperatures and for a sufficiently strong magnetic field the quantization of conduction electrons orbital movement becomes possible, which results in the oscillation of electron density of states on the Fermi surface. Kossov and Tsarevskii[23] have considered the effect of this factor on the NMR shift due to conduction electrons and have found that the amplitude of the oscillations must be $\tilde{\alpha} = \alpha(\hbar/E_F\tau)^{1/2}$ and the period $\tilde{H} = H \times \hbar\omega_0/E_F$. Here E_F is the Fermi energy, H is the external magnetic field, τ is the conduction electrons' spin-lattice relaxation time, $\omega_0 = eH/m^*c$ is the cyclotron frequency, e and m^* are the charge and the effective mass of a conduction electron. These formulae are valid if $\omega_0\tau \gg 1$ and $\hbar\omega_0 > kT$. Setting $E_F \sim 10^{-12}$ erg, $\tau \sim 10^{-10}$-10^{-9} s, $m^* \sim m_e$ (electron mass), $H \sim 5 \times 10^4$ Oe, we find $\tilde{\alpha}/\alpha \sim 10^{-3}$ and $\tilde{H}/H \sim 10^{-3}$. The measurements of the NMR shift oscillations could provide valuable data on conduction electron parameters. However, as the estimates show, such measurements are difficult to perform especially in view of the fact that ^{141}Pr and ^{169}Tm NMR linewidth in intermetallic compounds at liquid helium temperatures has been found to be proportional to the external magnetic field[19,20], this proportionality factor being larger than 10^{-2}.

NUCLEAR SPIN-LATTICE RELAXATION

The nuclear spin-lattice relaxation of paramagnetic ions in a singlet ground state must exhibit some specific features. In fact hyperfine and spin-orbital interactions couple the nuclear moment with electrons' orbital moment and, as a result, the reorientation of the nuclear spin may occur due to an electrostatic interaction between the paramagnetic ion and its environment modulated by lattice vibrations. The very first experiments[15,24] carried out at liquid helium temperatures on a thulium ethyl sulphate single crystal have demonstrated that the nuclear spin-lattice relaxation rate of ^{169}Tm nuclei is close in the order of magnitude to the electron spin-lattice relaxation rate and changes drastically with temperature variations obeying the dependence $\exp(-\Delta/\kappa T)$, where Δ is the first excited state energy. It was shown[25] that in TmES the Orbach process must be really very effective and that thulium nuclear spin-lattice relaxation determined by it must possess a very sharp angular dependence (in particular, $T_I^{\parallel}$ at H_0 being parallel to the crystallographic c-axis must exceed $T_I^{\perp}$ approximately by a factor of 10^4). The detailed analysis of relaxation processes in a TmES single crystal was done by Vaisfeld[26]. The direct and Raman processes were found to be ineffective while the theoretical estimates of the Orbach relaxation rate proved to be in very good agreement with the experiment.

Are the Orbach processes always effective in the nuclear spin-lattice relaxation of rare-earth singlet ground state ions ? One can hardly come to a definite conclusion on the basis of a single experimental fact (one could assume, for instance, that the Orbach relaxation proved to be effective in TmES owing to a special form of excited doublet wave functions). Only a series of measurements[12] on praseodymium salts made it possible to give an unequivocal answer to this question. Measurements have shown that the nuclear spin-lattice relaxation rate of rare-earth ions in all studied crystals is described by a sum of two terms

$$T_1^{-1} = AT + B\exp(-1.44\Delta/T),$$

the first term most probably describing nuclear relaxation via the paramagnetic impurities (according to EPR data the samples have always contained small amounts of Kramers ions) and the second term describing a two-phonon relaxation via the excited state of the 4f-shell with the energy Δ . The measurement results are listed in Table 3.

Table 3. Spin-lattice relaxation data on ^{141}Pr and ^{169}Tm nuclei in Van Vleck paramagnets.

Sample	T,K	Relaxation parameters		Excited state energy, cm^{-1}
		A, $s^{-1}K^{-1}$	B, s^{-1}	
TmES [15]	1.6-5.0	0.2	1.76×10^8	32.5
$Pr_2(SO_4)_3\ 8H_2O$ [12]	1.6-9.2	3.7	8.36×10^7	55
$Pr(NO_3)_3\ 6H_2O$ [12]	1.6-5.5	2.4	4.0×10^6	27
$Pr(ReO_4)_3\ 4H_2O$ [12]	1.6-5.2	3.35	2.21×10^6	19.7

All the attempts to observe the pulsed NMR of ^{141}Pr and ^{169}Tm nuclei in intermetallic compounds have failed. The cause of these failures apparently lies in the fact that even at liquid helium temperatures the nuclear relaxation time (T_2 and/or T_1) is too short. Tsarevskii[27] has obtained for the nuclear spin-lattice relaxation rate the following formula

$$T_1^{-1} = \frac{16\pi}{\hbar}\left(K_0\frac{\gamma\hbar}{g_J\mu_B}\right)^2\left(\frac{a_J}{a_S}\Gamma\frac{g_J-1}{g_J}\;\frac{\chi^{ion}}{2\mu_B^2}\right)^2 kT.$$

Assuming $a_s \sim a_J$, $K_0 = 3\times10^{-3}$, $\Gamma = -0.7$ eV [19,20] we find that in Pr-pnictides at T=1K the ^{141}Pr spin-lattice relaxation time is 10^{-6}s, which seems to be too short for the presently available pulsed NMR equipment.

CONCLUSION

The experimental data on the nuclear magnetic resonance of rare-earth ions in Van Vleck paramagnets obtained so far - some 20 substances have been studied - enable one to claim that the NMR method applied to singlet ground state systems is as effective as the EPR method for ordinary paramagnets. The possibilities afforded by NMR in the investigation of Van Vleck paramagnets can be summarized as follows:

(i) The NMR method provides data on the CEF symmetry.

(ii) The continuous wave method makes possible highly precise measurements of Van Vleck susceptibility components. This is particularly important when a unit cell contains magnetically unequivalent paramagnetic centers and any non-resonance method gives only average susceptibility values.

(iii) If the CEF symmetry is higher than rhombic, the continuous wave method enables one to determine the energy of the first excited state or (if the energy levels are known) the wave functions of the paramagnetic ion.

(iv) The pulsed NMR method makes it possible to determine the energy of the excited state for any CEF symmetry provided that the nuclear transverse relaxation time T_2 is not shorter than say 3 microseconds.

REFERENCES

I. S.A.Al'tshuler, Zh.Eksp.Teor.Fiz., Pis'ma Red., 3, I77 (I966)
2. K.Andres and E.Bucher, Phys.Rev.Lett., 2I, I22I (I968); J.Appl.Phys., 42, I522 (I97I); J.Low Temp. Phys., 9, 267 (I972)
3. M.M.Zaripov, Izvest.Akad.Nauk SSSR, Ser.Fiz., 22, I220 (I956)
4. R.J.Elliott, Proc.Phys.Soc.(London), 70, II9 (I957)
5. R.M.Mineeva, Fiz.Tverd.Tela, 5, I403 (I963)
6. S.A.Al'tshuler and R.M.Mineeva, Fiz.Tverd.Tela, 7, 3I0 (I965)
7. R.M.Mineeva, Thesis, Kazan State University (I966)
8. L.Ya.Shekun, Fiz.Tverd.Tela, 8, 2929 (I966)
9. Of the substances containing 3d-ions only V^{3+}-doped corundum Al_2O_3 has been studied (S.A.Al'tshuler and V.N.Yastrebov, Zh.Eksp.Teor.Fiz., 47, 382(I964)
I0. A.Abragam and B.Bleaney, Electron Paramagnetic Resonance of Transition Ions (Clarendon, Oxford, I970), Chap.I8, p.7I5
II. M.A.Teplov, Zh.Eksp.Teor.Fiz., 53, I5I0 (I967)
I2. I.S.Konov and M.A.Teplov, Fiz.Tverd.Tela, I8, 853 (I976)
I3. I.S.Konov and M.A.Teplov (to be published)
I4. I.G.Bol'shakov and M.A.Teplov (unpublished)
I5. F.L.Aukhadeyev, I.I.Valeyev, I.S.Konov, V.A.Skrebnev and M.A.Teplov, Fiz.Tverd.Tela, I5, 235 (I973)
I6. E.D.Jones, J.Phys.Chem.Solids, 29, I305 (I968)
I7. E.D.Jones and V.H.Schmidt, J.Appl.Phys., 40, I406 (I969)
I8. I.S.Konov and M.A.Teplov, Fiz.Tverd.Tela, I8, III4 (I976)

I9. E.D.Jones, Phys.Rev.Letters, I9, 432 (I967)
20. E.D.Jones, Colloq.Int.Cent.Natl.Rech.Sci.,2, 495 (I970)
2I. E.Bucher, K.Andres, F.J.di Salvo, J.P.Maita, A.C.Gossard, A.S.Cooper and G.W.Hull, Jr., Phys.Rev.B II, 500 (I975)
22. S.L.Tsarevskii, Fiz.Tverd.Tela, I3, 87 (I97I)
23. A.A.Kossov and S.L.Tsarevskii, Fiz.Tverd.Tela, I7, 2306 (I975)
24. S.A.Al'tshuler, F.L.Aukhadeyev and M.A.Teplov, Zh.Eksp.Teor.Fiz., Pis'ma Red., 9, 46 (I969)
25. M.A.Teplov, Thesis, Kazan State University (I969)
26. M.P.Vaisfeld, Fiz.Tverd.Tela, I4, 737 (I972)
27. S.L.Tsarevskii, Fiz.Tverd.Tela, I2, 2047 (I970)

LOW TEMPERATURE NMR KNIGHT SHIFT STUDIES IN $SmSn_3$

S.K. Malik and R. Vijayaraghavan

Tata Institute of Fundamental Research

Homi Bhabha Road, Bombay 400 005, India

ABSTRACT

The previous ^{119}Sn Knight shift measurements in $SmSn_3$ have been extended down to 15 K and the 4f-contribution to the ^{119}Sn Knight shift does not show a sign reversal in the temperature interval 300 to 15 K. The range of crystal field parameters for which simultaneous fit to the magnetic susceptibility and the extended Knight shift results in $SmSn_3$ is obtained can be substantially narrowed from that reported earlier.

INTRODUCTION

De Wijn et al[1] have performed a detailed analysis of the magnetic susceptibility and the ^{119}Sn Knight shift results in $SmSn_3$. These authors have obtained a range of crystal field parameters (reproduced in Fig. 1) for which a simultaneous fit is obtained to the magnetic susceptibility results from the magnetic ordering temperature to 850 K and to the Knight shift results from 77 to 300 K. Because of the limited temperature interval over which the ^{119}Sn Knight shift results were then available, the fit could be obtained for a wide range of crystal field parameters. However, the excess Knight shift K_f or $\langle S_z \rangle_{av}/H$ values predicted (defined in Ref. 1) at low temperatures by various combinations of crystal field parameter taken from the above mentioned range are strikingly different. Figure 2 shows a plot of $\langle S_z \rangle_{av}/H$ of Sm^{3+} ion versus the temperature

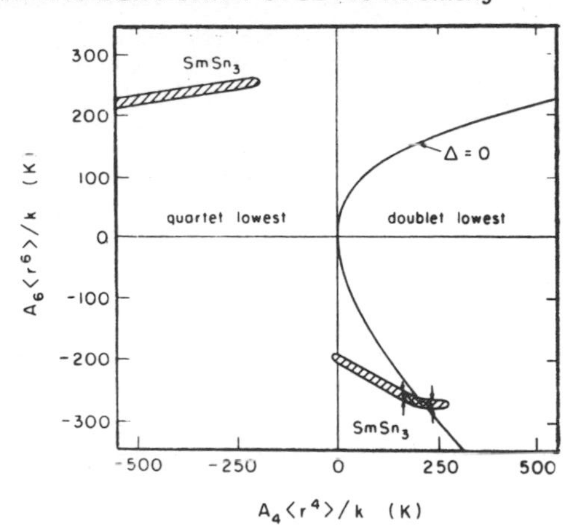

Fig. 1: Crystal field parameters $A_4\langle r^4\rangle$ and $A_6\langle r^6\rangle$ for Sm^{3+} in $SmSn_3$. For the combinations lying in the shaded region Knight shift in the temperature range 77-300 K and magnetic susceptibility could simultaneously be fitted to experiment (Ref. 1). The present Knight shift results in the temperature range 15-300 K (see text) could be fitted only for combinations lying in the region enclosed by the arrows.

calculated for certain combination of crystal field parameters which fit the susceptibility and Knight shift results in $SmSn_3$ (and keeping other parameters the same as in Ref. 1). It is to be noted that although all the curves show essentially the same behaviour in the temperature interval 77 to 300 K, the low temperature behaviour of $\langle S_z\rangle_{av}/H$ depends strongly on the crystal field parameters, and in some cases a crossover in $\langle S_z\rangle_{av}/H$ also appears. Therefore in order to establish the low temperature behaviour, and to narrow down the range of crystal field parameters, we have extended[2] the ^{119}Sn Knight shift measurements in $SmSn_3$ down to 15 K.

II. EXPERIMENTAL DETAILS AND RESULTS

Stoichiometric amounts of samarium metal of 99.9% purity and tin metal of 99.999% purity were sealed in a quartz tube under reduced argon pressure and then heated in an induction furnace. The ingot so obtained was annealed in vacuum at 800°C for one week. The resulting compound $SmSn_3$ has the f.c.c. Cu_3Au type of crystal structure with four formula units per unit cell. Each samarium is surrounded by twelve nearest

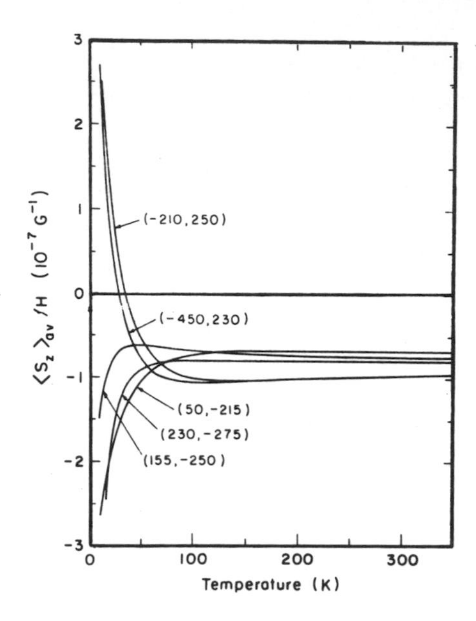

Fig. 2: $\langle S_z \rangle_{av}/H$ of Sm^{3+} versus the temperature for some combinations of crystal field parameters taken from the shaded region of Fig. 1. The curves are labelled with the values of $A_4\langle r^4 \rangle /k$ and $A_6\langle r^6 \rangle /k$ in kelvin, and throughout λ/k = 410 K and J_{ff}/k = - 25 K.

neighbours tin at equal distances. The point symmetry at samarium site is cubic, but is only axial or tetragonal at tin site.

The magnetic susceptibility of $SmSn_3$ measured by Tsuchida and Wallace[3] and by De Wijn et al indicate a Neel temperature close to 11 K. The ^{119}Sn Knight shift in $SmSn_3$ was measured earlier Rao[4], Malik[5] and others[7] upto 77 K. The present Knight shift measurements in the temperature interval 300 to 15 K were carried out using a Brucker B-KR 323s pulsed NMR spectrometer operating at a frequency of 16 MHz.

The Knight shift K_0 due to Pauli paramagnetism has been measured in $LaSn_3$, isostructural to $SmSn_3$, and found to have a value of 0.63%. The results of the present measurements of ^{119}Sn Knight shift in $SmSn_3$ in the temperature interval 15-300 K along with the results of earlier measurements in the temperature interval 77-300 K are shown in Figure 3, and it is to be noted that the crossover in the Knight shift is absent down to 15 K. No ^{119}Sn NMR could be observed between 10 and 4.2 K.

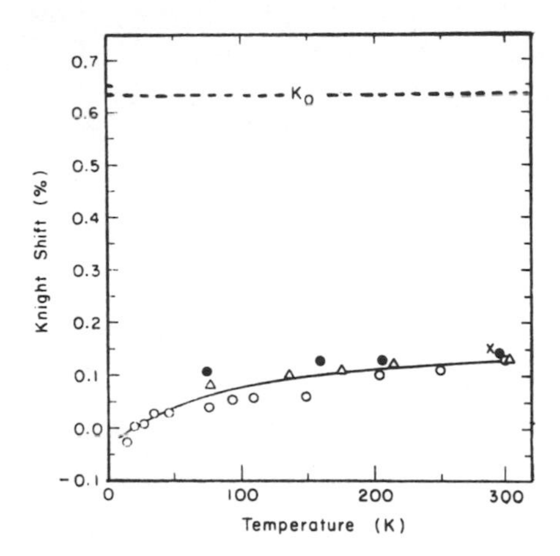

Fig. 3: The ^{119}Sn Knight shift K(T) in $SmSn_3$ versus the temperature. The present measurements are represented by open circles. The black dots and open triangles are data taken from refs. 7 and 5 respectively and cross is a data point taken from ref. 1.

III. DISCUSSION

The Sm^{3+} ions in $SmSn_3$ experience crystal fields of cubic symmetry the Hamiltonian for which can be written as in Ref. 1 where $A_4\langle r^4\rangle$ and $A_6\langle r^6\rangle$ are respectively the crystal field parameters for the fourth and the sixth degree terms in the crystal field Hamiltonian, and $C_m(n)$ $(\theta, \phi) = [4\pi/2n + 1]^{\frac{1}{2}} Y_m^{(n)}(\theta,\phi)$ are the tensor operators. The matrix elements of the operators $C_m(n)$, including those nondiagonal in J, can be easily determined by the tensor operator techniques[6]. The calculation of $\langle S_z\rangle_{av}/H$ of Sm^{3+} proceeds in the following manner: The Hamiltonian consisting of the spin-orbit term L.S. and the crystal field term $\mathcal{H}$ C is diagonalized between the M_J states arising from the lowest three multiplet levels to obtain the energy eigenvalues $E_m(0)$ and $|m\rangle$. The expectation value of the operator S_z is calculated and its Boltzmann average taken over all the levels to obtain $\langle S_z\rangle_{av}/H$. The 4f-electron susceptibility $X_f = N\mu_B (L_z + 2S_z{}_{av}/H$ can also be calculated in a similar manner.

The range of values of the crystal field parameters $A_4\langle r^4\rangle$ and $A_6\langle r^6\rangle$ obtained by De Wijn et al[1] which simultaneously fit the magnetic susceptibility results of $SmSn_3$ from the magnetic ordering temperature upto 850 K, and the ^{119}Sn Knight shift results in the temperature range 77 to 300 K is shown in Fig. 1. It is obvious that the range of $A_4 \langle r^4 \rangle$ and $A_6 \langle r^6 \rangle$ values which explains the present Knight shift results down to 15 K is also contained in the range shown in Fig. 1, and we tried to obtain a fit to the Knight shift results. As in Ref. 1, throughout the calculations we assumed the spin-orbit interaction parameters λ/k= 410 K and the exchange interaction constant (representing the exchange interaction between Sm^{3+} ions) J_{ff}/k = -25 K. It was found that a crossover in $\langle S_z \rangle_{av}/H$ appears when $A_4\langle r^4 \rangle$ is positive and $A_6\langle r^6 \rangle$ negative, the crossover temperature ranging from about 26 K to 40 K. Thus in view of the fact that no crossover temperature in K_f appears, all the values in the upper-left quadrant of Fig. 1 are inappropriate to $SmSn_3$. Proceeding in this way a fit to the Knight shift results was obtained only for a narrow range of crystal field parameters. This range is marked with arrows in Fig. 1. The range of values of $K_0 J_{sf}$ obtained from the present measurements remains the same as in Ref. 1.

We thank Dr. S.K. Garg and J. Ripmeester of National Research Council, Ottawa, for experimental assistance and discussions.

REFERENCES

1. H.W. De Wijn, A.M. Van Diepen and K.H.J. Buschow, Phys. Rev. B7, 524 (1973).

2. S.K. Malik, R. Vijayaraghavan, S.K. Garg and R.J. Ripmeester, to be published.

3. T. Tsuchida and W.E. Wallace, J. Chem. Phys. 43, 3811 (1965).

4. V.U.S. Rao and R. Vijayaraghavan, Phys. Lett. 19, 168 (1965).

5. S.K. Malik and R. Vijayaraghavan, J. Phys. (Paris) 32, C1-1028 (1971) and references therein.

6. B.G. Wybourne, Spectroscopic Properties of Rare Earths (Interscience Publishers, New York, 1965) p. 112, 164.

7. F. Borsa, R.G. Barnes and R.A. Reese, Phys. Status Solidi 19, 359 (1967).

SPIN REORIENTATIONS IN THE RARE EARTH-COBALT LAVES PHASES

D. Gignoux, F. Givord and R. Lemaire

Laboratoire de Magnétisme, C.N.R.S.

166X, 38042-Grenoble-Cedex, France

ABSTRACT

The cubic Laves phases $NdCo_2$ and $HoCo_2$ exhibit a spin reorientation, respectively at 42 and 14 K. For $HoCo_2$ this property has been evidenced from magnetization measurements on a single crystal, and for $NdCo_2$ from neutron diffraction experiments on powder oriented in a magnetic field. We analyse this behavior in terms of crystal field effects on Nd^{3+} and Ho^{3+}. The transition is of first order, the entropy discontinuity leads to a sharp specific heat anomaly.

INTRODUCTION

Among the RCo_2 Laves phases between rare earth and cobalt, $NdCo_2$ and $HoCo_2$ exhibit a peculiar magnetic behavior below their ordering temperature. The thermal variation of the magnetization measured in low fields on a polycrystalline sample shows a maximum at 42 and 14 K respectively, for $NdCo_2$ and $HoCo_2$[1] (figure 1). At the vicinity of these temperatures the thermal variation of the specific heat has a sharp peak[2,3] (figure 2). In order to understand these anomalies, we have performed magnetization measurements on a single crystal of $HoCo_2$ that we have prepared. For $NdCo_2$ we did not get a single crystal and the analysis was made by neutron diffraction in a magnetic field on a fine powdered sample.

EXPERIMENTAL RESULTS

Magnetization measurements were performed at the Service National des Champs Intenses of Grenoble. Neutron diffraction experiments were performed at the Laboratoire de Diffraction Neutronique of the Nuclear Center of Grenoble.

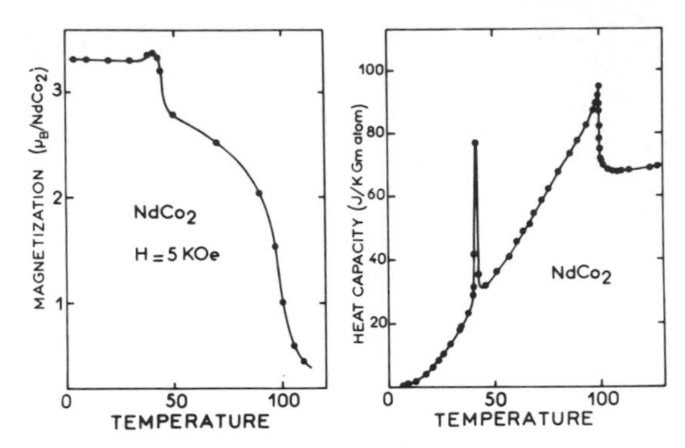

Fig. 1 : Thermal variation of the magnetization for $NdCo_2$

Fig. 2 : Heat capacity versus temperature for $NdCo_2$ from ref. 2

$HoCo_2$[4] : Magnetization measurements performed on a single crystal evidence for a change of the easy magnetization direction at 14 K. The spontaneous magnetization, which is parallel to the [110] axis at very low temperatures, becomes parallel to the [100] axis when temperature increases. It should be noted that below 14 K, the easy magnetization direction does not correspond to the direction for which the magnetization has the highest value. From the magnetization variation along [110] and [100] axes it has been possible to determine at each temperature the free energy difference between these two axes. Experimental points are reported on figure 3a.

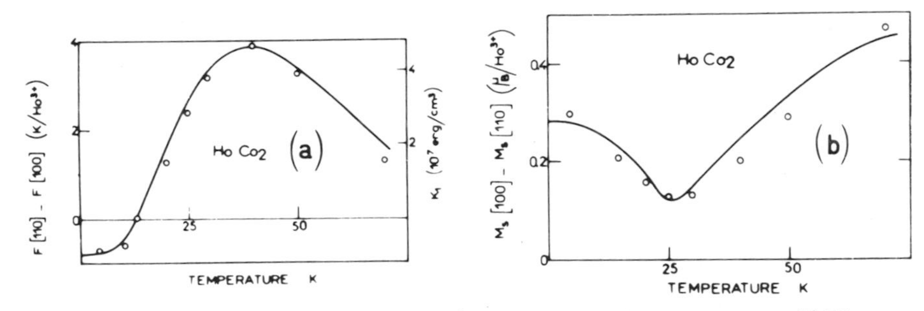

Fig. 3 : $HoCo_2$ - thermal variation of the free-energy differences F [110] - F [100] (a), and of the magnetization anisotropy M [100] - M [110] in a 100 kOe applied field (b). Round points represent the experimental values. Full lines represent the thermal variations calculated with $x = -0.469$, $W = 0.6$ K.

$NdCo_2$: In order to determine the moment direction below and above 42 K, we have performed neutron diffraction patterns on a sample reduced into fine powder. The first pattern was made at 4.2 K without an applied field. For the other patterns performed at 4.2

and 50 K, a field of 10 kOe was applied vertically, i.e. perpendicular to the scattering plane, before starting the experiment, and then reduced to zero. The three patterns are reported on figure 4. In the lower part of figure 4, we have presented the variation of the intensity of each peak depending on the easy magnetization direction. From these patterns we see that the spontaneous magnetization is parallel to the [110] axis at 4.2 K and parallel to the [100] axis at 50K.

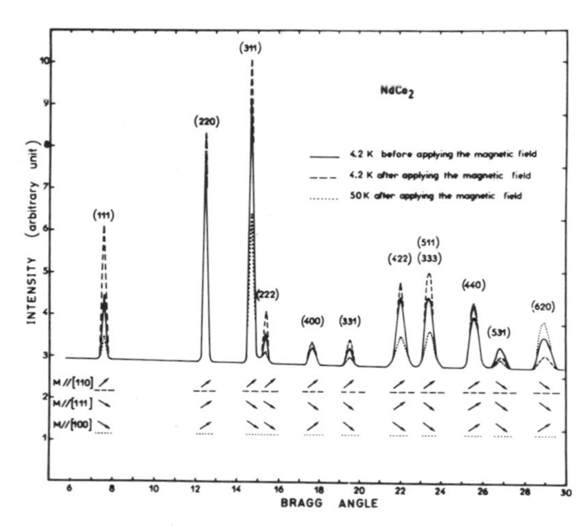

Fig. 4 : $NdCo_2$: neutron diffraction patterns performed in different conditions on a fine powder

ANALYSIS

The cobalt anisotropy measured in $GdCo_2$ being negligible[4], the changes of easy magnetization direction observed in $NdCo_2$ and $HoCo_2$ can be attributed to the crystal field effects on the 4f electrons of the Nd^{3+} and Ho^{3+} ions. We have analysed the experimental results using the perturbing Hamiltonian which can be written in the molecular field H_m model :

$$\mathcal{H} = W\left[x\, O_4/F_4 + (1 - |x|)\, O_6/F_6\right] + g_J \mu_B\, \vec{J}.(\vec{H}_m + \vec{H}_{ext})$$

$HoCo_2$: The parameters W, x and H_m have been obtained by fitting the thermal variation of the free energy difference $F[110] - F[100]$ determined from the magnetization curves along the [110] and [100] axes. The best agreement was obtained for W = 0.6 K, x = -0.469 and H_m = 340 kOe at 4.2 K. The calculated variation is compared to the experimental points on figure 3a. The thermal variation of the calculated magnetization anisotropy $M[100] - M[110]$ in a 100 kOe external field is compared to the experimental values on figure 3b. Using the calculation and the magnetization curve, we obtain $\mu_{Ho} = 9.3\ \mu_B$ and $\mu_{Co} = 0.8\ \mu_B$ at 4.2 K, values in good agreement with those determined by neutron diffraction[5]. A polarized neutron study has shown that the paramagnetic cobalt susceptibility is almost temperature independent[6]. Thus, the cobalt magnetic behaviour has to be described in the collective electron model. The interpretation of the paramagnetic susceptibility of $HoCo_2$ in this model allows the determination of the molecular field coefficients which lead to H_m = 318 kOe at 4.2 K in agreement with the previous result.

$NdCo_2$: The determination of the crystal field parameters was led as following : the molecular field value at 4.2 K was estimated from neutron diffraction experiments and from the thermal variation of the paramagnetic susceptibility : H_m = 1100 kOe. W and x were determined taking into account that the rotation occurs at 42 K and that μ_{Nd} = 2.6 μ_B at 4.2 K, value deduced from the neutron diffraction pattern. We have obtained W = 2.9 K and x = -0.18.

DISCUSSION AND CONCLUSION

The change of easy magnetization direction observed in $NdCo_2$ and $HoCo_2$ corresponds to a change due to the temperature from one energy level scheme to another (figure 5). Calculation of the free energy along intermediate directions between the [100] and [110] axes has shown that the magnetization passes at once from the [110] direction to the [100] one. The transition is of first order. In $HoCo_2$, the calculated entropy discontinuity ΔS (1.6 J K^{-1} $mole^{-1}$) at 14 K is in good agreement with that (1.4 J K^{-1} $mole^{-1}$) deduced from the specific heat measurements[3]. In $NdCo_2$ the calculated entropy gap is 0.6 J K^{-1} $mole^{-1}$. From the results previously published[2], it is difficult to give an estimation of ΔS at 42 K. The crystal field parameters determined experimentally and calculated in a point charge model assuming no charge on Co and +3e on Ho and Nd, are compared on Table I. The signs of the parameters are the same but the experimental values are much higher. Especially x is much weaker than the value given by the point charge model. This given evidence for the preponderant influence of the conduction electrons; because of their d character, their distribution is non uniform and aspherical.

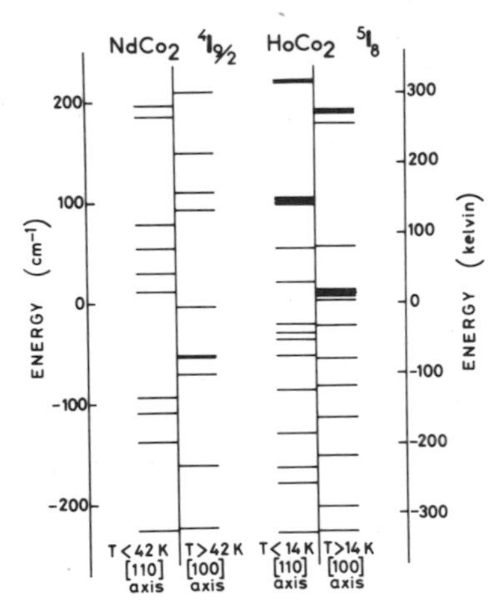

Fig. 5 : Splitting of the ground state multiplets of the Nd^{3+} and Ho^{3+} ions in $NdCo_2$ and $HoCo_2$ on both sides of the change of easy magnetization direction

The bump observed on the thermal variation of the magnetization of a polycrystalline sample in a constant field is associated with the drastic variation of the K_1 anisotropy constant which is null at the transition (Figure 3).

In $PrCo_2$ the specific heat anomaly observed below T_c[2] cannot be attributed to a spin reorientation. A neutron diffraction experiment on a powder oriented in a magnetic field has shown that at all temperatures below T_c, the magnetization is along the [001] axis.

Table I : Crystal field parameters of $NdCo_2$ and $HoCo_2$. Δ is the total splitting of the ground multiplet by the crystal field. $B_4 = \frac{Wx}{F_4}$; $B_6 = \frac{W(1 - |x|)}{F_6}$

		W(K)	x	$10^{+4}B_4$ (K)	$10^{+6}B_6$ (K)	Δ (K)
$NdCo_2$	Point-charge model	1.26	-0.81	-170	94	90.3
	Experimental results	2.9	-0.18	- 87	944	292.9
$HoCo_2$	Point-charge model	0.082	-0.785	- 10.7	1.3	49.4
	Experimental results	0.6	-0.469	- 46.9	23.0	267.8

REFERENCES

1. R. Lemaire, Cobalt, 33, 201 (1966)

2. C. Deenadas, R.S. Craig, N. Marzouk and W.E. Wallace, J. Sol. Stat. Chem., 4, 1 (1972)

3. J. Voiron, A. Berton and J. Chaussy, Phys. Lett., 50A, 17 (1974)

4. D. Gignoux, F. Givord and R. Lemaire, Phys. Rev. B, 12, 3878 (1975)

5. R.M. Moon, W.C. Koehler and J. Farrell, J. Appl. Phys., 36, 978 (1965)

6. D. Gignoux, F. Givord and W.C. Koehler, Intern. Conf. Magnetism, Amsterdam (1976) to appear.

EXCHANGE VS CRYSTAL FIELDS IN Gd-Pr SINGLE CRYSTAL ALLOYS

S. Legvold, P. Burgardt and B. J. Beaudry

Ames Laboratory-ERDA and Department of Physics

Iowa State University, Ames, Iowa 50011

ABSTRACT

We have measured the saturation magnetic moment of a hcp single crystal of $Gd_{0.91}Pr_{0.09}$ and obtain a value of 6.63 Bohr magnetons per atom. If we assume a magnetic moment of 3.2 μ_B per Pr atom and 7.55 μ_B per Gd atom and normal spin exchange coupling so as to have antiparallel moments we obtain a theoretical value of 6.60 Bohr magnetons in excellent agreement with the observed value. This demonstrates that the exchange field dominates the hexagonal crystal field in this sample. The sample exhibited strong magnetic anisotropy which corroborates our interpretation of the results.

INTRODUCTION

The light rare earth metal Pr has two electrons in the 4f shell and has the ABAC (dhcp) stacking of atoms. As a consequence, half the atom sites have cubic symmetry and half exhibit hexagonal symmetry. Apparently the ground state for either site is a singlet state according to a number of investigations[1-4]. The magnetic susceptibility studies on Pr have shown that it is strongly paramagnetic and the results of Sakamoto *et al.*[1] fit the theoretical calculations of Rainford[5] for Pr very well. Neutron studies of the magnetic excitations in Pr by Houmann *et al.*[2] lead them to believe that exchange effects are almost high enough to lead to magnetic ordering. In the work reported here we have explored exchange versus crystal fields for Pr. We subject Pr to strong exchange fields by alloying Pr into Gd which has high magnetic moment. By staying below fifteen atomic percent Pr we have maintained the hcp structure of Gd so the Pr atoms experience only hexagonal crystal fields.

EXPERIMENTAL

Samples of Pr-Gd were prepared by arc melting together the constituents over a copper hearth in an argon atmosphere. Several meltings followed by annealing procedures assured uniform mixing of the constituents. The first samples studied were polycrystalline bars 1x1x6 mm in size of $Gd_{0.9}Pr_{0.1}$ and $Gd_{0.85}Pr_{0.15}$. From x-ray diffraction studies the latter showed some Sm structure and was discarded but the former was found to be hcp as desired. Results of magnetic measurements at 4.5 K in fields up to 30 kOe were used for the magnetic moment versus 1/H plot shown as the curve marked poly $Gd_{0.9}Pr_{0.1}$ in Fig. 1. The extrapolated saturation moment indicated that Pr had 3.2 μ_B per atom which was ordered antiparallel to the Gd moment. However, the sample was far from saturation at 30 kOe making the extrapolation questionable and suggesting that strong magnetic anisotropy was present. This was also evident when samples cut from the same button in three

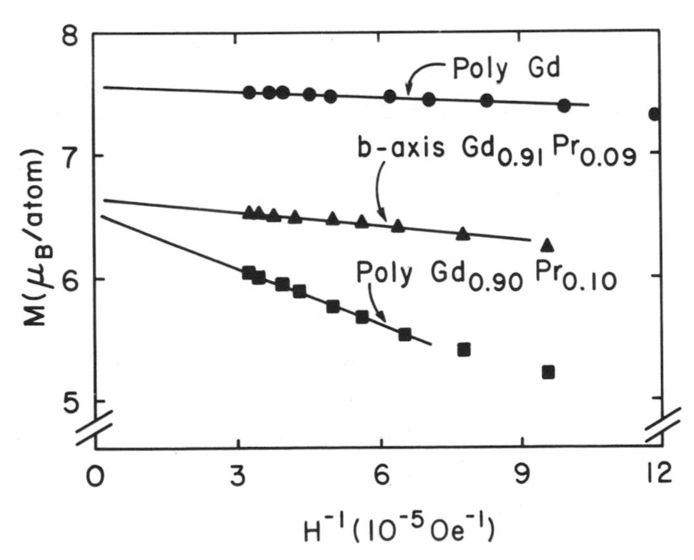

Fig. 1. Magnetic moment vs 1/H for polycrystalline Gd, for polycrystalline $Gd_{0.9}Pr_{0.1}$ and for single crystal $Gd_{0.91}Pr_{0.09}$ with H along a basal plane b-axis.

orthogonal directions showed different slopes. We then surmised single crystals were required for a proper investigation.

An ingot 1x2x6 cm of $Gd_{0.91}Pr_{0.09}$ was prepared by arc melting as described above (the actual concentration of the alloy is 8.8% of Pr content). This ingot was annealed for twelve hours in the strong temperature gradient region of an electric furnace at a mean temperature of 1095 C. Several single crystals of about 5 mm on an edge were visible in the ingot after this annealing procedure. A crystal was removed from the ingot and aligned with x-rays such that a 1x1x5 mm sample with a c-axis along the long axis of the sample was produced. The orientation of the sample is known to about $\pm 3^{\circ}$. Another crystal was removed from the ingot and etched down to a rounded cube 2 mm on an edge. This latter crystal was placed in the magnetometer and rotated to find the easy direction of magnetization. The easy direction was found to lie along a basal plane b-axis. Magnetization versus 1/H results for the 2 mm cube with the field along the easy direction are shown in Fig. 1. Also shown, for comparison, are similar data for a polycrystalline sample of pure Gd of the same quality as that used in this work.

Magnetization results for $Gd_{0.91}Pr_{0.09}$ as a function of field obtained at 4.5 K with the field parallel to the b and c axes are shown in Fig. 2. These data show that there is a large magnetic anisotropy in this alloy. The c-axis data show an interesting sharp increase in magnetization above 15 ± 1 kOe. Apparently this large change in magnetization indicates the critical external field necessary to overcome the internal anisotropy fields and allow the atomic moments to rotate into the c-axis direction. The remarkable c-axis data confirm our conclusion that the high field data for polycrystalline samples should be viewed with caution.

The saturation magnetization of the $Gd_{0.91}Pr_{0.09}$ sample is 6.63 μ_B/atom as read from Fig. 1. If we assume spin-spin coupling between the Gd and Pr atoms, the two angular momenta should align opposed to one another. Assuming this antiparallel alignment of moments we obtain a calculated moment of 6.60 μ_B per sample atom with the moment per Pr atom given by

$$J = L-S; \; M_{Pr} = \mu_B(g-1)J = 3.2 \; \mu_B$$

and the moment per Gd atom given by $M_{Gd} = 7.55 \; \mu_B$ (see the discussion below). The excellent correlation between the experimental and calculated moments indicates that the exchange fields are of the expected form.

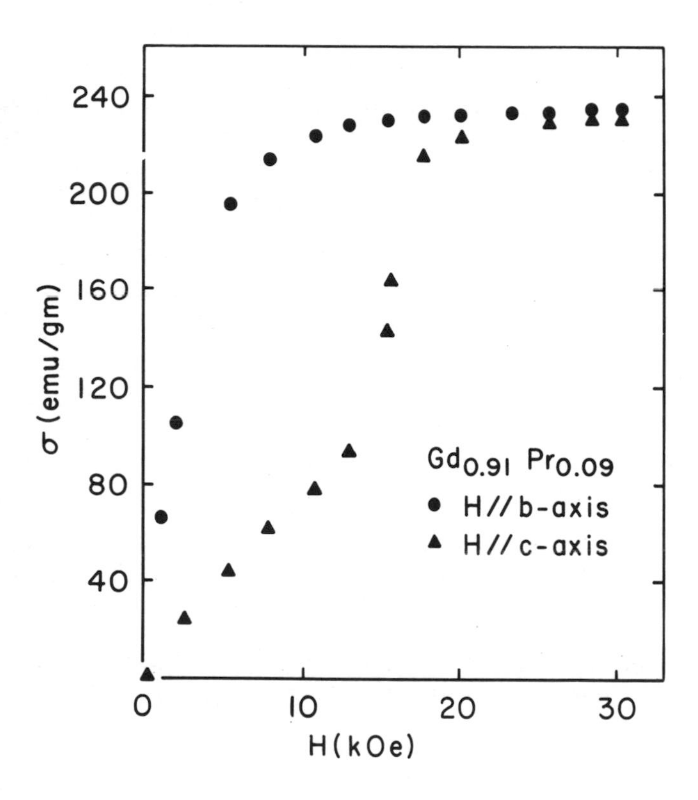

Fig. 2. Magnetization vs H for single crystal $Gd_{0.91}Pr_{0.09}$ with H along a basal plane b-axis and with H along a c-axis.

DISCUSSION

The use of 7.55 μ_B for Gd in the treatment above is taken from the saturation moment of Gd with the 0.55 μ_B attributed to the conduction electron polarization. We probably should use a slightly lower number to include the effect of the reduced internal field which brings about this polarization. However, such correction falls within the range of experimental error. In recent work[6,7] on high purity single crystal Gd it has been found that the saturation moment of Gd is 7.63 μ_B. Our use of 7.55 μ_B is based on the fact that the mother Gd we used exhibits this value.

It would be interesting to do work on the Pr rich dhcp end of the alloy system. Our experience in the present work indicates that single crystals would again be required. Methods for growing such crystals are under investigation at this laboratory.

The authors acknowledge the assistance of Harlan Baker in etching the single crystal. The assistance of J. H. Queen in taking the data is gratefully acknowledged.

REFERENCES

1. S. Sakamoto, J. Nakai and Y. Nakagawa, J. Phys. Soc. Jap. 40, 686 (1976).

2. J. G. Houmann, M. Chapellier, A. R. Mackintosh, P. Pak, O. D. McMasters and K. A. Gschneidner, Jr., Phys. Rev. Letters 34, 587 (1975).

3. T. Johansson, B. Lebech, M. Nielsen, H. B. Moller and A. R. Mackintosh, Phys. Rev. Letters 25, 524 (1970).

4. P. E. Lindelof, I. E. Miller, and G. R. Pickett, Phys. Rev. Letters 35, 1297 (1975).

5. B. D. Rainford, AIP Conf. Proc. No. 5, 591 (1972).

6. L. W. Roeland, G. J. Cock, F. A. Muller, A. C. Moleman, K. A. McEwen, R. G. Jordan, and D. W. Jones, J. Phys. F 5, L233 (1975).

7. H. W. White, B. J. Beaudry, P. Burgardt, S. Legvold and B. N. Harmon, AIP Conf. Proc. 29, 329 (1976).

CONFERENCE SUMMARY

Bernard R. Cooper

Department of Physics, West Virginia University

Morgantown, West Virginia 26506, U.S.A.

ABSTRACT

The Conference Summary tries to point out remaining questions and promising possibilities for further work. This is done by categorizing the work discussed at the Conference in several different ways, and discussing the situation in each category in turn.

INTRODUCTION

Professor Stevens' opening talk made explicit a point that is to a great extent a built-in theme of this Conference, the fact that when we talk about crystal-field effects, we mean crystal-field effects for local moment systems. That is why we have heard so much about rare earths and rather little about transition metal systems. Dr. Lander's talk has raised very fundamental questions with regard to this question of localization effects and the presence, or I should say so far at least largely the absence, of crystal-field effects in metallic actinide systems.

We do not have a single theme in a conference like this, because, of course, each author has his own theme, and his own area and work that he finds interesting. However, with that disclaimer, I am going to try to categorize in several different ways the distinctions between the different work that was discussed. I will then use that categorization to discuss what I regard as remaining questions and promising possibilities for further work.

We can first make a distinction between theoretical and experimental questions and problems. Second, we can make distinctions on the basis of type of phenomena, or if you like, type of measurement. In this regard, the characterizations that Professor Luthi made are very valuable. We can think of three classes of phenomena:

I. What Professor Luthi has called resonance properties, or what we might call spectroscopy of excited states, i.e. neutron scattering, optical absorption, microwave resonance, and Mössbauer effect studies. In the oral Summary I noted that nothing had been heard about optical studies. Several workers then remarked that they had unsuccessfully attempted transmission or reflection experiments trying to see crystal-field effects in metallic systems of interest, typically using lasers as light sources. However, this should not preclude the possibility of obtaining valuable information in the future by some optical technique.

II. Thermal, or macroscopic thermodynamic, properties: specific heat, magnetic susceptibility, elastic constants, thermal expansion, magnetostriction, among others.

III. Transport properties: electrical conductivity, magnetoresistance, thermal conductivity, thermoelectric power.

A third way of categorizing distinctions is on the basis of materials. Clearly the situation is quite different for actinides, as summarized by Dr. Lander, compared to rare earths.

Finally, we can distinguish several types of work that are relevant to important applications, either in the technological or the scientific sense. Here among topics that come to mind are Dr. Atzmony's discussion of anisotropy in rare earth-iron and cobalt compounds. This is a topic that is closely related to the importance of some of these materials, especially the non-cubic cobalt-rare earth compounds, for permanent magnets[1,2]; and also is related to the giant magnetostriction[3-5] (a topic that I believe was not discussed during the Conference) that makes some of the cubic materials candidates for important applications involving magnetoelastic effects. Another topic related to applications is the study of hydrides, which is potentially relevant to the question of hydrogen storage. An additional study related to application, but in the purely scientific sense, is that of the effective hyperfine field acting on the nucleus, as discussed by Dr. Teplov, for singlet ground state systems. This relates to the hyperfine enhanced nucleur adiabatic demagnetization cooling technique.[6-8]

THEORETICAL QUESTIONS

Referring back to Professor Stevens' talk, we have the question of the significance of the "crystal-field" parameters as presently introduced operationally into the analysis of experiments. Outside of Professor Stevens' talk, very little was said about developing a first principles theory for these "crystal-field", or as perhaps Stevens would prefer to say "spin Hamiltonian", parameters. Obviously the fundamental theory of crystal-field effects in metals, developed on a first principles basis, is quite complex. It involves the essential difficulty of combining localized and band electron effects in one theory. Perhaps, however, this question has to be faced up to by the theorists, at least to make clear the restrictions on the phenomenology that must be introduced in analyzing experimental results. Stevens has emphasized, in the discussion of the oral Summary, the fact that any such theory has to include strain and phonon effects, i.e. spin-phonon interaction.

Schmitt, Morin, and Pierre have at this Conference reported theoretical calculations that attempt to combine localized and band electron behavior. Dr. Brooks in the discussion of the oral Summary commented on the difficulty of treating many-electron 4f localized states together with single electron conduction band states in the same energy level scheme. The band aspects of a theory of the type of Schmitt et al might be best treated using the muffin-tin-orbital technique.[9-12] Certainly experimental crystal-field parameter behavior such as that found in the cerium monopnictides by Heer, Furrer, Halg, and Vogt requires a new fundamental theory. At present fundamental theoretical treatments for the size of crystal-field parameters and the variation in a related group of materials, or with temperature or pressure, consist largely of pointing to point charge model results; and in a rather embarrassed tone saying the point charge model works (disclaiming any fundamental significance to this), or that it doesn't work.

A great deal of the theory discussed in the Conference involved use of phenomenological spin Hamiltonians, including crystal-field effects, to treat the observed dynamic magnetic properties of rare earth systems, i.e. the excitation spectra. In considering the role of crystal-field effects in dynamic behavior, we are of necessity drawn into consideration of anisotropic exchange and elastic effects, and also of the mode softening or central mode divergence behavior as discussed by Elliott, Holden, Kjems, and others at the Conference. The nature of the divergence of the zero frequence response at phase transitions, particularly the central mode divergence question, warrants considerable additional thought.

At the beginning of the Conference, Dr. Lindgard's talk raised what is perhaps the central operational question in treating crystal-field effects on the dynamic properties of rare earth systems. Very good experimental results are available, especially from neutron scattering, giving very detailed information on the excitation behavior. One is then faced with an operational question in analyzing this wealth of detail. When faced with a discrepancy between experiment and an existing theory, say the theory for spin waves in heavy rare earth metals including single-ion anisotropy, i.e. crystal-field effects, and magnetoelastic effects, there are two obvious ways in which the discrepancy might be resolved. One is via the introduction of new interactions, such as two-ion anisotropy, i.e. anisotropic exchange or some variant thereon. The other possibility is that the approximations involved in the formalism of an existing theory may not be good enough. Lindgard has said that this latter is the case for spin wave behavior in the heavy rare earth metals, where he has used the behavior of erbium metal to illustrate this point. (Dr. Lindgard did point out that the question of the need for introduction of a new interaction is clear if the new interaction reduces the symmetry, and therefore would split some degeneracy in the excitation spectrum.)

This raises an important general question for analyzing experimental data on excitation behavior. When performing the quite detailed analyses that are often done currently, is one really building on an uncertain foundation when using the statistical treatments that typically have been used. Dr. Buyers, in the discussion of the oral Summary, expressed concern that as the neutron people study more complicated systems, the analysis seems to be on the basis of preconceived notions.

Another difficulty in analyses of experimental results at present is that problems in the correctness of the theory complicate the use of macroscopically measured parameters in analyzing excitation spectra and vice versa.

QUESTIONS AND NEED FOR FURTHER STUDIES IN THE VARIOUS EXPERIMENTAL AREAS

We now discuss questions and needs for further work in the three experimental areas, as classified in the Introduction:

I. Resonance Properties - The results reported for $PrSn_3$, by Murasik and Furrer and by Gross, Knorr, Murani, and Buschow, show that even in a "straightforward" case the fact that one often does not have a clearly resolved spectrum from neutron inelastic scattering makes for considerable difficulties in defining crystal-field level schemes. It is important to distinguish between phonons and crystal-field excitations.

Deducing crystal-field information from electron paramagnetic resonance (EPR) experiments has not achieved the level of refinement of the neutron inelastic scattering technique. However, EPR is potentially very valuable as a source of information on the crystal-field excitation spectrum, including information on crystal-field parameters and anisotropic exchange. Excited state EPR, as discussed by Dr. Huang, is an especially difficult, but extraordinarily sensitive, probe for such effects. For EPR experiments the effects of static strains in the polycrystalline samples typically used must be considered. (Phonon effects may be important for the anisotropic coupling observed in some experiments.)

II. Macroscopic Thermodynamic Properties - As discussed by Professor Luthi, understanding of crystal-field effects on macroscopic properties in the regime with no magnetic or distortional ordering is good. However, compatibility with understanding of the excitation spectra, and indeed achieving compatibility between different macroscopic measurements, requires more work. However, this is probably theoretical, or at least analytical, rather than experimental per se'. Study and understanding of the changes in macroscopic thermodynamic properties when magnetic or distortional ordering occurs requires further theoretical and experimental work. Study of Jahn-Teller distortional effects in metallic systems, as reported by Ott, Andres, Wang, Wong, and Luthi for $PrCu_2$, particularly warrants further study.

III. Transport Properties - This area of work must still be largely regarded as a frontier area so far as crystal-field effects are concerned. Much of the anomalous results that most stir one's interest unfortunately were not presented during the program of the Conference. In the paper of Dr. Lethuillier, as presented by Dr. Pierre, we did hear a brief report of the anomalous electrical conductivity behavior[13,14] of the system $Pr_cLa_{1-c}Sn_3$. An anomalous maximum appears in the resistivity at close to the same temperature across the entire concentration range. The behavior is suggestive of the anomaly being associated with scattering involving a crystal-field excited state, but it appears unlikely that the behavior can be understood from first order scattering effects. The Kondo side band mechanism[15] might offer an explanation for the behavior. However, in the discussion at the Conference there seemed to be some difference of opinion as to the degree of pathological behavior for the Fermi surface that would be required. The resistivity behavior of the terbium monopnictides when diluted by yttrium increases the puzzle offered by this type of anomaly. The very complete study for $Tb_cY_{1-c}Sb$ reported by Hessel Andersen,

Lindelhof, Smith, Splittorff, and Vogt shows quite well understood normal behavior. Normal behavior is also found[16] in $Tb_cY_{1-c}P$; but[16] $Tb_cY_{1-c}As$ shows anomalous behavior similar to that of $Pr_cLa_{1-c}Sn_3$ across the entire concentration range. Since the antiferromagnetic ordering temperatures and crystal-field splittings of TbP and TbSb bracket those of TbAs, it is very difficult to understand this behavior.

There are some questions and suggestions for further study common to all three areas of experimental technique. Foremost among these is the relationship, or question of compatibility, between crystal-field parameters of dilute and concentrated systems. This is particularly important because a device often used to separate single-ion (crystal-field) effects from two-ion interaction effects, particularly with regard to anisotropic properties, is to dilute a magnetic rare earth ion with a nonmagnetic ion, i.e. Y, Lu, La, or a combination of these. Depending on the rare earth being diluted and the diluent used, operationally sometimes there is a question of a change in the crystal-field parameters on dilution.

With regard to dilution experiments, the trade-off is between the complications of introducing spatial statistical problems associated with alloying and of introducing possible changes in electronic properties, and the simplification of diluting two-ion effects, while leaving single-ion effects (in the ideal situation) unchanged. Also, as pointed out by Dr. Buyers, in practice exchange striction, i.e. the distortional effect on exchange, has to be taken into account in deducing the behavior for concentrated systems from that for diluted systems. Dr. Lindgard pointed out the value of using alloys of two different magnetic rare earths to study effects of competing crystal fields.

Pressure effects, as discussed by Professor Guertin, are a means of varying crystal-field and interaction effects in a spatially uniform way, that has still seen only relatively little use in the area of interest of the Conference. This lack of use of high hydrostatic pressure effects is especially true of neutron scattering experiments. Also, in the context of his own theoretical ideas, Professor Stevens pointed out the possible use of pressure effects to push localized electrons into the conduction bands.

More work is necessary on the role of distortions, particularly internal distortions[17], or lattice internal rearrangement modes[18-21], in the correct interpretation of crystal-field effects. These internal lattice modes can change crystal-field properties substantially with no macroscopic strain.

In the discussion, Professor Stevens raised the question of using neutron scattering on isolated pairs of magnetic ions, analogous to EPR experiments on magnetic pairs in nonmagnetic insulating hosts, to study the nature of the interactions. Dr. Buyers pointed out that the practical difficulty for metals is to isolate pairs because of the long range of the interaction.

AREAS OF TECHNOLOGICAL AND SCIENTIFIC APPLICATION

Dr. Atzmony has clearly demonstrated the key role played by single-ion crystal-field anisotropy in determining the anisotropy properties of rare earth-transition metal intermetallic compounds. However, it would be interesting and valuable also fully to understand the role of the transition metal, particularly that of cobalt in the noncubic materials used in practical permanent magnet applications.

Study of hydrides is likely to be of increasing practical interest because of the relevance to the hydrogen storage problem. Crystal-field studies offer a valuable tool in seeing how hydrogen enters a lattice and its effect on a crystal. In this connection one should bear in mind the interesting fundamental effects on magnetic and superconducting properties associated with hydriding. A very fundamental question is what adding hydrogen does to the conduction electron concentration in the rare earth host material.

Dr. Ott and Professor Bucher, in the discussion, have brought us up to date on the state of the art for the induced hyperfine field cooling technique using crystal-field singlet-ground-state materials. There has been success in cooling He_3, where $PrNi_5$ has been a successful cooling material. The technique is very powerful for obtaining low temperatures, and then doing experiments. Professor Bucher pointed out that the difficulties in practice are obtaining a good thermal contact between the cooling material and the He_3, and that nuclear ordering interferes with nuclear cooling. Professor Bucher listed several promising cooling materials, $PrBe_{13}$, $PrIn_3$, PrS, the latter offering stoichiometry problems

Dr. Purwins in the discussion raised the question of crystal-field effects on chemical properties, particularly with regard to reaction kinetics. Professor Stevens pointed out that the small size of crystal-field splitting for rare earth materials on the scale of chemical effects decreased the likeliness of such effects being important.

ACTINIDE MATERIALS

So far as classes of materials are concerned, metallic actinide systems represent the frontier in crystal-field studies. A key problem is to sort out crystal-field effects from actinide-actinide anisotropic interaction effects. The behavior of cerium compounds may be useful[21] for guidance in doing this. Also new types[21] of magnetoelastic effects may be important as the source of actinide-actinide anisotropic interactions.

We must await the spectroscopic observation of excitations, whether crystal-field levels or spin waves, before we can proceed much further in understanding crystal-field effects in metallic actinide systems. Presumably this will be done by neutron scattering. Dilution of suitable actinide metallic materials with thorium or yttrium should be valuable in the sorting out process. Paramagnetic resonance experiments as well as neutron scattering should play a valuable role.

Pressure experiments should be useful in studying actinide metallic systems, particularly those of the light actinides (which are, of course, the only ones available in significant quantities). Professor Guertin, in the discussion, has pointed to the work[22] on UPt as an example indicating the potential value of pressure studies for metallic actinide materials.

In closing, I would like to emphasize that the greatest mystery in the actinide area is the lack of observation by neutron scattering of magnetic excitations.

REFERENCES

1. H. Stablein, A.I.P. Conf. Proc. No.5, 950 (1972).

2. D. L. Martin and M. G. Benz, A.I.P. Conf. Proc. No.5, 970 (1972).

3. A. E. Clark and H. S. Belson, A.I.P. Conf. Proc. No.5, 1498 (1972).

4. A. E. Clark and H. S. Belson, Phys. Rev. B5, 3642 (1972).

5. A. E. Clark and H. S. Belson, A.I.P. Conf. Proc. No.10, 749 (1973).

6. S. A. Altshuler, JETP Letters 3, 112 (1966).

7. K. Andres and E. Bucher, Phys. Rev. Letters 21, 1221 (1968).

8. K. Andres and E. Bucher, Phys. Rev. Letters 22, 600 (1969).

9. O. K. Andersen in "Computational Methods in Band Theory", page 178 (Ed. P. M. Marcus, J. F. Janak, and A. R. Williams, Plenum Press, 1971).

10. O. K. Andersen and R. V. Kasowski, Phys. Rev. B 4, 1064 (1971).

11. O. K. Andersen, Phys. Rev. B 12, 3060 (1975).

12. O. Jepsen, O. K. Andersen and A. R. Mackintosh, Phys. Rev. B 12, 3084 (1975).

13. A. I. Abou Aly, S. Bakanowski, N. F. Berk, J. E. Crow, and T. Mihalisin, Phys. Rev. Letters 35, 1387 (1975).

14. P. Lethuillier and P. Haen, Phys. Rev. Letters 35, 1391 (1975).

15. F. E. Maranzana, Phys. Rev. Letters 25, 329 (1970); F. E. Maranzana and P. Bianchessi, Phys. Stat. Sol. (b) 43, 601 (1971).

16. K. Sugawara, C. Y. Huang, and B. R. Cooper, paper SU6 at the Amsterdam International Magnetism Conference, 1976, to appear in the Proceedings of the Conference in Physics B.

17. S. J. Allen, Phys. Rev. 166, 530 (1968); Phys. Rev. 167, 492 (1968).

18. K. W. H. Stevens and E. Pytte, Solid State Commun. 13, 101 (1973).

19. J. Faber, Jr., G. H. Lander, and B. R. Cooper, Phys. Rev. Lett. 35, 1770 (1975).

20. J. Faber, Jr., G. H. Lander, and B. R. Cooper, A.I.P. Conf. Proc. No.29, 379 (1976).

21. B. R. Cooper, to appear in the Proceedings of the Second International Conference on the Electronic Structure of the Actinides, Wroclaw, Poland, 1976.

22. J. G. Huber, M. B. Maple, and D. Wohlleben, J. of Magnetism and Magnetic Materials 1, 58 (1975).

LIST OF PARTICIPANTS

Atzmony, U., Nuclear Research Centre-Negev, Beer Sheva, Israel

Baberschke, K., Freie Universität, Berlin, BRD
Barnes, S.E., Trinity College, Dublin, Ireland
Becker, K., Max-Planck-Institut, Stuttgart, BRD
Begum, R.J., Bhabha Atomic Research Centre CIRUS, Bombay, India
Bergsma, J., Reactor Centrum Nederland, Petten, The Netherlands
Boucherle, J.-X., Centre d'Etudes Nucléaires, Grenoble, France
Brooks, M.S.S., Kernforschungszentrum, Karlsruhe, BRD
Bucher, E., Universität Konstanz, BRD
Buschow, K.H.J., Philips Research Labs., Eindhoven, The Netherlands
Buyers, W.J.L. Atomic Energy of Canada Ltd., Chalk River, Canada

Carlile, C.J., Institut Laue-Langevin, Grenoble, France
Chappert, J., Centre d'Etudes Nucléaires, Grenoble, France
Chiu, J.C.H., University, Brandon, Canada
Cooper, B.R., West Virginia University, Morgantown, USA
Coqblin, B., Université de Paris-Sud, Orsay, France
Cornut, B., CNRS-CRTBT, Grenoble, France

Descouts, P., Université, Genève, Switzerland
Dixon, J.M., University of Warwick, Coventry, England

Eagles, D.M., National Measurement Labs., Chippendale, Australia
Elliott, R.J., University, Oxford, England

Fert, A., Université de Paris-Sud, Orsay, France
Fischer,P., Institut für Reaktortechnik ETHZ, Würenlingen, Switzerland
Fulde, P., Max-Planck-Institut, Stuttgart, BRD
Furrer, A., Institut für Reaktortechnik ETHZ, Würenlingen, Switzerland

Gamari-Seale, H., N.R.C. Democritos, Athens, Greece
Gardner, J.A., Oregon State University, Corvallis, USA
Genicon, J.L., CNRS-CRTBT, Grenoble, France
Gignoux, D., CNRS, Grenoble, France
Gross, W., Universität, Frankfurt, BRD
Guertin, R.P., Tufts University, Medford, USA

Hälg, W., Institut für Reaktortechnik ETH, Zürich, Switzerland
Happel, H., Universität, Frankfurt, BRD
Hardiman, M., Université, Genève, Switzerland
Hauck, J., Kernforschungsanlage, Jülich, BRD
Hedgcock, F.T., McGill University, Montréal, Canada
Heer, H., Institut für Reaktortechnik ETHZ, Würenlingen, Switzerland
Heidemann, A., Institut Laue-Langevin, Grenoble, France
Hessel Andersen, N., H.C. Ørsted Institute, Copenhagen, Denmark
Hoenig, H.E., Universität, Frankfurt, BRD
Holden T.M., Atomic Energy of Canada Ltd., Chalk River, Canada
Holland-Moritz, E., Kernforschungsanlage, Jülich, BRD
Huang, C.Y. Los Alamos Scientific Laboratory, Los Alamos, USA
Hulliger, F., Lab. für Festkörperphysik ETH, Zürich, Switzerland

Ishikawa, Y., Tohoku University, Sendai, Japan

Karlsson, E., Institute of Physics, Uppsala, Sweden
Keller, J., Universität, Regensburg, BRD
Kjems, J., A.E.K. Risø, Roskilde, Denmark
Knorr, K., Universität, Mainz, BRD
Koon, N.C., Naval Research Laboratory, Washington, USA

Lander, G.H., Argonne National Laboratory, Argonne, USA
Laves, F., Institut für Kristallographie ETH, Zürich, Switzerland
Leciejewicz, J., Institute of Nuclear Research, Otwock, Poland
Legvold, S., Iowa State University, Ames, USA
Lindgård, P.-A., A.E.K. Risø, Roskilde, Denmark
Loewenhaupt, M., Kernforschungsanlage, Jülich, BRD
Lüthi, B., Rutgers University, New Brunswick, USA

Matthies, S., JINR, Dubna, USSR
Meier, G., Institut für Reaktortechnik ETHZ, Würenlingen,Switzerland
Millhouse, A., Hahn-Meitner Institut, Berlin, BRD
Morin, P., CNRS, Grenoble, France
Muir, W.B., McGill University, Montréal, Canada
Murani, A.P., Institut Laue-Langevin, Grenoble, France
Murasik, A., Institute of Nuclear Research, Otwock, Poland

Nagel, J., Freie Universität, Berlin, BRD
Nicaud, A., Commissariat à l'Energie Atomique, Sevran, France

Olsen, J.L., Lab. für Festkörperphysik ETH, Zürich, Switzerland
Ott, H.R., Lab. für Festkörperphysik ETH, Zürich, Switzerland

Pellisson, J., Université, Genève, Switzerland
Perkins, R.S., Brown Boveri Research Centre, Dättwil, Switzerland
Pierre, J., CNRS, Grenoble, France

Pink, D.A., St. Francis Xavier University, Antigonish, Canada
Poirier, M., Université, Montréal, Canada
Purwins, H.-G., Lab. für Festkörperphysik ETH, Zürich, Switzerland

Rainford, B.D., Imperial College, London, England
Rhyne, J.J., National Bureau of Standards, Washington, USA
Roeland, L.W., University, Amsterdam, The Netherlands
Rossat-Mignod, J., Centre d'Etudes Nucléaires, Grenoble, France

Sarkissian, B.V.B., Imperial College, London, England
Savage, H.T. Naval Surface Weapons Center, Silver Spring, USA
Schmitt, D., CNRS, Grenoble, France
Seipler, D., Institut II für Festkörperphysik, Darmstadt, BRD
Shaltiel, D., Hebrew University, Jerusalem, Israel
Shamir, N., Nuclear Research Centre-Negev, Beer-Sheva, Israel
Sherrington, D., Institut Laue-Langevin, Grenoble, France
Siegel, E., Public Service Electric & Gas Co., Maplewood, USA
Southern, B.W., Institut Laue-Langevin, Grenoble, France
Steglich, F., Universität, Köln, BRD
Stevens, K.W.H., University, Nottingham, England
Suss, J.T., Soreq Nuclear Research Center, Yavne, Israel

Tcheou F., Centre d'Etudes Nucléaires, Grenoble, France
Tellenbach, U., Inst. für Reaktortechnik ETHZ, Würenlingen, Switzerland
Temple, A., University, Salford, England
Teplov, M.A., State University, Kazan, USSR
Thomas, H., Universität, Basel, Switzerland
Tiwari, M.D., Garhwal University, Srinagar, India

Umlauf, E., Zentralinstitut für Tieftemperaturforschung, Garching, BRD
Urban, P., Institut II für Festkörperphysik, Darmstadt, BRD

van Stapele, R.P., Philips Research Labs., Eindhoven, The Netherlands
Vijayaraghavan, R., Tata Institute of Fundamental Research, Bombay
Vogt, O., Lab. für Festkörperphysik ETH, Zürich, Switzerland
von Blanckenhagen, P., Ges. für Kernforschung, Karlsruhe, BRD
Vorderwisch, P., Hahn-Meitner Institut, Berlin, BRD

Waeber, W.B., Université, Lausanne-Dorigny, Switzerland
Wallace, W.E., University, Pittsburgh, USA
Wassermann, W.E., Technische Hochschule, Aachen, BRD
Wiesinger, G., Technische Universität, Wien, Austria

Yang, D.H.Y., Max-Planck-Institut, Stuttgart, BRD

Zevin, V., The Hebrew University, Jersualem, Israel
Zygmunt, A., Inst. of Low Temp. and Structure Research, Wroclaw, Poland

INDEX